12

Springer Series in Chemical Physics

Edited by J. Peter Toennies

Springer Series in Chemical Physics

Editors: V. I. Goldanskii R. Gomer F. P. Schäfer J. P. Toennies

S. A. Losev

Gasdynamic Laser

With 100 Figures

Springer-Verlag Berlin Heidelberg New York 1981

Professor Dr. Stalii A. Losev
Institute of Mechanics, Moscow State University,
Moscow SU-117192, USSR

Title of the original Russian edition:
Gazodinamicheskie lazary

ISBN 3-540-10503-4 Springer-Verlag Berlin Heidelberg New York
ISBN 0-387-10503-4 Springer-Verlag New York Heidelberg Berlin

Library of Congress Cataloging in Publication Data. Losev, S. A. (Staliĭ Andreevich). Gasdynamic laser. (Springer Series in Chemical Physics ; v. 12) Translation of Gazodinamicheskie lazery. Bibliography: p. Includes index. 1. Gasdynamic lasers. I. Title. II. Series. TA1695.L6713 621.36'63 81-1574 AACR2

Offset printing: Beltz Offsetdruck, Hemsbach/Bergstr. Bookbinding: J. Schäffer oHG, Grünstadt.
2153/3130/543210

In fond memory of
Rem Khokhlov –
a man of men,
a good friend,
and a great scientist.

Preface

This book deals with gasdynamic lasers (GDL)—high-energy sources of coherent directional radiation. The theory and practice of gasdynamic lasers are based on three fields of modern science: quantum electronics, physicochemical kinetics, and gasdynamics. This circumstance has determined the content of this book, which can be divided into two parts. The first four chapters, which occupy a considerable part of the book, prepare the reader for discussing the theoretical and experimental results of up-to-date GDL studies. The necessity of such a presentation is dictated by the fact that gasdynamic lasers can be a subject of interest for readers who are experts in only one of the mentioned fields, or those who have just begun studying GDLs. Such a representation is based on the experience gained by the author over several years in delivering lectures on this subject to students of Moscow State University. These lectures formed the basis of the book, which can also be used as a text.

The basic information on gasdynamic lasers is contained in the last two chapters of the book. These chapters give mathematical models of processes occurring in existing high-energy infrared GDLs (CO_2 GDLs, carbon monoxide GDLs, etc.), experimental results for these lasers, and also consider some possible new designs of GDLs. Of greatest interest in this respect are the attempts to create recombination electronic GDLs and plasmadynamic lasers.

The number of studies devoted to the development of high-energy GDLs is continuously increasing. Several reviews on this subject have been published. The first monograph about gasdynamic lasers, the book by Anderson, which was published in the USA in 1976, is a very good introduction to the field. A year later our book, with an extensive bibliography of the work of Soviet scientists, was published in the USSR.

In preparing the present English edition of the book we have taken into account recent advances in the development of gasdynamic lasers. In particular a new section devoted to GDLs on twin modes of the CO_2 molecule vibrations was added. We also omitted the references to papers not easily accessible to English readers but retained the English references. Most of the Soviet scientific journals and books are published in the USA in English translation. In those cases where the author had no information about the availability of a translation, the original Russian references are given.

The author wishes to give special acknowledgement to his colleagues N.A. Fomin, A.I. Demin, N.V. Evtyukhin, V.K. Konyukhov, G.A. Lukyanov, V.A. Kochelap, and V.A. Khokhryakov, who helped him greatly in preparing the English translation of the book. The author also gives thanks to the translator N.V. Deineko, who worked laboriously to present this book in English. We both hope that this work will be helpful and will contribute to better understanding between scientists of different countries.

Moscow, January 1981 *S.A. Losev*

Contents

1. Introduction

In a gasdynamic laser (GDL), radiation is produced by rapidly cooling a preheated gas. This can be achieved by accelerating the gas to supersonic velocities. However, not every laser with a rapidly moving working medium can be called a gasdynamic laser. The utilization of moving gases in electric-discharge lasers and other types of lasers have made it possible to increase the output power considerably but have not altered the basic principle of creating an active medium. The specific feature of gasdynamic lasers considered in this book is that the laser effect results from the motion of a gas that leads to a disturbance of thermal equilibrium in the system and creates the population inversion of energy levels in gas molecules. Thus, gasdynamic lasers are devices in which thermal energy is directly transformed into coherent directional radiation.

At present most high-power gasdynamic lasers operate in the infrared spectral region on optical transitions between vibrational energy levels of CO_2 molecules. Carbon monoxide, nitrous oxide, and carbon sulphide lasers have also been developed. Attempts are being made to create high-power GDLs operating in other spectral regions—visible and ultraviolet, that is, employing transitions between electronic energy levels of atoms and molecules.

A schematic diagram of a gasdynamic laser is given in Fig.1.1. Preheated working gas is delivered to a nozzle where it is accelerated and cooled. During rapid cooling, the different deactivation rates of molecules possessing different energies lead to a population inversion of energy levels and the concentration of more highly excited molecules exceeds the concentration of less highly excited molecules. Then the gas enters a cavity, which consists of two mirrors (one of them is semitransparent) parallel to the flow. In the cavity, a part of the energy associated with population inversion is transformed into directional coherent radiation that passes through the semitransparent mirror, to form a laser beam.

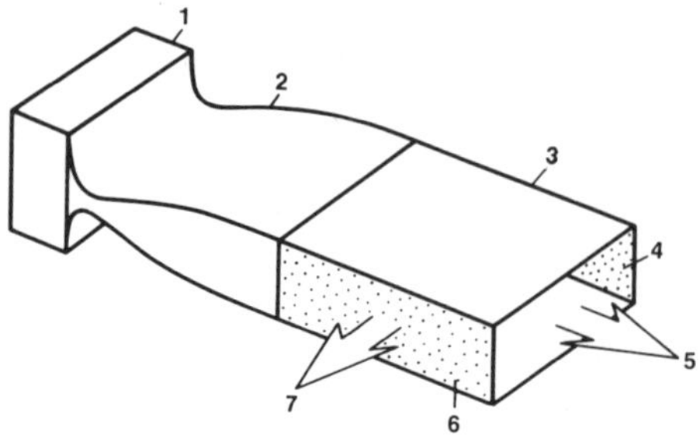

Fig.1.1. Schematic of a gasdynamic laser. (*1*) heater; (*2*) nozzle; (*3*) cavity; (*4*) reflecting mirror; (*5*) gas flow; (*6*) semitransparent mirror; (*7*) output laser beam

The development of gasdynamic lasers has begun quite recently. Several reviews have been published [1.1], and the monograph by ANDERSON [1.2] is a very good introduction to the field. But before considering the subject matter of the book let us mention some historical background which shows the development of ideas underlying the gasdynamic laser concept.

First works suggesting that thermal excitation of a quantum system could be used to produce coherent electromagnetic radiation appeared in 1959-1963 [1.3]. These works considered a three-level system with different rates of establishing thermal equilibrium, in which population inversion formed during cooling. In [1.4] an analysis of a concrete scheme of obtaining population inversion of energy levels of CO_2 molecules by creating a temperature gradient in a gas contained between hot and cold walls was given. In [1.5] the authors put forward an idea of possible population inversion of electronic energy levels during rapid cooling of a gas in a supersonic nozzle. The authors of [1.6] suggested that cooling a plasma can also serve as a GDL active medium.

The possibility of a considerable distortion of local thermodynamic equilibrium in a rapidly accelerated gas, accompanied by its heating (during compression) or cooling (during expansion), is well known in gas dynamics (see [1.7,8]). Among the first experiments in which a delay of deactivation of vibrational degrees of freedom of molecules was observed were [1.9] (CO_2 cooling in a flow) and [1.10] (nitrogen flow in a nozzle of a shock tube). On the other hand, still earlier in [1.11] a highly effective energy transfer from nitrogen to CO_2 molecules was observed in an electric discharge. In [1.12,13] this phenomenon was used for the first time for selective excitation of the upper laser level in creating a CO_2 electric-discharge laser (see [1.14]).

The first proposal for creating an infrared gasdynamic laser operating on a CO_2-N_2 mixture with supersonic nozzle cooling appeared in 1966 [1.15,16].

In [1.17,18] the authors considered the possibility and methods of thermal pumping of vibrational energy levels of polyatomic molecules. In [1.19] it was suggested that carbon monoxide burning should be utilized to obtain excited CO_2 molecules and to create population inversion during subsequent cooling.

Operating conditions of GDLs were experimentally studied in [1.20]. At last, in 1970 first reports about CO_2 gasdynamic laser performance appeared. Experiments were repeated using shock tubes with nozzles [1.21-23], continuously operating setups with electrical heating at a nozzle entrance [1.24], with fuel mixtures [1.22], and then, in 1971, using detonation of solid explosives [1.25]. Subsequently many other reports about not only CO_2 GDLs but also carbon monoxide GDLs (first in [1.26]), nitrous oxide (first reports [1.27,28]), and carbon sulphide [1.29,30] GDLs appeared. The population inversion of electronic energy levels of hydrogen atoms was first obtained in [1.31] in a cooling plasma jet.

To date gasdynamic lasers provide 20 to 25 J of coherent radiation energy per gram of a working gas [1.32,33] (in [1.34] a specific power of 50 J per gram of nitrogen was reported). This is inferior to specific power output from rapid-flow electric-discharge or chemical lasers. The GDL efficiency, about 1%, is also low so far. This, however, does not narrow the prospects of gasdynamic lasers, since these devices provide a direct conversion of thermal energy into the energy of a laser beam.

Advantages of gasdynamic lasers become especially noticeable in creating continuously operating high-power lasers, since the efficiency of these devices grows with a scale of a setup due to the decrease of heat losses. It should be noted that the development of high-power chemical lasers is complicated because of high toxicity and chemical activity of their working mixtures. An important advantage of gasdynamic lasers, especially of CO_2 lasers, is the possibility of obtaining laser radiation using combustion products of common fuels, which greatly increases their field of application.

High-power continuously operating GDLs will eventually have a great impact on many fields of science and technology. For example, lasers can be utilized in high-quality processing (deep welding, precision cutting, heating, spraying, and so on), in the development of a new trend in laser chemistry and chemical technology (selective control of chemical reactions, in particular isotope separation and obtaining perfectly pure substances), in wireless transmission of high energy over large distances (to orbital stations, spacecraft, and so on) and in weather control (elimination of mist and clouds in airport zones). The development of high-power industrial lasers has begun quite recently, and they have many applications. The role of gasdynamic lasers in this respect may turn out to be very significant.

In the present book we consider mainly the theoretical foundations of gasdynamic lasers. In this respect the book has a somewhat wider scope than the known monograph by ANDERSON [1.2]. However, this shouldn't discouvage an unprepared reader since we start from a simple presentation of the fundamentals of those branches of science which are required for the development of gasdynamic lasers and which are quite comprehensible for a wide circle of readers including theoreticians and experimenters. These are quantum electronics, physico-chemical kinetics, and gas dynamics. The next three chapters of the book deal with these fields and contain the information and examples required for understanding the theory of gasdynamic lasers. If the reader has a desire to go deeper in these fields of science, we refer him to the textbooks and monographs contained in References.

Chapter 5 is the main part of the book as it gives the concept of the infrared CO_2 gasdynamic laser — the most widespread type of lasers which has already found practical applications in technology. Other types of gasdynamic lasers are described in Chap.6. The most interesting here are the promising developments of high-power gasdynamic lasers for a visible and ultraviolet spectral regions.

The main problem considered in this book is that of obtaining an active medium by gasdynamic methods and describing its properties using various mathematical models. The technical problems concerning the creation of gasdynamic lasers (for example, the development of heaters, combustion chambers, high-power optical cavities, and diffusers) are beyond the scope of the book and are considered in detail in references. Therefore, this book cannot be regarded as a manual for the designers, although it is impossible to develop a gasdynamic laser without understanding the theoretical principles of its operation.

So, let's start!

2. Basic Concepts of Quantum Electronics

This chapter considers the rotation and vibration of molecules. The concepts of spontaneous and induced transitions and the Einstein coefficients determining the optical transition probability are introduced along with a description of spectral line contours in a gas and the mechanism of line broadening. Introductions to the concepts of light absorption and amplification coefficients in a gas, the population inversion of energy levels and pumping are given. Building on these principles the creation of population inversion in gasdynamic lasers is then discussed. The chapter continues with a discussion of resonator performance, in which laser radiation is generated; threshold conditions for generation and types of losses; and the concepts of useful radiation power and resonator efficiency. The problem of radiation extraction from a cavity, and the formation of a directed beam are also discussed and various types of resonators (stable and unstable) and multipass amplifiers are considered. The chapter concludes with a consideration of the problem of a rapid active medium flow in a cavity, losses due to gas inhomogeneities, and the presence of condensed particles in a flow.

2.1 Energy Levels and Quantum Transitions in Atoms and Molecules

Quantum electronics, which is a specific combination of atomic and molecular physics and optics, actually appeared about twenty-five years ago. It has, however, an earlier history that is closely related to the development of 20th century physics, specifically the combination of ideas concerning the wave and corpuscular nature of matter known as quantum mechanics. In 1954 Basov and Prokhorov in the Soviet Union and Townes in the USA created the first lasers—NH_3 molecular generators of radiation at about 1 cm wavelength. The first ruby laser operating in the optical range was developed in 1960, and the first gas lasers were reported in 1961[1]. (Footnote see next page.)

Quantum electronics studies processes occurring in atoms and molecules. Apart from translational motion of a molecule as a whole, the simplest motion of a molecule is its rotation. To a first approximation the rotation is described by the model of a *rigid rotator*. The quantization of its rotational energy yields

$$E_{rot}(J) = (h^2/8\pi^2 I)J(J + 1) \quad , \tag{2.1.1}$$

where h is Planck's constant, I is the moment of inertia of a molecule, and J is the rotational quantum number. The separation of rotational energy levels $E_{rot}(J)$ increases with J.

To describe vibrations of atoms in a molecule the concept of the *harmonic oscillator* is used, in which an elastic force, returning the system to equilibrium, is proportional to the amplitude of deflection. The potential energy of a harmonic oscillator is (Fig.2.1a)

$$U = k(r - r_e)^2/2 \quad , \tag{2.1.2}$$

where r is the distance between nuclei in a molecule and r_e is that distance at equilibrium.

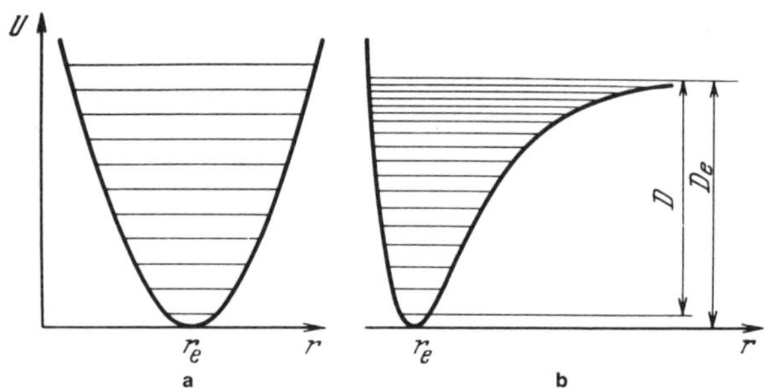

Fig.2.1a,b. Potential energy curves for a harmonic oscillator (a) and a Morse anharmonic oscillator (b). D is the energy of dissociation

Quantum mechanics permits for a harmonic oscillator (2.1.2) only those states in which its energy is

$$E_{vib}(v) = h\nu_0(v + 1/2) \quad , \quad v = 0, 1, 2, \ldots \quad , \tag{2.1.3}$$

where ν_0 is the fundamental frequency of vibrations. Energy levels in a harmonic oscillator are equidistant. This model is a good approximation to re-

1 Fundamentals of quantum electronics and laser physics are given in detail in [1.14, 2.1-4].

ality when amplitudes of vibrations of atoms in a molecule are small. In the next approximation we must account for the anharmonicity of vibrations using, for example, the *Morse oscillator* model (Fig.2.1b),

$$U = D_e\{1 - \exp[-\alpha(r - r_e)]\}^2 \quad , \tag{2.1.4}$$

where D_e is the potential well depth and α is a parameter related to the frequency of vibrations and D_e. The limit at $r \to \infty$ corresponds to dissociation of a molecule. The expression for energy levels of the Morse oscillator corresponding to potential (2.1.4) is

$$E_{vib}(v) = h\nu_0(v + 1/2) - (h^2\nu_0^2/4D_e)(v + 1/2)^2 \quad , \tag{2.1.5}$$

where ν_0 is the fundamental frequency of vibrations of the respective harmonic oscillator. From (2.1.5) it is clear that for (2.1.4) energy levels get closer as v increases.

The motion of electrons in fields of nuclei of atoms and molecules is described by more complicated equations. The potentials of this type of motion and the respective energy levels E_{el} are given, for example, in [2.5-7].

The spacing of vibrational energy levels, ΔE_{vib}, is usually much greater than the spacing or rotational energy levels, ΔE_{rot}, and in turn $\Delta E_{vib} \ll \Delta E_{el}$. For this reason we say that to every electronic state of a molecule there corresponds a system of vibrational levels, and to each of these levels, a system of rotational levels.

Sometimes for one value of energy there can be two or more states of a system, which are indistinguishable for macroscopic observations in ordinary conditions. Deformation (bending) vibrations or a linear CO_2 molecule can serve as an example (Fig.2.2). Here two independent directions of vibrations at the same frequency correspond to one value of vibrational energy. One of these vibrations proceeds in the plane of the figure, the other, in the perpendicular direction. In this case it is said that deformation vibrations of the CO_2 molecule are doubly degenerate. The number of different states corresponding to one value of energy is called the degree (or multiplicity) of degeneration, or the *statistical weight* of the energy level, and is usually denoted by g.

When a state of the system under consideration changes, a quantum transition between energy levels occurs. Two kinds of transitions are distinguished for atoms and molecules: *collisional* transitions, occurring as a

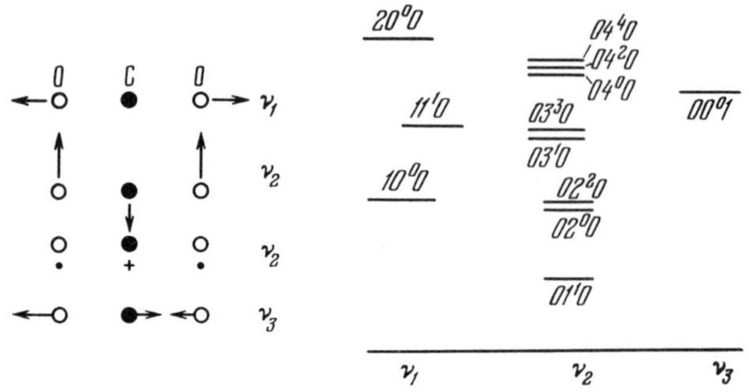

Fig.2.2. Vibrational modes and lower vibrational energy levels for CO_2

result of collisions with other atoms and molecules, and *radiative* transitions, accompanied by radiation or absorption of light. For radiative transitions $E_m - E_n = h\nu$, where ν is the frequency of radiated or absorbed light. Here and below the subscript m designates the upper and n the lower energy level. To every transition there corresponds a line in the electromagnetic spectrum. The transition energy ΔE is related to frequency ν, wavelength λ, and wave number ν' as follows: $\Delta E = h\nu$, $\lambda\nu = c$, $\nu' = 1/\lambda$, where c is the speed of light.

In atoms only transitions between separate electronic energy levels are possible; a set of spectral lines corresponding to them is called the *line spectrum*. This spectrum is observed in visible and ultraviolet regions. In molecules transitions between rotational energy levels correspond to far infrared and millimeter spectral ranges, provided that other types of energy remain unchanged. In the infrared range vibration-rotational transitions are observed in which both kinds of energy change. Finally, in the visible and ultraviolet spectral ranges there occur transitions in which electronic, vibrational, and rotational energies change simultaneously. The corresponding spectra are of a band nature owing to the close location of the lines of the vibrational structure (vibrational lines). Such spectra are called the *band spectra*. The appearance of certain lines is determined by different selection rules (see, for example, [2.5-7]). It should be noted that in the infrared region the transitions with distortion of symmetry (dipole transitions) have an appreciable intensity. For instance, among diatomic molecules sufficiently intensive infrared spectra are observed for CO, NO, CN, etc. molecules, but not for O_2, H_2, and N_2 molecules.

Let us consider some examples which are interesting from the point of view of the subject matter of the book. The first concerns vibrational and rotational energy levels of the CO_2 molecule and its infrared spectrum.

In a ground electronic state CO_2 molecules are linear with axes of symmetry and symmetry planes perpendicular to them (see Fig.2.2). There are three typical modes of vibrations for CO_2 molecules: symmetric vibrations of oxygen atoms with respect to a carbon atom (the first mode ν_1), doubly degenerate symmetric deformation (bending) vibrations of oxygen atoms with respect to a carbon atom (the second mode ν_2), and antisymmetric valence vibrations (the third mode ν_3). To describe these vibrations three quantum numbers, v_1, v_2, v_3, respectively, are introduced. For doubly degenerate deformation vibrations one additional quantum number ℓ describes the quantization of the angular momentum associated with deformation vibrations and is directed along the axis of a molecule [2.7]. ℓ acquires the values $\ell = v_2, v_2 - 2, \ldots, 0$ for even v_2, and $\ell = v_2, v_2 - 2, \ldots, 1$ for odd v_2. The levels with $\ell = 0$ are nondegenerate, and those with $\ell > 0$ are doubly degenerate. Thus, the statistical weight of a level with a given v_2 is $g_{v_2} = v_2 + 1$. The remarkable feature of the first two vibrational modes is that the fundamental frequency of the first mode (ν_1) and the double frequency of the second mode ($2\nu_2$) are almost equal (Fermi resonance). As a result of such a strong interaction, degeneracy is removed and splitting of the levels of the second vibrational mode is observed (see [2.7]). Vibrational energy levels of CO_2 are designated by the notation v_1, v_2^{ℓ}, v_3; a schematic diagram of lower vibrational levels of this molecule is shown in Fig.2.2. To every level there corresponds a whole set of rotational levels which are not shown in the figure. Complete equations describing the positions of all vibrational levels of CO_2 taking account of interactions between vibrations and rotation can be found, for instance, in [2.7-9].

The Fermi resonance in the CO_2 molecule results in energy level shifts even for the lowest vibrational states. Under certain circumstances the resulting deviation from the harmonic oscillator model must be taken into account. In this case it is essential to realize that, owing to the Fermi resonance, the strict separation of symmetrical valence and symmetrical deformation modes of the CO_2 vibrations becomes impossible: these modes are mixed together. The levels of these mixed modes form a system which can be characterized by a quantum number $v = 2v_1 + v_2$, in the harmonic approximation each level of this system is doubly degenerate with a statistical weight $g_v = (v + 2)^2/4$ for even v, and $g_v = (v + 1)(v + 3)/4$ for odd v [2.10].

The selection rules for infrared transitions of CO_2 and other molecules are described, for example, in [2.7-9]. It should be noted that only those transitions, in which deformation or antisymmetric vibrations participate,

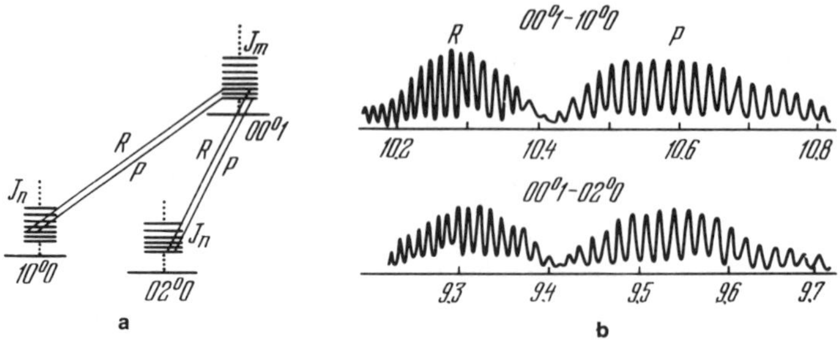

Fig.2.3a,b. R- and P-branch radiative transitions between rotational levels of the 00^01, 10^00, and 02^00 vibrational states (a) and spectrum for these transitions (b) in CO_2

contribute to this spectrum. The rotational quantum number J either changes by ±1 or does not change at all for certain bands ($\Delta J = 0$). Of great interest for us are the transitions $00^01 \rightleftharpoons 10^00$ and $00^01 \rightleftharpoons 02^00$, on which laser radiation of the highest power has been obtained in CO_2 (at a wavelength of 10.6 and 9.6 μm, respectively). A more detailed scheme of transitions between rotational lines for these bands is given in Fig.2.3a; here $\Delta J = -1$ (P branch) and $\Delta J = +1$ (R branch), other transitions being forbidden. A general view of a spectrum for these transitions is shown in Fig.2.3b.

In the infrared spectrum of CO molecules the most important for us is its anharmonicity, that is, a decrease of the vibrational quantum with an increase of the level number v according to (2.1.5). The most probable transitions here are between neighboring vibrational levels ($v \rightleftharpoons v-1$), while higher harmonics $\Delta v \geqq 2$ are less intense. The bands corresponding to these transitions are in a wavelength range of 4.7 μm.

Let us consider now one of the fundamental concepts of quantum electronics—*induced* and *spontaneous* transitions between energy levels. Suppose that an atom or a molecule is in an excited state m and has energy E_m. This excited state is unstable; in a certain time the atom will spontaneously pass over to a lower state n, emitting a quantum $h\nu$. This is a spontaneous transition. All spontaneous transitions obey statistical laws owing to the random nature of each act of spontaneous transition. The uncertainty of spontaneous transition results in the uncertainty of phase, direction of radiation, and polarization plane, since atoms emit independently and not at the same time. Hence spontaneous emission is incoherent and nonpolarized. Time dependence of a number of excited atoms N_m for spontaneous emission is expressed by an exponential law[2] (footnote see next page)

$$N_m = (N_m)_0 \exp(-t/\tau_{mn}) \quad , \qquad (2.1.6)$$

where τ_{mn} is the so-called *optical lifetime* of the mth level with respect to the $m \to n$ transition. The quantity $A_{mn} = 1/\tau_{mn}$ determines the average number of quanta emitted as a result of the $m \to n$ transition per atom per unit time. This quantity is called the spontaneous emission probability. Relating N_m to a unit volume, we find that the number of photons (quanta) Z^s_{mn} spontaneously emitted per unit volume per unit time is

$$Z^s_{mn} = A_{mn} N_m \quad . \qquad (2.1.7)$$

Note that the process of spontaneous absorption is meaningless from the physical point of view, since in this case there is an external effect of radiation of an atom. This effect results in induced transitions. Thus if an atom is in the lower level N_n, then, under the effect of external electromagnetic radiation at a frequency equal to the frequency ν of the transitions under consideration, its transition to the upper level m is possible. This is an induced process, since it occurs only as a result of the external effect. Let ρ be the density of radiation, that is, light energy per unit volume. The number of absorbed quanta Z^{in}_{nm} is proportional to the density of radiation ρ and concentration N_n of atoms in the nth state,

$$Z^{in}_{nm} = B_{nm} N_n \rho \quad . \qquad (2.1.8)$$

Here B_{nm} is the probability of induced absorption, equal to the number of photons absorbed on the average by an atom per unit time with unit density of radiation.

When an atom occupying the upper energy level m is acted upon by external radiation, secondary (induced) emission can be observed. Induced emission was established for the first time by Einstein in 1916, when he analyzed the heat balance between emitting and absorbing media (equilibrium between radiation and substance). Since spontaneous emission is determined only by the properties of atoms, and absorption depends also on the intensity of radiation acting on the atom, the balance can set in only when induced emission is present. The number of photons Z^{in}_{mn} appearing with induced emission, as in the case of absorption, is proportional to the density of radiation ρ

---

2 Concentration N_m, N_n, ... of atoms or molecules in a certain energy state m, n, ... is often called the *population* of the level m, n, ...

and concentration, but now it is the concentration N_m of atoms in the upper state m,

$$Z_{mn}^{in} = B_{mn}N_m\rho \quad , \qquad (2.1.9)$$

where B_{mn} is the induced emission probability, equal to the number of photons emitted on the average by an atom per unit time per unit density of radiation. Induced emission has the same frequency, polarization, and direction of propagation as the acting radiation owing to strong interaction between radiation and the excited atom. Thus, a secondary photon is indistinguishable from a primary one. The quantities A_{mn}, B_{nm}, and B_{mn} are called the *Einstein coefficients* for the m ⇌ n transition.

It is well known that the optical transition m ⇌ n is not strictly monochromatic, that is, spectral lines have finite widths. Spectral line broadening is, due to the nature of the atoms themselves, the effect of external forces on them and the translatory motion of atoms in a gas. With line broadening the transition probability varies depending on frequency within the limits of the line. The *Einstein spectral coefficients* denoted by $a_{mn}(\nu)$, $b_{nm}(\nu)$, $b_{mn}(\nu)$, after integrating them over frequency ν, naturally yield the above-mentioned total (integral) Einstein coefficients for the m ⇌ n transition,

$$A_{mn} = \int a_{mn}(\nu)d\nu \quad ,$$

$$B_{nm}\rho = \int b_{nm}(\nu)\rho_\nu d\nu \quad ,$$

$$B_{mn}\rho = \int b_{mn}(\nu)\rho_\nu d\nu \quad , \qquad (2.1.10)$$

where ρ_ν is the spectral density of radiation per unit frequency interval. If, as usual, frequency ν varies within the limits of the line insignificantly, the Einstein coefficients, both spectral and total, are interrelated (g_n and g_m are the statistical weights of the levels n and m),

$$g_n b_{nm}(\nu) = g_m b_{mn}(\nu) \quad , \quad a_{mn}(\nu) = (8\pi h\nu^3/c^3)b_{mn}(\nu) \quad . \qquad (2.1.11)$$

These relations are derived in quantum electrodynamics (see, for example, [2.1,11]). They also can be obtained from the principle of detailed balancing, according to which in thermodynamic equilibrium the numbers of transitions (collisional or radiative) in the forward and reverse directions via any channel are equal, so that the number of particles arriving at a certain level is equal to the number of particles leaving this level. This principle makes it possible to obtain equations for such a balance of particles. If coefficients in these equations are independent of the particle concentration, then the relations between the coefficients will be valid for any concentration, including nonequilibrium. In the example under consideration the balance of the number of photons emitted and absorbed during radiative transitions between the levels m and n per unit frequency interval yields

$$a_{mn}(\nu)\bar{N}_m + b_{mn}(\nu)\bar{N}_m\bar{\rho}_\nu = b_{nm}(\nu)\bar{N}_n\bar{\rho}_\nu \quad . \tag{2.1.12}$$

Here the bar above a letter denotes equilibrium values of concentrations N_m, N_n and of spectral density of radiation ρ_ν. Using the Boltzmann equation for $\bar{N}_m$, $\bar{N}_n$ and the Planck equation for $\bar{\rho}_\nu$, we obtain (2.1.11), which are valid also for other, nonequilibrium values of N_m, N_n, and ρ_ν; this derivation is given in Appendix A.

The frequency dependence of probabilities $a_{mn}(\nu)$, $b_{mn}(\nu)$, and $b_{nm}(\nu)$ determines the shape of the spectral line $g(\nu)$. It is convenient to relate $g(\nu)$ to the center of the line $\nu_0 = (E_m - E_n)/h$, so that[3]

$$a_{mn}(\nu) = a_{mn}(\nu_0)g(\nu) \quad ,$$

$$b_{nm}(\nu) = b_{nm}(\nu_0)g(\nu) \quad ,$$

$$b_{mn}(\nu) = b_{mn}(\nu_0)g(\nu) \quad . \tag{2.1.13}$$

According to (2.1.10), after integration over ν and normalization, we get

$$a_{mn}(\nu) = A_{mn}g(\nu)/\overline{\Delta\nu} \quad ,$$

$$b_{nm}(\nu) = B_{nm}g(\nu)/\overline{\Delta\nu} \quad ,$$

$$b_{mn}(\nu) = B_{mn}g(\nu)/\overline{\Delta\nu} \quad , \tag{2.1.14}$$

where $\overline{\Delta\nu} = \int g(\nu)d\nu$. The intensity of radiation I (or the absorbed energy) is proportional to the probability and, hence, varies within the line as $g(\nu)$. The respective curve is usually bell-shaped (Fig.2.4). The line *width* can be characterized by $\Delta\nu$, which is equal to the frequency difference for the points where the spectral probability of transition (or radiation intensity), is decreased to one half of its value at the center of the line.

---

3 Here it is assumed that the contours of spectral lines are the same for both emission and absorption, which is true only for separate lines and not for wide bands, in which the frequency varies significantly (see [2.7])

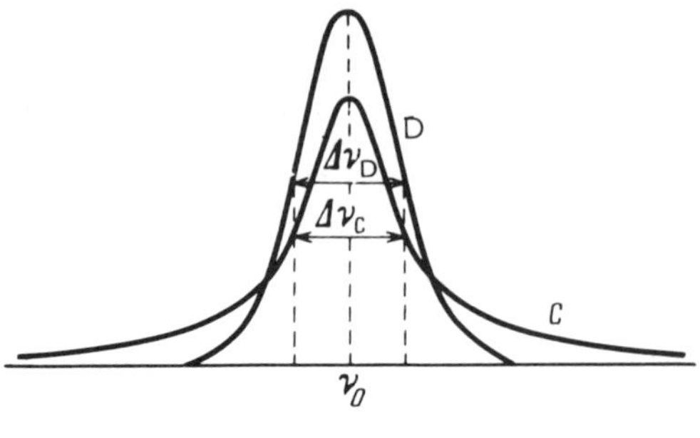

Fig.2.4. Shape of a spectral line for the Doppler (D) and collisional (c) broadening

There are several types of spectral line broadening, among which the most important are the *collisional* broadening and the *Doppler* broadening. The collisional broadening is due to the effects of other particles on an atom during collisions, since in each collision the phase of radiation changes, thus disturbing monochromaticity. In this case

$$g_c(\nu) = \frac{(1/\tau_0)^2}{4\pi^2(\nu - \nu_0)^2 + (1/\tau_0)^2} , \qquad (2.1.15)$$

and the collisional width of a line is[4]

$$\Delta\nu_c = \frac{1}{\pi}\tau_0 , \qquad (2.1.16)$$

where τ_0 is the mean free time of an atom. It is related to the mean free path ℓ_0 and the average velocity of particles $\bar{v}$, so that for a gas composed of similar particles $\tau_0 = \ell_0/\bar{v}$, and

$$\ell_0 = 1/\sqrt{2}\pi N\sigma^2 , \quad \bar{v} = \sqrt{8RT/\pi\mu} , \quad \Delta\nu_c = 4\sigma^2 N\sqrt{RT/\pi\mu} , \qquad (2.1.17)$$

where σ is the so-called optical diameter of collisions resulting in the broadening of the line, R is the gas constant, μ is the molecular mass of the gas, and N is the concentration of particles (the number of particles in a unit volume). Expressing N in terms of pressure p and temperature T($N = \mathcal{N}_A p/RT$, $\mathcal{N}_A$ is the Avogadro number), we find that the collisional width for a given pressure is inversely proportional to the square root of temperature. Actually, however, σ also depends on gas temperature, so the dependence $\Delta_c \propto 1/\sqrt{T}$ is valid only approximately. For collisional broadening, all the particles participating in the interaction with radiation have the same

4 We must distinguish between the total width of the line and a quantity half as large called the half width (see the next section).

frequency dependence of probability. In quantum electronics this broadening is called *uniform* broadening, and the line shape given by (2.1.15) is called the *dispersion contour*.

Atoms in a gas are in thermal motion. According to the Doppler effect, the frequency emitted by an atom moving along the x axis is $\nu = \nu_0[1+(v_x/c)]$, where v_x is the velocity of thermal motion along the x axis. Proceeding from the well-known Maxwellian velocity distribution for thermal motion of particles in a gas, we get, upon averaging over all the velocity values and the direction of molecular motion, the expression for $g_D(\nu)$

$$g_D = \exp\left[-\frac{\mu c^2}{2RT}\left(\frac{\nu-\nu_0}{\nu_0}\right)^2\right] \quad . \qquad (2.1.18)$$

This is the *Gaussian* distribution, which drops more sharply to the "wings" of the line compared to the dispersion distribution (see Fig.2.4). According to (2.1.18), the Doppler line width is

$$\Delta\nu_D = \frac{2\nu_0}{c}\sqrt{\frac{RT\ 2\ \ln 2}{\mu}} \quad . \qquad (2.1.19)$$

Thus Doppler broadening is essentially determined by the resonance frequencies of transition for individual particles. For this reason the contribution to the probability of emission at a frequency ν does not depend on the nature of interaction of radiation with atoms at other frequencies. This broadening is called *nonuniform*.

As the gas temperature increases at constant pressure, the nature of the broadening changes from collisional broadening to Doppler broadening. The pressure and temperature ranges employed in gasdynamic lasers are such that there can be cases when the difference between the collisional and the Doppler line widths is so great that one of the broadenings can be neglected. In other cases both types of broadening are observed simultaneously. To obtain the analytic expression for $g(\nu)$ with a joint action of uniform and nonuniform broadening, we must consider a dispersion distribution of intensity in a line and then take account of the contribution of the Doppler broadening, with the center at a certain frequency ν_*, to the value of the probability of emission at a frequency ν. After that we must integrate over ν_* to obtain the relation (see [2.11,12]) $g(\nu) = H(a,\omega)$, where

$$H(a,\omega) = \frac{a}{\sqrt{\pi}}\int_{-\infty}^{\infty}\frac{\exp(-y^2)dy}{(\omega-y)^2+a^2} \quad , \qquad (2.1.20)$$

$$a = \frac{\Delta\nu_c}{\Delta\nu_D}\sqrt{\ln 2} \quad , \quad \omega = \frac{\nu_0 - \nu}{\Delta\nu_D} 2\sqrt{\ln 2} \quad , \quad y = \frac{\nu_* - \nu}{\Delta\nu_D} 2\sqrt{\ln 2} \quad . \tag{2.1.21}$$

The quantity a is an important parameter in the theory of line broadening; it is equal to the ratio of the collisional and the Doppler line widths. The quantity ω shows the difference between the center of the line ν_0 and the frequency ν under consideration. The integral H (a, ω), which determines the shape of the line in this case, is called the *Voigt function*. Experiments confirm the existence of such a line shape (for the lines of the CO_2 infrared spectrum see [2.13a]). The integral H (a, ω) cannot be taken in the finite form; asymptotic and other approximate analytical expressions for the Voigt functions are given in Appendix B. As we pass over from the Doppler broadening (a = 0) to the collisional broadening ($a \to \infty$), the intensity at the center of the line continuously decreases from H(0, 0) = 1 to H(∞, 0) = 0 and grows in the wings. Thus, the line shape is determined by the Doppler broadening at the center and by the collisional broadening at the wings.

Note that as a result of normalization we obtain

$$H(a, \omega)d\nu = \sqrt{\pi}\Delta\nu_0/2\sqrt{\ln 2} \quad , \tag{2.1.22}$$

so that according to (2.1.14)

$$a_{mn}(\nu) = \frac{2\sqrt{\ln 2}}{\sqrt{\pi}} A_{mn} \frac{H(a, \omega)}{\Delta\nu_D} \quad , \tag{2.1.23}$$

and similar expressions can be obtained for other Einstein coefficients.

2.2 Absorption and Amplification of Light in a Gas: Pumping and the Gain Coefficient

Let us consider the passage of light through a gas, neglecting for the moment dispersion. Suppose monochromatic radiation of frequency ν is incident on a certain area perpendicular to the direction of its propagation. For an infinitely thin gas layer dx the change of the energy flux is proportional to the thickness of the layer and the intensity of incident radiation, so that per unit area we have

$$dI_\nu = k_\nu I_\nu dx \quad , \tag{2.2.1}$$

where I_ν is the spectral intensity of radiation propagating in the x direction per unit frequency interval[5]. The law $dI_\nu \propto dx$ was established by Lambert, and Burger found that the ratio dI_ν/I_ν does not depend on intensity. The Lambert and Burger law (2.2.1) characterizes the so-called *linear optics* range; here k_ν does not depend on the radiation intensity. The proportionality factor k_ν is called the *absorption coefficient* or the *gain coefficient*, depending on the result of the interaction of radiation with a gas.

On the other hand, the variation of light energy dI_ν in the layer dx per unit cross-sectional area of a light beam is equal to the difference of the numbers of photons emitted and absorbed in a layer dx per unit frequency interval multiplied by the photon energy $h\nu$,

$$dI_\nu = (z_{mn}^{in} - z_{nm}^{in})h\nu\ dx = (b_{mn}N_m\rho_\nu - b_{nm}N_n\rho_\nu)h\nu\ dx \quad , \qquad (2.2.2)$$

where b_{mn} and b_{nm} are the Einstein spectral coefficients for frequency ν, and z_{mn}^{in}, z_{nm}^{in}, being distinct from (2.1.8,9), are calculated per unit frequency interval. Then, bearing in mind that $I_\nu = \rho_\nu c$ and taking into account (2.2.1), we obtain the expression for k_ν

$$k_\nu = (h\nu/c)(b_{mn}N_m - b_{nm}N_n) \quad , \qquad (2.2.3)$$

or, using (2.1.11),

$$k_\nu = (h\nu/c)b_{nm}[(g_n/g_m)N_m - N_n] \quad . \qquad (2.2.4)$$

Consequently, the result of interaction of radiation with a gas (absorption or amplification) depends on the sign of the expression in the square brackets in (2.2.4), that is, on the relation between populations of the levels under consideration. In equilibrium the population of the upper level is always much lower than the population of the lower level; in this case k_ν is negative and absorption is observed. Moreover, if the difference between N_m and N_n is so large that induced emission can be ignored, we have

$$k_\nu = -(h\nu/c)b_{nm}N_n \quad . \qquad (2.2.5)$$

This formula is the expression of the law of Beer, who established for a monochromatic radiation the proportionality between the absorption coefficient k_ν and the concentration of the absorbed substance.

---

5 The total (or integral) values of the density of radiation ρ, the intensity I, and other energy characteristics can be obtained by integrating spectral quantities over the frequency range under consideration: $\rho = \int \rho_\nu\, d\nu$, $I = \int I_\nu d\nu$, $Z_{mn} = \int z_{mn}\, d\nu$, etc.

Thus, for the light amplification in a gas the following inequality must hold:

$$g_n N_m > g_m N_n \quad , \tag{2.2.6}$$

which implies population inversion, i.e., such a distribution of atoms and molecules over energy levels that the number of molecules on the upper level is greater than on the lower level (accounting for the statistical weights). In the case of population inversion $k_\nu > 0$, and after integration over the path length in a medium, we have

$$I_\nu(x) = I_\nu(0)\exp(k_\nu x) \quad , \tag{2.2.7}$$

which implies light amplification. Hence the amplification is due to the contribution from induced emission. In quantum electronics the inversion medium capable of amplification of radiative intensity is called the *active medium*.

Thus, to create an active medium we must obtain population inversion, that is, supply (or pump) so many atoms to the upper level that inequality (2.2.6) becomes valid. There can be different methods of pumping. Inversion level population can be created in a medium as a result of action of light (optical pumping), chemical reactions occurring in the medium (chemical pumping), and thermal effects in the medium with a subsequent disturbance of equilibrium (thermal excitation). Optical pumping is the basic method of creating population inversion in solid and liquid lasers. In this case, as a result of high-power radiation, so many excited particles are formed in crystals and liquids owing to absorption of light that population inversion is created. Electric discharges are typically used to activate a gaseous medium as a result of collisions of electrons with atoms, molecules, and ions. Electric current creates the population inversion in semiconductor lasers (there are semiconductor lasers employing optical pumping from a ruby laser radiation). Chemical lasers with gaseous active media are being intensively developed, in which excited products of chemical reactions create a population inversion. In photodissociation lasers a primary act resulting in the formation of excited particles occurs owing to high-power optical radiation of various compounds.

In gasdynamic lasers thermal excitation is employed, which can be realized in different ways: with the help of electrical heaters and discharges, by combustion of fuels and explosions, in shock waves, etc. In this case population inversion in a gas is not obtained, but excitatation takes place,

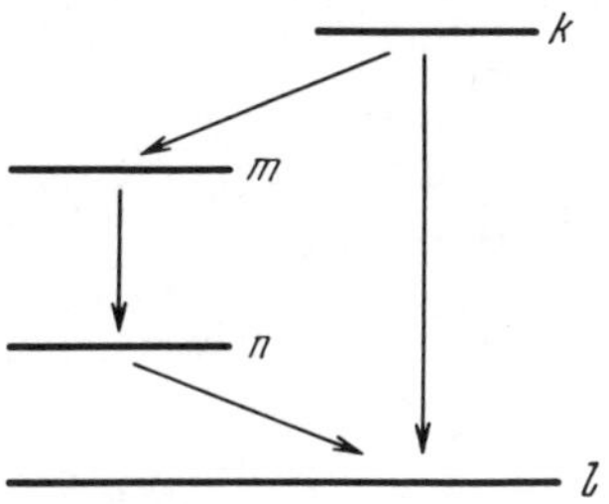

Fig.2.5. Schematic diagram of transitions between conventional energy levels

which most often results in a new state of equilibrium with increased temperature. As stated in Chap.1, in a typical scheme of a gasdynamic laser the inversion appears at sufficiently rapid cooling of a preheated gas. To do this, the heated gas is set in motion and is accelerated in a nozzle to high supersonic speeds. The principle of pumping and creating inversion in a gasdynamic laser is illustrated by a schematic diagram of energy levels given in Fig.2.5, where for simplicity it is assumed that statistical weights of all the levels are identical. After heating at equilibrium the population of the levels corresponds to the Boltzmann distribution. Since in this case the gas temperature is sufficiently high, a considerable number of atoms (or molecules) N_k occupy the upper level k, but so that $N_k < N_m < N_n < N_\ell$. If the gas is cooled, the noncompensated collisional transitions of particles between the levels in the downward direction occur in it, the population of the levels being determined by the rate of transitions of the particles from one level to another. In the case of slow cooling the inequality $N_k < N_m < N_n < N_\ell$ can remain in force. In the case of rapid cooling the equilibrium is disturbed to a greater extent. If in this process the rate of the $k \to m$, $n \to \ell$ transitions becomes much higher than the rate of the $m \to n$ transition and other channels populating the kth and m^{th} levels, then the population of the mth level will increase and that of the nth level will decrease. This can result in population inversion of the nth and m^{th} levels ($N_m > N_n$) and thus in the positive value of the gain coefficient. Hence, to create a gas dynamic laser we must choose a gas possessing the appropriate rates of transitions and provide the conditions for its heating and rapid cooling. The questions relating to the kinetics of physico-chemical processes in a gas, relaxation phenomena, and gas dynamics are treated in Chaps.3 and 4.

The basic characteristic of an active medium is the gain coefficient. Let us transform (2.2.4) to express k_ν in terms of the quantities known from theory or experiments. Passing over, according to (2.1.11), to the spontaneous emission probability and taking into account (2.1.23), we get

(omitting the subscript ν)

$$k = \frac{\sqrt{\ln 2}}{4\pi^{3/2}} \cdot \frac{c^2}{\nu^2} \cdot \frac{A_{mn}}{\Delta\nu_D} \left(N_m - \frac{g_m}{g_n} N_n \right) H(a, \omega) \quad . \tag{2.2.8}$$

Sometimes, accounting for (2.1.21), (2.2.8) is expressed in the form

$$k = \frac{c^2}{4\pi^{3/2}\nu^2} A_{mn} \left(N_m - \frac{g_m}{g_n} N_n \right) \frac{a}{\Delta\nu_c} H(a, \omega) \quad , \tag{2.2.9}$$

where $\Delta\nu_c$ is the collisional line width. The quantities $\Delta\nu_D$ and $\Delta\nu_c$ in these expressions correspond to the total width of the line. In practice the half-widths of lines that are smaller than a factor of two are frequently used. If $\Delta\nu_D$ and $\Delta\nu_c$ designate the half-widths, then the right-hand side of (2.2.9) must be reduced by a factor of two. To describe the infrared spectral range the scale of wave numbers is often used, $\nu' = \nu/c$. In this case we have

$$k = \frac{\lambda^2 A_{mn}}{8\pi^{3/2} c} \left(N_m - \frac{g_m}{g_n} N_n \right) \frac{a}{\Delta'_c} H(a, \omega) \quad , \tag{2.2.10}$$

where λ is the wavelength and Δ'_c is the collisional half-width of the line in terms of wave numbers (cm^{-1}). The expression

$$\Delta N = N_m - (g_m/g_n) N_n \tag{2.2.11}$$

in (2.2.9,10) determines the absolute population inversion. Sometimes it is convenient to represent the expression for k in the form of the sum of two terms, $k = k^+ + k^-$, where

$$k^+ = \sigma_{mn} N_m \quad , \quad k^- = \sigma_{nm} N_n \quad . \tag{2.2.12}$$

Here σ_{mn} is the so-called induced optical transition cross section of $m \to n$, and σ_{nm} is the same for the reverse transition. The value of σ_{mn} equals the gain coefficient per particle in the m state with no account of reverse transitions and is the measure of the induced transition probability. The explicit expressions for σ_{mn} and σ_{nm} are directly obtained from (2.2.9,10). Clearly, k^+ is the upper limit for the k values. If $N_m \gg N_n$ (as is usually the case with sufficient pumping), then $k \approx k^+$.

2.3 Laser Radiation in Resonators: Radiation Power

Until now we have considered amplification of light during its propagation in an active medium. As in radio engineering, the transition from amplification to generation is made by introducing a feedback. To do this it is sufficient to make a light beam repeatedly pass through an active medium, using a system of mirrors. If the increase in the radiation intensity exceeds inevitable losses, then a light amplifier becomes an oscillator which is called a *laser*. A system of mirrors utilized here is called a *resonator* (or optical *cavity*). To begin the oscillation process there is no need for an extraneous radiation source of the required frequency, since the initial level of spontaneous emission in the gas itself is sufficient for this purpose.

In general, a resonator is an oscillatory system in which the energy of oscillations (here electromagnetic oscillations in the optical range) can be stored. In this case the wavelength is much less than the dimensions of the cavity, which makes it possible to utilize so-called *open resonators*. An open resonator consists of a system of mirrors (in the simplest case, of two mirrors) which is not bounded by side surfaces. Basically there are two types of resonators—stable and unstable—distinguished by the way the light energy is stored in them. The problem of stability of oscillations and the characteristic features of both types of resonators will be treated in the next section. Here, using the simplest examples, we shall consider the principle of operation of optical cavities. Let us consider a plane-parallel resonator (often called the Fabry-Perot resonator[6]) which consists of two parallel mirrors placed at a distance d from each other (Fig.2.6)[7]. The mirrors used in a resonator can be either circular or rectangular with sides a and b, as in Fig.2.6. Let us assume so far that such a resonator is perfect from the geometrical viewpoint, the medium filling it is homogeneous, the losses are absent and, in particular, the mirror material has infinite conductivity. Then after a certain, rather small, time from the moment the oscillations are excited, if there is some initial store of electromagnetic energy, a system of standing waves will appear in the resonator. It is known

---

6 The design of such a resonator is based on the Fabry-Perot interferometer with plane-parallel mirrors, which is well known in physical optics [2.13b, 13c,32].

7 In quantum electronics the quantity d is usually called the resonator length. In the analysis of the processes in gasdynamic lasers, however, it is convenient to call it the resonator width.

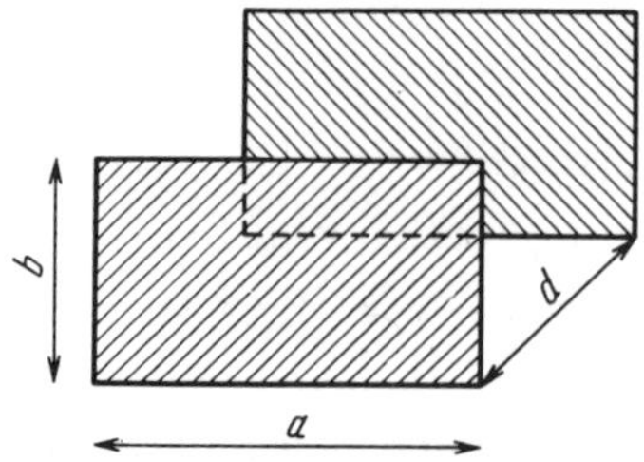

Fig.2.6.
Plane-parallel Fabry-Perot resonator

that this system can be described by the Maxwell wave equation, which for the electric field strength $\underset{\sim}{E}$ has the form

$$\Delta E + \bar{k}^2 \underset{\sim}{E} = 0 \quad , \tag{2.3.1}$$

where $\bar{k}$ is the wave number. Since the medium in the resonator is homogeneous and energy losses are absent, we can write another equation for $\underset{\sim}{E}$,

$$\mathrm{div}\ \underset{\sim}{E} = 0 \quad . \tag{2.3.2}$$

For the boundary conditions we assume that at the surface of the mirrors the amplitude of E_x and E_y components is zero (ideal conductivity of the mirror material) and the electric field strength at the surface of the mirrors is also close to zero. The solution of (2.3.1,2) in this case is

$$E_x = A \cos(m\pi x/a) \sin(n\pi y/b) \sin(q\pi z/d) \quad ,$$

$$E_y = B \sin(m\pi x/a) \cos(n\pi y/b) \sin(q\pi z/d) \quad ,$$

$$E_z = C \sin(m\pi x/a) \sin(n\pi y/b) \cos(q\pi z/d) \quad , \tag{2.3.3}$$

where A, B, C are complex constants depending on the boundary conditions, the quantities m, n, q take the values of 0, 1, 2, ..., and the wave number $\bar{k}$ is

$$\bar{k} = \pi[(m/a)^2 + (n/b)^2 + (q/d)^2]^{\frac{1}{2}} \quad , \tag{2.3.4}$$

so that the resonant frequencies of oscillations for the given resonator are

$$\nu_{mnq} = (c/2)[(m/a)^2 + (n/b)^2 + (q/d)^2]^{\frac{1}{2}} \quad . \tag{2.3.5}$$

Thus, to each set of numbers m, n, q there corresponds a definite frequency ν_{mnq} and a specific pattern of distribution of electromagnetic waves in the resonator. The pattern corresponding to a given set of values of m, n, q is called the *mode of oscillations*.

According to (2.3.3), the value of q is equal to the number of half-waves which can fit into the distance d along the z axis in the resonator, that is,

$$q = 2d/\lambda \quad . \tag{2.3.6}$$

For typical resonators (d is of the order of tens of centimeters) and for the optical wavelength range the values of q are very large ($q \sim 10^6$), that is, we always have $q \gg 1$. If in this case $m = n = 0$, then all such modes with different q are called *longitudinal*, or *axial*. If $m \neq 0$ and/or $n \neq 0$, the modes are called transverse, or *angular*. The field distribution for longitudinal modes in an optical cavity can be easily understood if we compare it with the pattern of standing waves on a string, water surface, and so on. For transverse modes the pattern is not so simple. Some idea of it is given in Fig.2.7, where field configurations in a certain cross section of a rectangular resonator are given for different values of m and n.

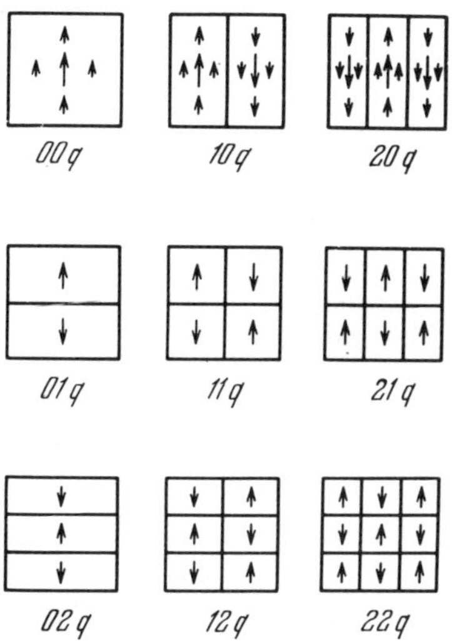

Fig.2.7. Field configuration in a plane-parallel Fabry-Perot resonator with rectangular mirrors

Since the value of q is very high, in developing resonators for the optical range we can hope that a resonance frequency corresponding to the frequency of transitions in an active medium can always be found. This results from the fact that the frequency difference between the two neighboring longitudinal modes is small compared to the fundamental frequency of oscillations. Indeed, from (2.3.5) it follows that

$$\Delta\nu = \nu_{00q+1} - \nu_{00q} = c/2d \quad . \tag{2.3.7}$$

For a numerical example take d = 50 cm; then $\Delta\nu$ = 300 MHz, and for λ = 10 µm we get $\Delta\nu/\nu = 10^{-5}$. Note that

$$\Delta\nu/\nu = 1/q \quad . \tag{2.3.8}$$

The difference between resonant frequencies for transverse modes is much less than for longitudinal modes. As the resonant frequencies are very close, oscillations in the cavity can appear at several frequencies simultaneously. Such operating conditions are called *multimode* as distinct from *single-mode* operation, when only one frequency is generated. Sometimes single-mode laser performance is preferable, since in this case oscillations are more stable and other characteristics are improved (higher coherence, lower divergence of the beam is observed when a laser is operating on a longitudinal mode, etc.).

In real resonators the wave pattern of radiation intensity distribution is in principle the same as in the case of an ideal resonator, but it is more complicated. Here different modes of oscillations are interrelated, so that the energy is pumped from one mode to another. The rate of this pumping depends on the manufacturing quality of a resonator and appears in the form of oscillation pulsations at a certain mode. If we take into account nonhomogeneity and motion of the active medium, the analysis of wave processes in the resonator becomes still more complicated. The mode distribution in the resonator in this case is considerably deformed.

Many difficulties can be avoided by using the geometrical optics approximations, that is, we assume that the wave front is locally flat and the amplitude and the direction of propagation of the wave change little over distances of the order of a wavelength. In this case the wave fronts can be treated as nearly straight lines, the tangents to which coincide with the direction of wave propagation [2.14]. This makes it possible to obtain sufficiently simple but important relations describing resonator performance.

Let us introduce the concept of a *Q factor*, which characterizes the design and manufacturing quality, as well as optical properties of a medium in a resonator. Qualitatively the Q factor is equal to the ratio of the electromagnetic energy stored in a system to the average energy lost in the resonator during a period of oscillations. More pictorially the rate of energy losses can be characterized by the *damping time* τ_d, that is, the time during which the energy (or intensity) stored in a resonator is decreased by a factor of e,

$$I(t) = I(0)\exp(-t/\tau_d) \quad . \tag{2.3.9}$$

Usually in resonators the damping is small. Then introducting the period $T = 1/2\pi\nu = 1/\omega$, we obtain

$$Q = \frac{I(0)}{I(0) - I(T)} = \frac{1}{1 - \exp(-T/\tau_d)} \approx \frac{\tau_d}{T} = \omega\tau_d \quad , \qquad (2.3.10)$$

since for small damping $T/\tau_d \ll 1$.

The quantity τ_d is often expressed in terms of the so-called *loss coefficient* α. The product αI characterizes the decrease in oscillation intensity per pass of a wave in a resonator from one mirror to the other, including reflection. The time of one such pass equals d/c. Then, using (2.3.9) for one pass, we obtain (again for small damping) $\tau_d = d/c\alpha$. Consequently, the Q factor can be written in the form

$$Q = \omega d/c\alpha \quad \text{or} \quad Q = 2\pi d/\alpha\lambda \quad . \qquad (2.3.11)$$

Hence Q factors of typical resonators are sufficiently large (for example, $Q \approx 3 \times 10^7$ for d= 50 cm, λ = 10 µm when α = 1%).

Energy in a resonator is lost owing to absorption of light during reflection, diffraction at mirror edges, and dispersion and absorption of light in a medium in the resonator and on equipment parts inside the resonator; there are also losses due to geometrical imperfections of resonators.

The losses upon reflection of light from the mirrors are accounted for by the reflection coefficient r; this part of the loss coefficient is

$$\alpha_r = 1 - r \quad . \qquad (2.3.12)$$

It is more difficult to take account of the losses due to diffraction at the mirror edges. As a result of diffraction, some part of the light waves that pass near the mirror edges propagates both inside the cavity and outside of it. The escape of radiation from the resonator leads to losses. Quantitative estimates of the diffraction losses are based on the concept of the Fresnel number, which is equal to the number of Fresnel zones seen in one mirror from the center of the other (see, for example, [2.15,16]). The Fresnel number $N_F = a^2/4\lambda d$, where a is the size (diameter) of a mirror and d is the distance between the mirrors. As the Fresnel number increases (with the increase in the mirror diameter and decrease in the resonator width), diffraction losses decrease. For a plane-parallel Fabry-Perot resonator we can approximately take

$$\alpha_d \approx 1/N_F = 4\lambda d/a^2 \quad . \qquad (2.3.13)$$

For $a = 5$ cm, $d = 50$ cm, $\lambda = 10\ \mu m$, the Fresnel number $N_F = 125$ and $\alpha_d = 8 \times 10^{-3}$, which is much less than the losses due to reflection.

Dispersion losses of light in a medium filling the resonator are very important. In gasdynamic lasers a gas flows in a cavity, which results in the appearance of optical inhomogeneities associated with shock waves, boundary layers, flow turbulence, dust content, etc. (see Sects.2.5 and 4.4).

Laser radiation lines have a finite width. The width of these lines $\Delta\nu$ is related to the damping time τ_d by the expression similar to (2.1.6),

$$\Delta\nu = 1/2\pi\tau_d \qquad (2.3.14)$$

or

$$\Delta\nu = \nu/Q \quad . \qquad (2.3.15)$$

Since the Q factor of a resonator is large, the width of a laser line is very small, much smaller than a gas line width.

Resonant frequencies of resonator oscillations determine the radiation spectrum. To obtain laser oscillations, however, it is essential that an active medium should spontaneously emit radiation at this frequency (to initiate oscillations) and that the gain coefficient for these frequencies should exceed losses. These conditions are fulfilled only for frequencies coinciding with radiation frequencies of the gas. This is illustrated by Fig.2.8, where the solid bell-shaped line depicts the shape of a gas spectral line, while dash horizontal lines show the loss levels for different modes of oscillations in a resonator. Sometimes several lines corresponding to longitudinal modes can fit into the width of the spectral line, when the frequency difference, according to (2.3.7), is less than the gas spectral line width. The lines of transverse modes are satellites of the longitudinal mode lines; the losses corresponding to them are usually higher. Vertical lines in Fig.2.8 correspond to spectral lines for which generation conditions are fulfilled. It is precisely the frequencies at which laser radiation occurs (Fig.2.8, lower diagram). Thus, the laser radiation spectrum is composed of a definite number of very narrow lines; the number of the lines is determined by the relation between the frequency difference for various resonator modes and gas line width, and also by the loss level for different modes.[8]

8 More complicated questions related to interaction between an active medium and a field in a resonator (for example, frequency pulling into a resonator, "burnout" of a spectral line, etc.) are beyond the scope of this book (see, for example, [1.14, 2.17]).

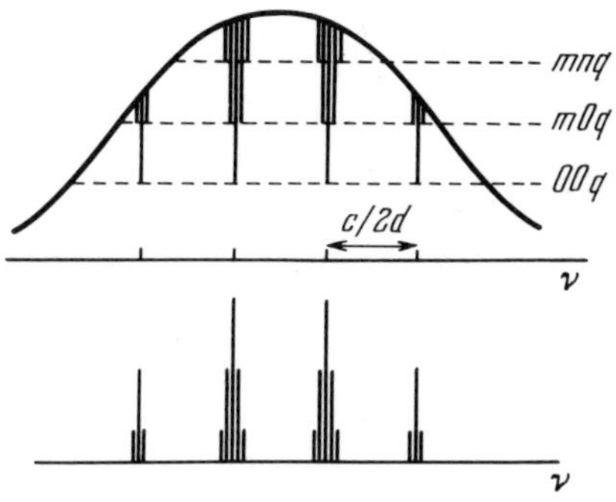

Fig.2.8.
Multimode oscillation spectrum

The mode selection in resonators provides for conditions in which radiation takes place only on quite definite modes or on a single mode. Methods of selection of longitudinal modes include the selection of the active medium with respect to composition, temperature, and gas pressure, so that spectral lines are sufficiently narrow. If the resonator width d is small, then, according to (2.3.7), the conditions of oscillations at only one longitudinal mode are met. This is observed for CO_2 lasers with low pressure and not very large d. Sometimes an induced rearrangement of the longitudinal mode pattern in a resonator is used; in this case absorbing cells are mounted across the resonator axis. The transverse mode selection in a resonator is achieved by mounting into it diaphragms with a central aperture. These procedures, however, decrease the output power and can hardly be used in gasdynamic lasers. The most expedient method of mode selection in gasdynamic lasers is the utilization of unstable and multipass resonators (see Sect.2.4). It is clear that other conditions being equal, the laser output power in single-mode operating conditions is lower than in multimode performance.

Thus, if the gain in a cavity exceeds all the energy losses, an amplifier becomes an oscillator. This enables us to formulate the *self-oscillation*, or *threshold condition* for laser radiation. In the framework of usual geometrical optics this can be conveniently related to the gain coefficient. Let us equate the increase in the radiation intensity per pass in a resonator of width d, which for the $(s+1)^{th}$ pass is

$$I_{s+1} - I_s = [\exp(kd) - 1]I_s \tag{2.3.16}$$

to the decrease in intensity due to losses, that is, αI_s. Then the self-oscillation relation becomes

$$\exp(kd) - 1 > \alpha \quad . \tag{2.3.17}$$

Usually near the threshold the gain is low; thus the threshold relation of oscillation becomes

$$k > \alpha/d \quad . \tag{2.3.18}$$

If the losses due to reflection play a major role, then this relation is

$$k > (1 - r)/d \quad . \tag{2.3.19}$$

When oscillations are initiated, the gain in each pass exceeds losses, and therefore the radiation intensity increases. This increase is limited by the energy which is transformed into electromagnetic oscillations. This limit is a result of a change in the state of an active medium in the cavity. Indeed, the increase in the radiation intensity is connected with the increase in the number of atoms which transfer from the upper to the lower laser level. This results in the population inversion decrease and in the drop of the gain coefficient. If this decrease is compensated for by some source of pumping the atoms to the upper level and removing them from the lower level, then a steady-state regime of laser performance sets in. Thus, the gain coefficient in an active medium is at maximum for a very low intensity of the amplified wave ("zero" intensity). In this case it is called the *small-signal gain*; let us denote it by k_0. The value of k_0 does not depend on the radiation intensity. The dependence $k(I)$ is the most important characteristic of an active medium. The idea of this dependence can be derived by considering an isotropic stationary active medium with a constant pumping rate and a uniform broadening of lines (when collisional broadening predominates). Then, according to Appendix C,

$$k = \frac{k_0}{1 + I/\bar{I}} \tag{2.3.20}$$

where $\bar{I}$ is a saturation parameter, equal to the radiation intensity for which $k = k_0/2$ [see (C.6)].

Quantitative characteristics of a steady-state laser intensity can be represented in different forms. One of them relates the consecutive values of radiation intensity before and after the sth pass of the beam in a resonator. In saturation the gain is compensated for by the losses, that is,

$$(1 - \alpha)I_s = I_{s-1} \quad \text{or} \quad (1 - \alpha)\exp(k_* d) = 1 \quad , \tag{2.3.21}$$

where, as before, k_* is the saturated gain coefficient. In addition to losses, let us take account of the useful output of radiation from a cavity. Let the mirrors of the Fabry-Perot resonator be semitransparent with the transmission factor t. Then for reflection of the beam from the surface of each mirror the equality $r + t + \alpha = 1$ holds, and (2.3.21) is replaced by $(1 - \alpha - t)\exp(k_* d) = 1$. In this case often two consecutive passes are considered, that is, the beam returns to the starting point with the initial propagation direction. Then

$$(1 - \alpha_1 - t_1)(1 - \alpha_2 - t_2)\exp(2k_* d) = 1 \quad ,$$

where α_1, t_1, α_2, t_2 are loss coefficients and transmission factors for the first and the second mirror, respectively. This relation can be written in the form

$$r_1 r_2 \exp(2k_* d) = 1 \quad , \tag{2.3.22}$$

where r_1, r_2 are reflection coefficients for the first and the second mirror. Usually radiation is extracted from a resonator only through one mirror. For simplicity we can calculate losses for a double pass, assuming that $\alpha' = 2\alpha$. Then to a good approximation we obtain

$$(1 - \alpha' - t)\exp(2k_* d) = 1 \quad . \tag{2.3.23}$$

If, as usual, $(\alpha' + t) \ll 1$, then

$$k_* = (\alpha' + t)/2d \quad . \tag{2.3.24}$$

Equations (2.3.21-24) correspond to a local steady-state laser intensity. Constraints on the gain coefficient k_* are imposed by the properties of the medium in which the beam travels and on the mirror quality at points of reflection. In contrast to this we can consider steady-state conditions which characterize the processes in some volume of a resonator. To do this we write the balance equation for a number of photons N in the volume V of the resonator. The change in the number of photons with time is determined by the rate Φ of appearance or disappearance of photons as a result of induced radiation and light absorption in a gas and by the rate of variation of the number of photons due to losses $(dN/dt)_\ell$,

$$\frac{dN}{dt} = \Phi + \left(\frac{dN}{dt}\right)_\ell \quad . \tag{2.3.25}$$

According to (2.1.8,9), we have

$$\Phi = \int (z_{mn}^{in} - z_{nm}^{in})dV = \iint [b_{mn}(\nu)N_m - b_{nm}(\nu)N_n]\rho_\nu d\nu dV \quad ,$$

where the integration is carried out over volume V and over laser line width. Using (2.2.3) and taking into account, while integrating over ν, that the gain coefficient changes little within the laser line width, we obtain, passing over to radiation intensity $I = \rho c$ along the optical axis of the resonator,

$$\Phi = (1/h\nu) \int kIdV \quad .$$

According to (2.3.9,10), $(dN/dt)_\ell = -N\omega/Q$, where Q is the Q factor of the resonator, and the number of photons N is

$$N = (1/h\nu c) \int IdV \quad .$$

For a steady-state radiation regime in (2.3.25), $dN/dt = 0$ and $k = k_*$; thus, bearing in mind that $\omega = 2\pi c/\lambda$, we obtain

$$\int k_* IdV = \frac{2\pi}{\lambda Q} \int IdV \quad . \tag{2.3.26}$$

If the intensity I changes little along the optical axis of the resonator (see Sect.2.5), from (2.3.26) it follows that

$$\int k_* IdS = \frac{2\pi}{\lambda Q} \int IdS \quad , \tag{2.3.27}$$

where S is the area of the mirror in the resonator. The radiation intensity is sometimes assumed constant over the entire volume of the resonator. Then, introducing the loss coefficient α per pass and using (2.3.11), we obtain from (2.3.27) the distribution of k_* in the resonator,

$$(1/S) \int k_* dS = \alpha/d \quad . \tag{2.3.28}$$

A similar equation for a steady-state radiation regime is obtained after integrating (2.3.22) over the mirror area S (see Sect.2.5),

$$\int [\ln(r_1 r_2) - 2k_* d] dS = 0 \quad . \tag{2.3.29}$$

These integral equations are used in calculations [2.18-20]. Equations (2.3.21-24) or (2.3.6-29) together with equations describing the saturation state of an active medium make it possible to determine k_* and I. The theory presented here does not take into account wave processes occurring in a resonator and their effect on radiation intensity distribution. Nevertheless, even such a simplified approach enables one to solve fundamental problems of computation of GDL radiation power.

One of the important laser characteristics is the radiation power extracted from the cavity, that is, the output power P. To determine P let us take into account the fact that radiation intensity in a resonator is the sum of radiation propagating in opposite directions from one mirror to the other: $I = I^+ + I^-$ (see Sect.2.5). If there is only one mirror of area S transmitting light, then only the radiation intensity corresponding to one (say, negative) direction of propagation contributes to the output power. In this case the intensity of extracted radiation is $I_t = tI^-$, and its power is

$$P = \int I_t dS = t \int I^- dS \quad ,$$

where I^- is measured near the surface of the mirror. Upon reflection $I^+ = rI^-$; consequently,

$$P = \frac{t}{1 + r} \int IdS \quad . \tag{2.3.30}$$

This relation holds for $r \approx 1$. A more general formula for the output power is given in Sect.2.5.

Note that the energy W transferred into radiation per unit resonator volume per unit time is

$$W = \int (z^{in}_{mn} - z^{in}_{nm}) h\nu d\nu = \int (b_{mn}N_m - b_{nm}N)\rho_\nu h\nu d\nu \quad ,$$

or, with the account of (2.1.10, 2.3) and the fact that a laser radiation line is narrow, we obtain

$$W = kI \quad . \tag{2.3.31}$$

The computation of radiation power can be carried out only by specifying the properties of an active medium filling a resonator.

For an isotropic and nonmoving active medium in a resonator with $\alpha' + t \ll 1$ and $r \approx 1$, we obtain a simple expression for the output power extracted from the resonator [2.21,22]. Thus, for uniform broadening, by equating expressions for k_* from (2.3.20) and (2.3.24), we arrive at

$$P = \frac{ab\bar{I}}{2} t \left(\frac{2k_0 d}{\alpha' + t} - 1 \right) \quad , \tag{2.3.32}$$

where a and b, as before, are the length and the height of the mirror, respectively, and $\bar{I}$ is determined according to (C.6). It is clear from (2.3.32) that the extracted power attains its maximum at a certain optimum value of transmission factor t_0. This value can be obtained, as usual, by equating to zero the derivative dP/dt, so that

$$t_0 = \sqrt{2\alpha' k_0 d} - \alpha' \quad . \tag{2.3.33}$$

Substituting (2.3.33) into (2.3.32), we find that for optimum t_0 and uniform broadening, P is proportional to $(1 - \sqrt{\alpha'/2k_0d})^2$.

Another important quantitative characteristic of a resonator is its *efficiency* η_r, which can be defined as the ratio of the energy actually obtained and extracted from the cavity to the total energy which can be transferred into radiation. In the simple case under consideration, efficiency η_r can be calculated from the ratio of power P and the product WV, where W is the energy per unit resonator volume according to (2.3.31) for a maximum gain coefficient ($k \to k_0$) and V is the resonator volume ($V = abd$). Then for the same saturation parameter $\bar{I}$ we get

$$\eta_r = \frac{t}{2k_0d}\left(\frac{2k_0d}{\alpha' + t} - 1\right) \quad , \tag{2.3.34}$$

or, for the optimum transmission factor t_0 (2.3.33), recalculating for losses for a single pass, we obtain

$$\eta_r = (1 - \sqrt{\alpha/k_0d})^2 \quad . \tag{2.3.35}$$

Even in an ideal resonator (with $\eta_r = 1$) it is impossible to transform into radiation the entire energy of molecular excitation. Some molecules populate the upper level m in the absence of inversion ($g_nN_m \leqq g_mN_n$), then it is impossible to obtain laser radiation even with zero resonator losses. By decreasing the population N_n of the lower level, we find that laser intensity is limited by the so-called *quantum efficiency* η_q. As is clear from Fig.2.5, η_q is equal to the ratio of the energy of the $m \to n$ transition to the energy required to deliver particles from the ground level 1, where the particles were before pumping, to the level m,

$$\eta_q = \frac{E_m - E_n}{E_m - E_1} \quad . \tag{2.3.36}$$

The total laser efficiency is not only composed of η_q and η_r, but also depends on the effectiveness of pumping and creating inversion. The analysis of the GDL efficiency is carried out in Sect.5.6 for a CO_2 laser.

To extract radiation from the Fabry-Perot resonator and stable resonators, sometimes use is made of opaque mirrors with holes through which radiation can pass. Diffraction of light waves on the edges of these holes provides the filling of the holes with radiation. Therefore the hole diameter must be small to make the filling complete and to prevent noticeable distortions introduced by the holes into the intensity distribution in the resonator.

For this reason the Fresnel number N_F' for each hole must be sufficiently small. Usually it is assumed that $N_F' = (\delta^2/4\lambda d) \leqq 1$, where δ is the hole diameter. Thus for resonators of ordinary size, hole diameters should not exceed 1 mm. If the holes are distributed uniformly over the mirror surface, the transmission factor of such a mirror is $t = nS_\delta/S$, where n is the number of holes, S_δ is the area of one hole, and S is the surface area of the mirror. Since the gas pressure of the surrounding medium p_{ex} and the resonator pressure p_r are different, the mirrors are usually covered from outside by transparent windows to prevent the gas from flowing through the holes.

At very high values of power (3 kW/cm^2 and higher) the radiation extraction through the mirror is hindered because of overheating and destruction of the mirror material as a result of absorption in it of even a small part of radiation power [2.23]. In this case, aerodynamic windows of high transparency are used to extract radiation. In these windows, to prevent the gas from flowing into the cavity, a gas stream is moving across the beam at a high velocity, thus creating a "gas screen" and maintaining the required pressure difference between p_{ex} and p_r using a system of shock waves or expansion waves (Fig.2.9). The aperture of these windows can reach tens of centimeters with an output beam deviation only of several degrees (see [1.2, 2.23,24]). Owing to the high transparency, aerodynamic windows are expedient in the schemes with unstable resonators and multipass amplifiers (see Sect.2.4).

Apart from the radiation power, one of the fundamental characteristics of laser performance is the quality of field distribution for radiation extracted from a cavity, since it is important not only to extract high power but also to supply it in the form required for a consumer, that is, laser *beam quality* must be high. A detailed discussion of this question is beyond the scope of this book. It should be noted that a laser radiation field is described differently for the regions close to a resonator exit (the near-field zone) and far from it (the far-field zone). This difference is connected with diffraction and is determined by the relation between the distance L from the resonator and the quantity $L_d = a^2/8\lambda$, where a is the effective diameter of the resonator exit. The inequality $L < L_d$ corresponds to the near-field zone of laser radiation (or to the Fresnel zone of microscopic diffraction), whereas for $L > L_d$ we have the far-field zone (the Fraunhofer zone of telescopic diffraction). In the analysis of the intensity distribution in the most important far-field zone, the waves can be considered plane when their superposition is taken into account, which considerably simplifies the problem (see [2.15,16]). In the best case the

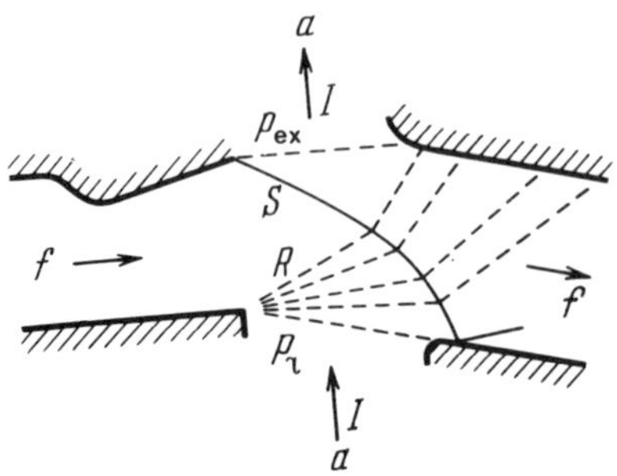

Fig.2.9. Schematic of a gasdynamic window for extracting high-energy laser radiation I from the cavity in the a-a direction across the f-f flow. (S) shock wave, (R) expansion waves

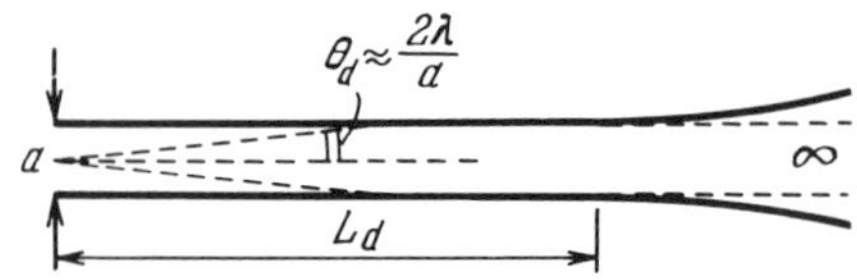

Fig.2.10. Far-field ($L > L_d$) and near-field ($L < L_d$) zones of laser beam; $L_d = a^2/8\lambda$. (a) effective exit diameter of the cavity; (λ) wavelength, (θ_d) diffraction-limited angle of divergence

divergence of a laser beam in the far-field zone is determined by the diffraction-limited angle $\theta_d \approx 2\lambda/a$ (Fig.2.10). Thus, the basic problem solved in developing laser systems is to "achieve the maximum output power in a beam with minimum divergence, limited only by diffraction" [2.25].

2.4 Unstable and Multipass Resonators

High-power gasdynamic as well as other types of lasers require a considerable volume of active medium and, hence, large resonators. The increase in the mirror size in the Fabry-Perot resonator results in a large number of transverse modes at the expense of quality of the laser beam. For this reason the so-called unstable resonators and multipass resonators and amplifiers are being used now in high-power GDLs. The concept of "stability" as applied to resonators should be understood from the point of view of geometrical optics, that is, in *stable* resonators the rays do not leave the cavity after repeated reflections from the mirrors, so that the resonator is stable with respect to conservation of light energy, and the output power is extracted through semitransparent mirrors. In contrast to this, in *unstable* resonators a beam after repeated reflections from the mirrors can eventually avoid the mirror and leave the resonator.

A stability relation can be obtained from the laws of geometrical optics. If we introduce the quantity

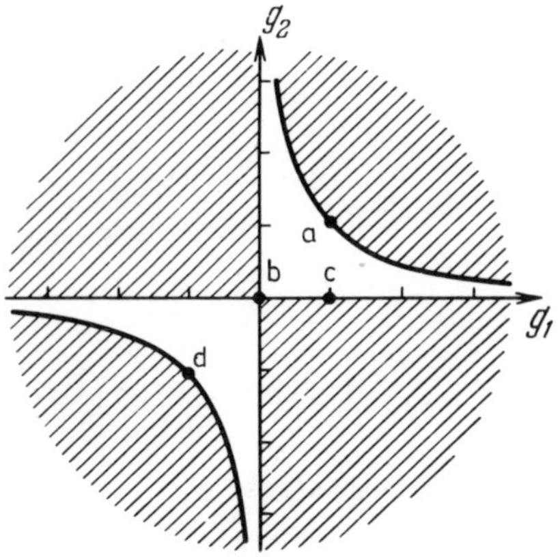

Fig.2.11

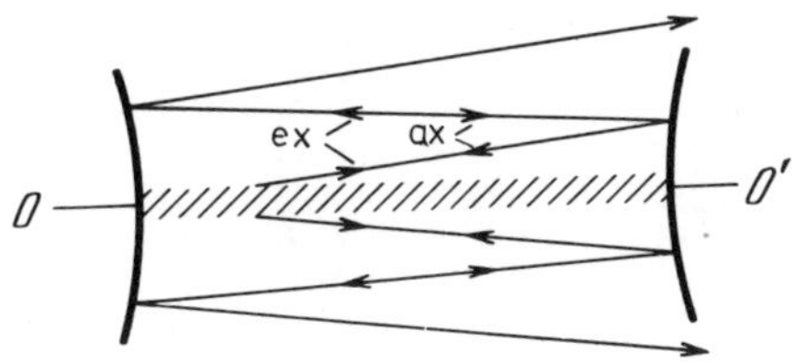

Fig.2.12

Fig.2.11. Diagram of oscillation stability in a resonator. (a) The plane-parallel Fabry-Perot resonator considered above ($R_1 = R_2 = \infty$, $g_1 = g_2 = 1$); (b) the confocal resonator, which consists of two spherical concave mirrors at a distance equal to the sum of their focal lengths. If the mirrors are identical, the distance between them is equal to the radius of the mirror curvature ($R_1 = R_2 = 0$, $g_1 = g_2 = 0$); (c) the semiconcentric resonator, that is, a system of a concave and a flat mirrors separated by a distance equal to the radius of curvature of the first mirror ($R_1 = \infty$, $R_2 = d$, $g_1 = 1$, $g_2 = 0$); (d) the concentric resonator with two concave mirrors of equal curvature, separated by a distance of a double the radius of curvature ($R_1 = R_2 = d/2$, $g_1 = g_2 = -1$)

Fig.2.12. Paths of the rays in an unstable resonator—to the axis (ax) and to the resonator exit (ex)

$$g_1 = 1 - d/R_1 \quad , \quad g_2 = 1 - d/R_2 \quad , \tag{2.4.1}$$

where d is the distance between the mirrors, R_1 and R_2 are the radii of curvature of the mirrors (the mirrors can be convex, that is, have negative radius of curvature), then the resonator is stable when

$$0 \leqq g_1 g_2 \leqq 1 \quad . \tag{2.4.2}$$

Relation (2.4.2) makes it possible to draw the so-called stability diagram (Fig.2.11), which shows instability zones (hatched) and stability zones (remaining parts of the plane). Some simplest types of resonators operate at the boundary between stable and unstable zones; these cases are depicted on the diagram.

If the diameter (or diagonal) a of a mirror is much larger than the resonator width d, so that the Fresnel number $N_F = a^2/4\lambda d$ is close to unity, then almost the entire energy of oscillations is concentrated in the ground

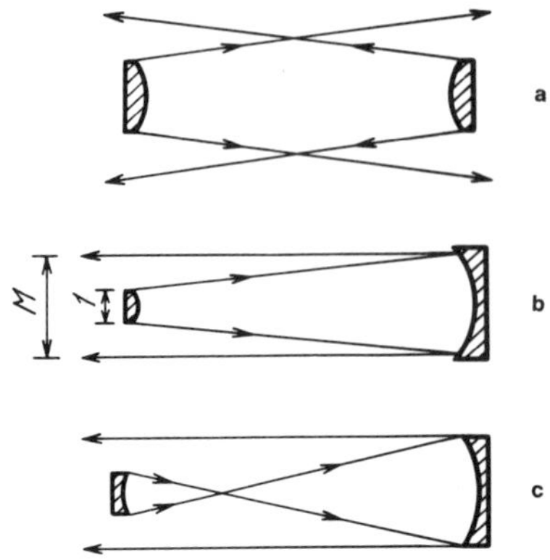

Fig.2.13a-c.
Types of unstable resonators

longitudinal mode. In stable resonators N_F grows with an increase of the mirror size, and an increasing part of energy is delivered to the higher-order modes (transverse modes), while the ground mode possesses only a small part of energy, approximately equal to $1/N_F$. The appearance of intensive transverse modes make the laser beam quality worse. For $N_F \approx 1$ diffraction effects are small, and all the modes in a resonator are weakly coupled, which is unfavorable for coherence and stability of oscillations. In an unstable resonator the rays propagating in the ax direction (Fig.2.12) are concentrated in the narrow central region, which in Fig.2.12 is shown by hatching. In this region $N_F \approx 1$, therefore practically single-mode oscillations appear. As a result of diffraction, some part of the rays deviates from the axis oo' and, propagating in the direction ex, covers the entire resonator volume. Their intensity constantly increases, and mode composition changes slightly. Eventually these rays leave the resonator, so that radiation output from unstable resonators is determined by diffraction losses. Consequently, a considerable advantage of unstable resonators is a large volume of the active medium occupied by a single mode, appropriate suppression of other modes, and a convenient extraction of energy, which makes it possible to utilize only reflecting mirrors [2.25-27].

The most typical examples of unstable resonators are given in Fig.2.13. In case a) a resonator is symmetrical, and radiation output is the same at both sides of the resonator. In practice, however, it is more convenient to use resonators with extraction of radiation in one direction [b) and c)]. Schemes b) and c) with confocal convex and concave mirrors provide for collimation of the output beam and are called *telescopic*. In case b) $g_1g_2 > 1$ (the so-called unstable resonators of the positive branch), and in case c) $g_1g_2 < 0$ (the unstable resonators of the negative branch). In the latter case the beam is focussed inside the resonator, which for a high-power radiation intensity can lead to increase in the active medium inhomogeneity and even to optical breakdown. For this reason unstable resonators of the positive branch are preferable.

The basic quantitative characteristic of an unstable resonator is the *linear magnification* M of the diameter of a light beam when it performs round trips from one mirror to the other (Fig.2.13b, where $M = R_2/R_1$). During each pass a definite part of energy escapes from the cavity,

$$t = 1 - (1/M)^n \quad , \tag{2.4.3}$$

where n = 1 for cylindrical mirrors and n = 2 for spherical mirrors. This quantity is similar to the transmission factor for stable resonators. After solving the Maxwell equation for unstable resonators, we obtain the value of t somewhat lower than it follows from (2.4.3), for example, 35 to 40% instead of 50% [2.28]. The distribution of radiation field in a resonator and parameters of the output beam to a good approximation agree with the theoretical results obtained using geometrical optics approximations [2.26, 27,29]. This considerably simplifies the analysis of laser operating conditions and determining the output power extracted from an unstable resonator.

The physical nature of basic processes (excitation of laser oscillations, saturation effects, etc.) in unstable resonator remains the same as in stable ones: oscillations are also excited when the gain in a pass exceeds losses, and a steady-state level of radiation intensity corresponds to the balance of the energy transferred into radiation and the energy leaving the cavity owing to power extraction and losses. Quantitative relations for unstable resonators are similar to the relations derived in the previous section; it must only be taken into account that the radiation intensity decreases by a factor of M^n during each pass in a resonator. Thus the threshold relation for laser oscillations now can be written as

$$k > (\ln M^n + \ln\alpha)/d \quad , \tag{2.4.4}$$

where α is the loss coefficient. The steady-state radiation regime in an unstable resonator can be achieved only in the central region (hatched in Fig.2.12). It is natural to formulate it in the local representation; then, by analogy with (2.3.22), for a homogeneous active medium we obtain

$$r_1 r_2 \exp[2k_*(0)d]/M^n = 1 \quad , \tag{2.4.5}$$

where $k_*(0)$ corresponds to the saturation gain coefficient at the resonator axis. Other parts of the active medium volume operate in the amplification regime.

The output power P can be calculated with the help of transfer equations for radiation density ρ, which take into account the increase in ρ during each sth pass from one mirror to the other and in the reverse direction,

$$\rho(s+1) = F(s)\rho(s) \quad , \tag{2.4.6}$$

where in the simplest case of constant k(s) for the sth pass we have

$$F(s) = r_1 r_2[\exp 2k(s)d]/M^n \quad . \tag{2.4.7}$$

Such a chain of equations can be conveniently solved by iterative methods, beginning with $\rho(0)$ at the resonator axis. The principal difficulty in solving this problem consists in determining $\rho(0)$ and k(s), since these quantities are results of interaction of an active medium and a radiation field. This dependence is determined by specific properties of the medium and the method of pumping [2.30,31]. Despite the difference in the processes in stable and unstable resonators, the optimum radiation output and, hence, the optimum amplification M in an unstable resonator, as well as the maximum efficiency η_r (with the mean values of k_0 and α), can be assessed using (2.3.34,35) [2.30], which were obtained for the plane-parallel Fabry-Perot resonator. Equations (2.3.26-29) can be used to estimate the integral equation for the steady-state radiation regime in an unstable resonator if the magnification M is small [2.18].

The field distribution over the cross section of the beam leaving an unstable resonator is also described well in the geometrical optics approximation [2.27]. In this case, in the far-field zone, this distribution corresponds to the Fraunhofer diffraction pattern obtained with uniform illumination of the ring (in the case of a circular output mirror) or two slits (for a rectangular mirror). The required concentration of energy in the central region is ensured when this ring (or slits) is sufficiently wide and diffraction effects are small. Hence, from this point of view, it is desirable to have unstable resonators with the highest possible value of M. Many questions related to the choice of the type and parameters of unstable resonators are treated in [2.26].

The highest values of extracted power are obtained in GDLs with multipass resonators (Fig.2.14). By using a multiple light beam reflection, we can achieve a considerable decrease in the Fresnel number (to $N_F \sim 1$) and thus obtain good characteristics of a laser device without decreasing an active volume of a resonator. Such resonators can operate both in stable (Fig.2.14a) and unstable (Fig.2.14b) conditions. To calculate the power of a multipass resonator, use should be made of the equations given in the preceding section

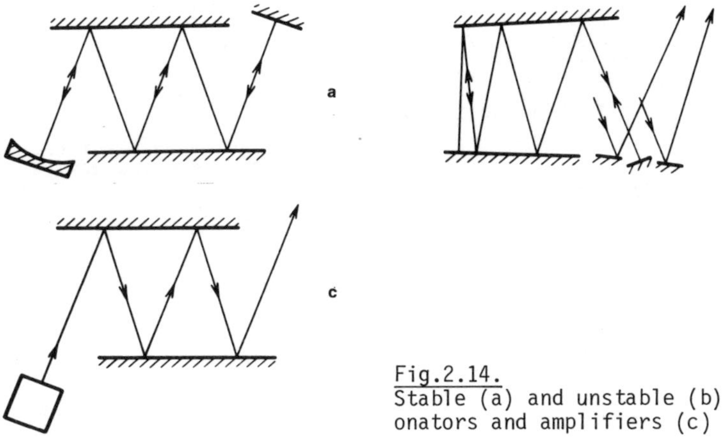

Fig.2.14.
Stable (a) and unstable (b) multipass resonators and amplifiers (c)

(stable resonators) and in this section (unstable resonators) taking account of the repeated reflection of radiation by the mirrors. The optimal number of passes is determined by operating conditions and active medium properties.

To obtain a high power of laser radiation, multipass *amplifiers* (MOPA) are used (Fig.2.14c), which are supplied with a master oscillator. The feedback between amplified radiation and input parameters of the amplifier is considerably weakened. The advantages of multipass amplifiers include the control of intensity, frequency, mode composition, and laser beam quality. An amplification regime close to self-oscillation provides for frequency synchronization and, hence, makes it possible to vary, within certain limits, the spectral composition of radiation. For this purpose a small external signal of appropriate frequency from the master oscillator is introduced into the cavity, which provides for oscillations at a required mode.

If the flat mirror in the Fabry-Perot resonator is inclined, the resonator becomes either unstable or turns into a multipass amplifier, depending on a state of the medium and the input signal. It is not difficult to compute the number of passes in such a resonator with the mirror length a and the angle on inclination β: upon each reflection the angle between the light beam and the axis of the system increases by 2β, and the beam is shifted from one edge of the mirror first by $d \tan 2\beta \approx 2\beta d$, and then by $6\beta d$, $10\beta d$, and so on, so that the total number of passes is $2\sqrt{2a/\beta d}$. It should be noted that in a similar way we can compute the losses in the Fabry-Perot resonator, connected with nonparallelism of the mirrors [2.32].

The computation of the power extracted from a multipass amplifier can be carried out using the system of equations (2.4.6,7) and assuming that M = 1.

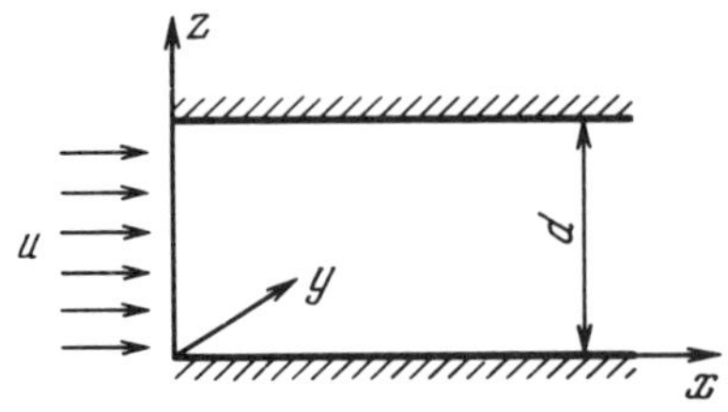

Fig.2.15. Schematic gas flow in the Fabry-Perot resonator. (d) flow width; (u) flow velocity

In high-power gasdynamic lasers combined schemes with multipass resonators and amplifiers are utilized to increase the effectiveness of extracting energy from an active medium (see, for example, [2.33]).

2.5 Active Medium Flow in a Cavity

Until now we have assumed that an active medium in a cavity is homogeneous and nonmoving. In gasdynamic lasers, however, an active medium is blown through a cavity at a high velocity, which causes considerable perturbations of the resonator performance.

The fast flow of an active medium in a cavity, which was used before the invention of gasdynamic lasers [2.34], promoted a decrease of undesirable heating in a gas, effective increase of the active medium mass, and, hence, a considerable increase of the laser power. For resonator performance in these conditions it is not important by which method pumping is achieved and the active medium is created—either by a gasdynamic method with thermal excitation, in an electric discharge, or as a result of chemical reactions. It is important that the flow of active medium through the cavity should occur at a high (in GDLs, supersonic) speed. Fresh active medium is supplied to the cavity entrance (for instance, to the Fabry-Perot plane-parallel resonator, Fig.2.15) and is used up in the resonator flow, so that with a maximum resonator efficiency the inversion at the exit falls to zero. Thus, forced inhomogenity of the active medium downstream from the resonator flow is created. There are some other sources of inhomogeneity in a gas flow through a resonator: turbulent pulsations of gas density, local expansions and compressions, shock waves, etc. During a GDL performance liquid and solid condensate particles can appear in a flow, which degrade the active medium homogeneity. All these factors affect to a certain extent the laser power and phase characteristics of radiation.

We can assess the role of these factors, introducing considerable simplifications, which make it possible to solve the problem analytically or using simple numerical methods. One such simplification is the solution in geometrical optics approximations and the utilization of equations of energy transfer along the beams. Let laser radiation propagate from one mirror to the other across the gasdynamic flow (or, as in the case of unstable resonators, at a small angle to the optical z axis connecting the centers of the mirrors). At a steady-state regime for every cross section of a resonator it is always possible to write the equation for variation of the radiation intensity along the axis, when a light wave propagates to the forward (I^+) and reverse (I^-) directions

$$\frac{\partial I^+}{\partial z} = (\sigma_{mn}N_m - \sigma_{nm}N_n)I^+ - \beta I^+ \quad ,$$
$$-\frac{\partial I^-}{\partial z} = (\sigma_{mn}N_m - \sigma_{nm}N_n)I^- - \beta I^- \quad , \tag{2.5.1}$$

where σ_{mn}, σ_{nm} is the cross section of the induced optical transition $m \rightleftharpoons n$ (2.2.12); β is the loss coefficient in the active volume per unit length along the optical axis, which we shall call the attenuation factor. The first term on the right-hand side of (2.5.1) corresponds to the increase of intensity as a result of induced optical transitions and can be obtained from (2.3.31); the second term is associated with losses in the medium. For a moving medium the dependence of intensity on the x coordinate along the flow enters into (2.5.1) implicitly through the dependence of populations N_m, N_n and coefficients σ_{mn}, σ_{nm}, and β. The boundary conditions for (2.5.1) are the relations which determine the change of intensity on the mirror surface upon reflection

$$r_1 I^-(0) = I^+(0) \quad \text{and} \quad r_2 I^+(d) = I^-(d) \quad , \tag{2.5.2}$$

where, as before (see Sect.2.3), the quantity r_i is connected with light absorption in the mirror material and with radiation output from the cavity ($r_i = 1 - \alpha_i - t_i$). Write (2.5.1) in the form

$$\frac{1}{I^+}\frac{\partial I^+}{\partial z} = -\frac{1}{I^-}\frac{\partial I^-}{\partial z} = k - \beta \quad , \tag{2.5.3}$$

from which it follows that I^+I^- = constant (for a given x). Taking into account (2.5.2), we obtain

$$I^+(d)/I^-(0) = \sqrt{r_1/r_2} \quad . \tag{2.5.4}$$

Integrating (2.5.3) with respect to z, we can find, using (2.5.2), the mean value of $k - \beta$ across the flow in the given cross section x

$$(1/d) \int (k - \beta)dz = -\ln(r_1 r_2)/2d \quad .$$

The total radiation intensity at a certain point of the resonator is obtained, after integrating (2.5.3), in the form

$$I = I^+(0)\exp\left[\int_0^z (k - \beta)dz\right] + I^-(d)\exp\left[-\int_0^z (k - \beta)dz\right] \quad .$$

Proceeding from the last two expressions and (2.5.2,4), we can assess the radiation field inhomogeneity across the flow in the form of the ratio of the minimum (I_{min}) and maximum (I_{max}) values of intensity on the z axis [2.35]

$$I_{min}/I_{max} = 2\sqrt{r_m}/(1 + r_m) \quad , \qquad (2.5.5)$$

where r_m is the smaller of the values r_1 and r_2. From (2.5.5) it is clear that for $r_m > 0.5$ the intensity change across the flow vanishes.

Let us determine now the value of the output power without limiting ourselves, as in Sect.2.3, to the assumption that r_1, r_2 are close to unity and, as in (2.5.5), that the intensity I along the z axis varies little. Let the reflection losses in each mirror be similar ($\alpha_1 = \alpha_2 = \alpha$), and radiation leave the cavity only through mirror 1 at the point $z = 0$ (to the negative direction), so that $t_1 = t$, $t_2 = 0$, $r_1 = 1 - \alpha - t$, $r_2 = 1 - \alpha$. The mean (over the width d of the resonator) value of the product $(k - \beta)I$, which, according to (2.3.31), determines the laser power, is

$$\begin{aligned}\langle(k - \beta)I\rangle &= (1/d)\int_0^d (k - \beta)I dz \\ &= (1/d)[I^+(d) - I^+(0) - I^-(d) + I^-(0)] \\ &= I^-(0)(1 + \sqrt{r_1/r_2})(1 - \sqrt{r_2 r_2}) \quad ,\end{aligned}$$

where we have used (2.5.2-4). Then, bearing in mind that the intensity I_t of radiation emerging from the cavity is $I_t = tI^-(0)$, we obtain

$$P = \int_S I_t dS = \int_S \frac{t\langle(k - \beta)I\rangle d}{(1 + \sqrt{r_1/r_2})(1 - \sqrt{r_1 r_2})} dS \quad , \qquad (2.5.6)$$

where S is the area of the output mirror. This expression is simplified when reflection losses are small ($\alpha \ll 1$, $r_2 \approx 1$) and coefficients k and β are constant over the entire resonator volume. If $t < 0.5$, then, according to (2.5.5), I varies little along the z axis. Then, using (2.3.22) and putting $r_1 = r$, $r_2 = 1$, we arrive at

$$P = \frac{td}{\alpha + t}(k - \beta) \int_S I dS = \frac{t \ln(1/r)}{2(\alpha + t)} \int_S I dS \quad . \qquad (2.5.7)$$

For $r \approx 1$ this expression coincides with (2.3.30).

To analyze the state of an active medium flowing in a cavity, we must find the distribution of the gain coefficient k and, hence, the distribution of laser level populations in the flow, which in turn depends on the field intensity I. This dependence can be obtained using (2.3.31): the decrease in the population of the upper laser level N_m and similar increase in the population of the lower level N_n are equal to the number of optical transitions $m \rightarrow n$ and, according to (2.3.31), amount to $kI/h\nu$ (per unit resonator volume per unit time). We denote the rate of variation of population of these levels due to collisions by $(dN_i/dt)_c$. Then, with (2.2.12), we can write for the population balance in the flow

$$u \frac{dN_m}{dx} = \left(\frac{dN_m}{dt}\right)_c - (\sigma_{mn}N_m - \sigma_{nm}N_n)\frac{I}{h\nu} \quad ,$$

$$u \frac{dN_n}{dx} = \left(\frac{dN_n}{dt}\right)_c + (\sigma_{mn}N_m - \sigma_{nm}N_n)\frac{I}{h\nu} \quad , \qquad (2.5.8)$$

where u is the flow velocity and $I = I^+ + I^-$. The system of (2.5.1,8) can be used to solve for the steady-state laser radiation for an active medium flow in geometrical optics approximation. The explicit expression for the first term on the right-hand side of (2.5.8), which determines the change of N_m and N_n upon collisions, can be obtained from the analysis of a specific mechanism of the physico-chemical processes in a gas. Examples of such analysis can be found in [2.35-39].

According to (2.5.5), for $r \approx 1$ the intensity I across the flow changes little, which permits us without solving (2.5.1) to assume with a good approximation that all the quantities are independent of the z coordinate. Then we must supplement (2.5.8) with a certain relation for a steady-state regime. Using integral equations (2.3.26-29), the intensity is often assumed either to be constant over the entire resonator volume or a different distribution of I is taken, proceeding from experimental results or from the analysis in the framework of the wave approximation. If local equations in the form (2.3.22-24) are used for describing a steady-state radiation regime, then in the analysis of processes in the Fabry-Perot resonator a peculiarity appears at the resonator entrance, connected with the change in the gain coefficient from k_0 (before the resonator) to k_* (in the resonator region). In resonators with sharp mirror edges this transition occurs very rapidly (owing to a high Q factor of the resonator) and is, in essence, a jump of intensity I, populations N_m and N_n, and coefficient k. The boundary conditions for such a jump can be easily obtained if we assume that in this case the populations N_m and N_n change only as a result of optical transitions $m \rightleftharpoons n$, and therefore the sum $N_m + N_n$ remains constant

$$N_m + N_n = \text{const.} \quad . \tag{2.5.9}$$

The value of k_* behind this jump is determined by (2.3.22) for known values of r_1 and r_2. Then for the state of the active medium at the beginning of the zone of steady-state oscillations we have

$$\sigma_{mn}N_m - \sigma_{nm}N_n = k_* \quad , \tag{2.5.10}$$

which together with (2.5.9) makes it possible to obtain the boundary conditions for system (2.5.8). The method described here does not take into account the part of the energy liberated in the transient zone of the resonator in the total energy balance [2.35]. To eliminate this shortcoming we must consider the continuous transition $k_0 \rightarrow k_*$. This can be done, for example, either by assuming artificially that the values of r_1 and r_2 in this region are variable [2.40] or using a kinetic equation for electromagnetic energy density for resonator excitation (see [2.1,41]).

Let us consider now an active medium flow through an unstable resonator of the telescopic type (Fig.2.16). Remaining within the framework of the geometrical optics approximation (see [2.30,42]) and proceeding from (2.4, 6,7), we can write the latter of these equations in the form

$$F(s) = r_1 r_2 \exp\left[\int_{\ell(s)} k(z)dz\right]/M^n \quad , \tag{2.5.11}$$

where $\ell(s)$ is the path length for the sth round trip. The gain coefficient k at any point of the resonator can be obtained, according to (2.2.12), after the system of equations (2.5.8) has been solved together with (2.4.6, 5.11). Since the state of the active medium (the values of N_m, N_n, and k) and the radiation intensity I (or the density of radiation ρ) are mutually dependent, the gain coefficient and $\rho(0)$ in the zone of steady-state oscillations at the resonator axis are determined by conditions for the entire path length of the medium to this region (see Fig.2.16). Indeed, this problem

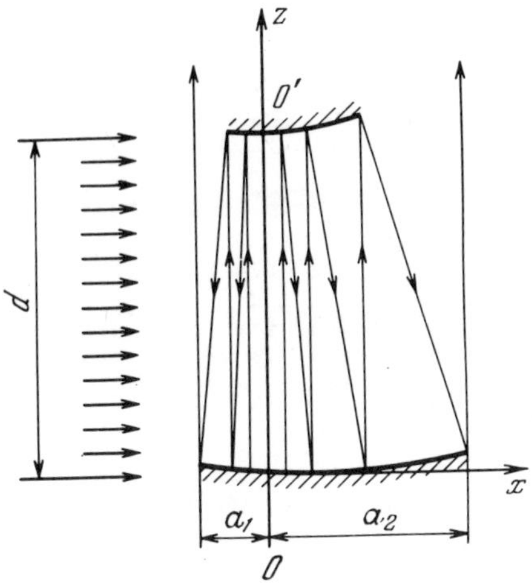

Fig.2.16. Schematic of the gas flow in an unstable resonator of the telescopic type. d - Flow width; a_1 - length of the oscillator; a_2 - length of the amplifier parts of the resonator

can be solved only numerically, using step-by-step approximations to a steady-state regime of the flow through the resonator [2.30]. Additional complications arise in describing the active mixture flow through an unstable resonator with spherical mirrors, which requires the analysis of three-dimensional fields of intensity and gain coefficient [2.42]. The results of the solution show that the entire flow in such a resonator can be devided by the axis 00' into two parts: "oscillator" of length a_1 at the resonator entrance and "amplifier" of length a_2 to the right of the axis 00', that is, at the resonator exit. In the oscillator part such operating conditions are created that the radiation intensity and its distribution provide for a decrease in the gain coefficient required for the steady-state regime at the axis 00'. Consequently, the transition from k_0 to k_* in the unstable telescopic resonator occurs gradually, over a distance a_1. The solution for the right-hand side of the resonator can be found in a simpler way since the value of $\rho(0)$ is already known. In fact, this part of the resonator is a multipass amplifier. In specific calculations it is possible to find the most advantageous relationship between dimensions a_1 and a_2 of both parts of the telescopic resonator (see Sect.5.3) [2.42]. An approximate solution for a continuously diverging unstable resonator was obtained in [2.43]. Clearly, unstable resonators can be used to obtain more uniform fields of radiation intensity. In [2.44] a simplified theory of multipass gasdynamic laser cavities is developed.

When an active medium flows through an unstable resonator, a pulsation regime can appear. This occurs when the active medium in the oscillator

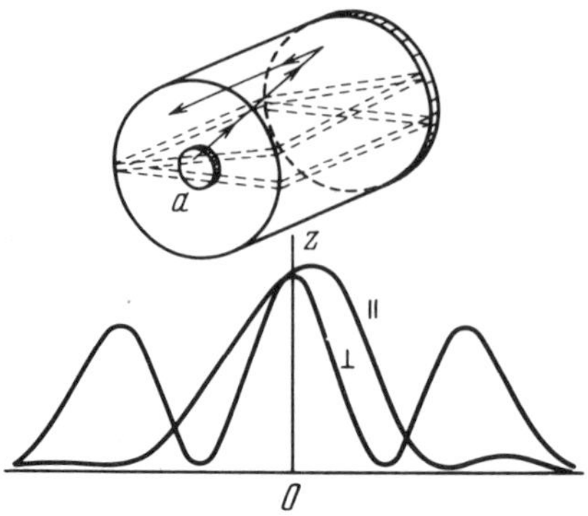

Fig.2.17. Far-field intensity distribution in an unstable resonator with annular aperture a in the direction parallel to the flow (||) and orthogonal to the flow (⊥) with oblique shock waves in the resonator (----); (0z) is the optical axis of the beam

part is destroyed for a short time. The oscillation breakdown occurs and, after a fresh mixture is delivered to the central region, oscillations set in again (see [2.45]). The period of these pulsations is close to the time of flight of gas particles through the resonator. To provide a steady-state oscillation regime, special measures should be taken (selection of resonator configuration, introduction of exciting radiation into the central part of the resonator, etc.) [2.42].

In the case of a supersonic gas flow the homogeneity of the active medium in the cavity may be considerably degraded due to shock waves and expansion waves created in the flow at the GDL nozzle exit (see Sect.4.4). In shock and expansion waves the gas density as well as the refractive index change considerably, so that the optical path in different parts of the flow varies, which results in distortions of the phase pattern of laser radiation. Figure 2.17 shows the results of an analysis of the far-field intensity distribution (after Fraunhofer) for a confocal unstable resonator, where the active medium is distorted by two oblique shock waves (shown in the figure by dashed lines), in which the gas density increases by 7% [2.25]. The effect of shock waves on the output beam quality can be decreased with such an arrangement of mirrors that provides for the mutual compensation in the change of the optical path length due to shock waves [1.2, 2.25,46]; this, however, is not always convenient because of construction difficulties.

Another important cause of active medium inhomogeneities is a random turbulence of gasdynamic flow. The sources of medium turbulence can be both the processes in the cavity before the GDL nozzle entrance and the interaction of the flow with nozzle walls, which results in turbulent wakes emanating from the exit edges of the nozzle and entering the cavity. An isotropic homogeneous field of a turbulent flow is determined by the so-called integral turbulence scale size Λ corresponding to the mean size of turbulent

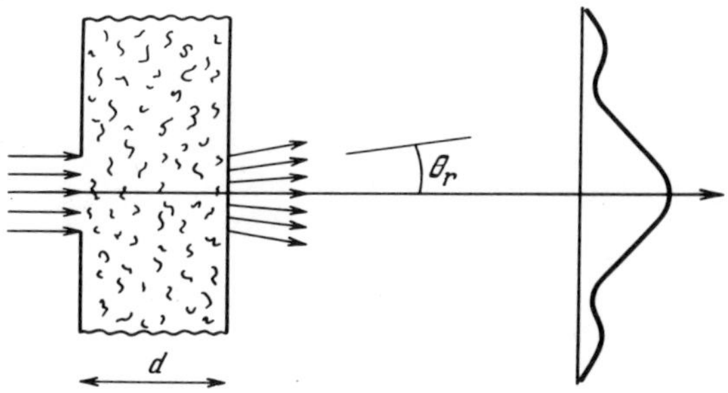

Fig.2.18. Plane wave passing through a turbulent flow. To the right — intensity distribution in the far-field zone

vortices. These vortices create pulsations of the gas density $\Delta\rho$ and, hence, pulsations of the refractive index Δn, which are related by the equation $\Delta n = K\Delta\rho$, where K is the Gladstone-Dale constant depending on the gas composition and wavelength. Passing through such a turbulized medium, light is attenuated as a result of dispersion and absorption. The coefficient β introduced into (2.5.1) can serve as a quantitative determination of this attenuation. This effect leads to a reduction of the gain coefficient and, besides, causes beam divergence and loss of coherence. The intensity I_r of radiation which thus becomes noncoherent and leaves the limits of the beam is determined not only by attenuation of the beam but also includes the energy spent for amplification of dispersed rays. In addition to (2.5.1) we must write

$$\frac{dI_r}{dz} = \beta I + kI_r \quad . \qquad (2.5.12)$$

For a flat layer of width d (Fig.2.18) [2.47] we obtain from (2.5.1,12)

$$I(d) = I(0)\exp[(k - \beta)d] \quad , \quad I_r/I = \exp(\beta d) - 1 \quad . \qquad (2.5.13)$$

Thus, for $\beta d \approx 1$ about half of the energy of coherent radiation is dissipated. Of interest is scattering, when $\beta d < 1$ (the opposite case is considered in [2.47]), that is, when radiation is scattered only at small angles to the initial direction—within the angle $\theta_r = 2/\bar{k}\Lambda$, where $\bar{k}$ is the wave number ($\bar{k} = 2\pi/\lambda$), Λ is the above-mentioned integral turbulence scale size. Usually this angle is larger than the diffraction-limited angle $\theta_d \approx 2\lambda/a$ (a is the diameter or diagonal of the output window). The expression for the scattering loss coefficient β, both in geometrical optics approximation and as a result of the solution of the wave equation [2.47], is obtained in the form

$$\beta = 2\bar{k}^2\langle\Delta n^2\rangle\Lambda \quad , \qquad (2.5.14)$$

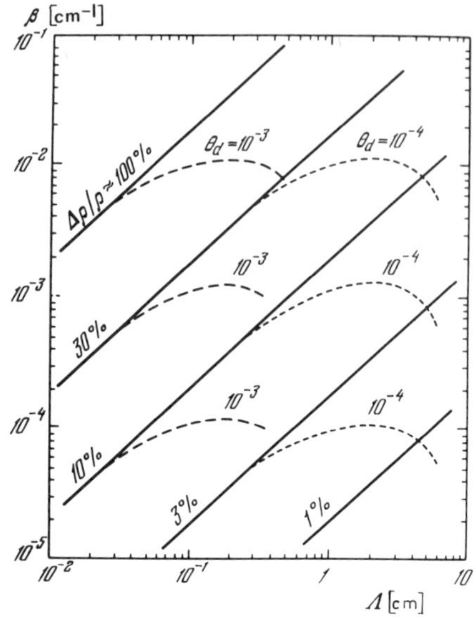

Fig.2.19. Scattering loss coefficient β in a turbulent CO_2 flow ($\lambda = 10.6$ μm for $p = 0.1$ atm, $T = 300$ K) as a function of the integral scale of turbulent pulsations Λ [2.25]

where $\langle\Delta n^2\rangle$ is the mean squared variation of the refractive index fluctuations. Figure 2.19 gives the values of β of various levels of the density fluctuations $\Delta\rho/\rho$ for λ = 10.6 μm (CO_2 laser) (the problem of determining Λ is treated in Sect.4.4). Dashed lines on the graph indicate the limit of decrease in β, which results from the diffraction-limited angle θ_d of the beam [2.25].

The computation of laser power in the resonator with account of losses due to turbulent fluctuation can be carried out using (2.5.12) for two opposite directions and the system of equations (2.5.1,8). To the right-hand side of the second equation of (2.5.8) we must add the term $(\sigma_{mn}N_m - \sigma_{nm}N_n)I_r/h\nu$, and the same term must be subtracted from the first equation to take account of the decrease in population inversion, when the active medium is radiated by scattered photons. The solution of this problem in a simplified form (with no account of the population variation as a result of molecular collisions and at a constant velocity of the flow through the resonator) can be obtained analytically [2.39].

According to (2.5.14), the effect of turbulence at a constant gas temperature in a cavity increases in proportion to the squared pressure and may become very strong for high gas pressure. The far-field diffraction pattern for the relative intensity distribution changes little in the presence of turbulence [2.47].

The analysis of the effect of small particles in a condensed phase on laser power and field distribution can be carried out in a similar manner. These particles get into a cavity either from a chamber at the GDL nozzle entrance (e.g., carbon particles due to incomplete combustion of fuel, or dust), or are formed during condensation as a result of gas cooling. Attenuation of the light intensity during its passage through such an inhomogeneous medium is also determined by scattering and absorption. We can describe this effect quantitatively if the particles are uniformly distributed over the volume, and their separation is much greater than the radiation wavelength and at least several times as great as the particle size. We simply introduce into (2.5.1) the attenuation coefficient γ in addition to the coefficient β, which corresponds to light attenuation due to scattering in the turbulent medium. For spherical particles of similar radii R (a monodisperse medium) we have

$$\gamma = \pi R^2 f_{at} n_c \quad , \tag{2.5.15}$$

where n_c is the number of condensed particles per unit volume and f_{at} is the attenuation factor equal to the part of the light flux that undergoes scattering or absorption. In a system of particles with different radii (a polydisperse medium) we have

$$\gamma = \pi \int f_{at} n_c(R) R^2 dR \quad , \tag{2.5.16}$$

where $n_c(R)$ is the distribution of the particle number with respect to radius R. The attenuation factor f_{at} is additively determined by scattering f_{sc} and absorption f_{ab}: $f_{at} = f_{sc} + f_{ab}$. These quantities depend on the complex refractive index of the particle material $m = n - \kappa i$, their shape, and the wavelength [2.48]. If $z = 2\pi R/\lambda$ is much smaller than unity, then

$$f_{sc} = \frac{8z^4}{3} \left| \frac{m^2 - 1}{m^2 + 2} \right|^2 \quad , \tag{2.5.17}$$

$$f_{ab} = -4z \, \mathrm{Im} \left| \frac{m^2 - 1}{m^2 + 2} \right| = \frac{24 z n \kappa}{(n^2 - \kappa^2 + 2)^2 + 4n^2\kappa^2} \quad . \tag{2.5.18}$$

In the general case f_{at} can be determined by numerical calculations. The results of a computation of f_{at} for carbon are given in [2.49]. Usually scattering losses are much lower than losses due to absorption, and the more so for long waves, for which the factor z^4 in (2.5.17) is small.

The attenuation coefficient γ can be conveniently expressed in terms of the mass fraction α_c of condensate in a mixture; then γ will not depend on

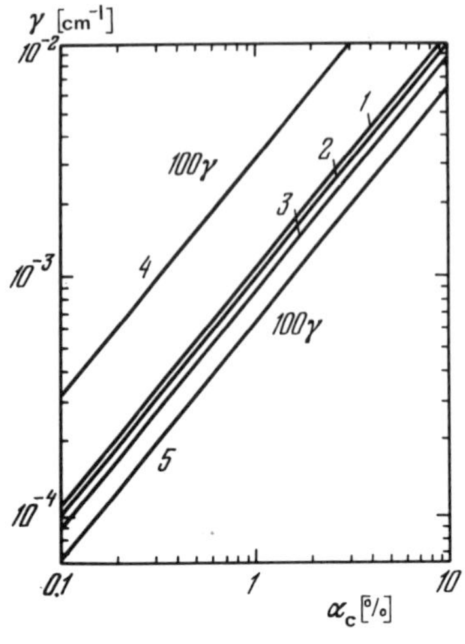

Fig.2.20. Effect of particle concentration α_c in the flow on the attenuation factor γ in a CO_2 active medium containing dust (λ = 10.6 μm for p = 0.1 atm, T = 300 K). (*1*) water drops; (*2*) carbon blacks; (*3*) quartz; (*4*) aluminum; (*5*) copper

the particle size if it is small ($z \ll 1$) and scattering in (5.17) can be neglected [2.50]. Proceeding from (2.5.18), we obtain

$$\gamma = \frac{36\pi\alpha_c}{\lambda} \cdot \frac{\rho}{\rho_c} \cdot \frac{n\kappa}{(n^2 - \kappa^2 + 2)^2 + 4n^2\kappa^2} \quad , \tag{2.5.19}$$

where ρ_c is the density of the particle material and ρ is the density of a gas in the mixture.

Figure 2.20 gives the dependence of the attenuation coefficient γ on the mass fraction α_c of particles in a gas for light with wavelength 10.6 μm passing through the medium containing particles of different substances. It is clear from the graph that the presence of absorbing particles in a flow leads to a considerable decrease in laser efficiency. The method of solving the problem of laser oscillations in the resonator through which such a heterogeneous mixture passes is the same as in the case of a flow with turbulent fluctuations described above.

To estimate losses due to scattering and absorption, we must add the quantity $(\beta + \gamma)d$ to the loss coefficient α in all the expressions derived earlier For example, for the resonator efficiency with a homogeneous medium for the optimum transmission coefficient, instead of (2.3.35) we can write

$$\eta_r = \{1 - \sqrt{[\alpha + (\beta + \gamma)d]/k_0 d}\}^2 \quad . \tag{2.5.20}$$

Similar estimates can be made in determining the threshold relation of laser oscillations and also for other cases considered in Sect.2.3.

It should be noted that in high-power lasers a cause of inhomogeneity of the radiation field may be the deformation of resonator mirrors due to mechanical (for instance, vibrations) and thermal effects (as a result of radiation absorption by the mirrors, dust appearing on their surfaces, and so on). These effects become especially noticeable for large field intensities in the cavity [2.51].

3. Physico-Chemical Gas Kinetics

This chapter introduces the concept of the Boltzmann distribution of molecules over energy levels and gives the basic laws for gases (the thermal and caloric equations of state, and Dalton's law for gas mixtures). Proceeding from the generalized Boltzmann equation, the balance equations for the number of molecules in different quantum states are obtained for the case of molecular collisions. The chapter then introduces the concept of the probability of inelastic energy exchange upon molecular collisions and gives the classification of the processes of energy exchange. A derivation for the vibrational transition probabilities is presented and specifies the type of the temperature dependence of these quantities. It also derives the relaxation equations for the vibrational energy with attention given to the peculiar shape of the molecular distribution curves for the model of anharmonic oscillators. Gives as an example the vibrational exchange kinetics in CO_2. The chapter concludes with a phenomenological description of nonequilibrium kinetics of chemical reactions in a gas and kinetics of the processes in heterogeneous mixtures.

3.1 Thermodynamic Properties of Gas Mixtures

The most important property of an active medium is the population inversion of energy levels. Thus the population of a level is determined by the processes in a gas whose state considerably differs from the equilibrium state. These processes are the subject matter of physico-chemical kinetics[1]. Let us begin with the description of complete and partial gas equilibrium, since even in nonequilibrium gas systems we shall assume the existence of subsystems in which partial equilibrium is realized.

1 For a review of physico-chemical kinetics, see [3.1-5].

Suppose that N_i particles (to be more specific, gas moleceles) have energy E_i, where i can take on all possible values. The set i, E_i, N_i [or the dependence N(E)] represents the energy distribution of gas molecules. Each of these distributions, which differ from one another, is called the *state of the system* (or subsystem). The state with the highest probability, or the equilibrium state, corresponds to a quite definite energy distribution; for an ideal gas it is the Boltzmann distribution (see [3.6])[2]

$$N_i = (N/Q)g_i \exp(-E_i/kT) \quad , \tag{3.1.1}$$

where g_i is the statistical weight of the energy level E_i, k is the Boltzmann constant, T is the temperature of the system, N is the total number of molecules in the system, and Q is the statistical sum (partition function)

$$Q = \sum_i g_i \exp(-E_i/kT) \quad . \tag{3.1.2}$$

The temperature T here is an important macroscopic parameter, which describes the equilibrium state of the system.

Let us consider a gas which consists of polyatomic molecules. Different kinds of energy in such a gas are in the subsystems of translational, rotational, vibrational, and electronic degrees of freedom of molecules. We shall assume here and below that in the subsystem of translational motion of molecules there is an equilibrium described by a Maxwellian velocity distribution for temperature T [3.6]. To describe the equilibrium in other subsystems we shall use the Boltzmann distribution (3.1.1), the temperature of the jth subsystem being T_j. This permits us to introduce the concept of a temperature to describe the distribution of molecules over rotational and vibrational degrees of freedom (rotational T_{rot} and vibrational T_{vib} temperatures). The term "electronic temperature" is usually used to describe the average translational energy of free electrons. At statistical equilibrium $T_{rot} = T_{vib} = T$.

To describe a subsystem, we must choose a model of motion and thus determine the arrangement of energy levels, that is, find the partition function and the expression for energy ε_j of the jth subsystem, depending on molecular constants and temperature. This analysis for simple models, with the assumption that each kind of motion in a molecule proceeds independently, is given in Appendix D. In this approximation the total energy of a molecule

---

2 An ideal gas is a gas in which the interaction between molecules consists only of collisions of short duration.

is the sum of all energies ε_j ($\varepsilon = \sum \varepsilon_j$), and the total partition function equals the product of partition functions Q_j ($Q = \Pi Q_j$). The approximation of separate motions of a molecule proves to be quite sufficient in most cases of considering processes in gas dynamic lasers. At the next approximation we must take into account the interaction between various kinds of motion in a molecule (see [2.8]).

In gasdynamic lasers mixtures of various gases are usually utilized. According to Dalton's law [3.6], we have

$$\sum p_j = p \quad , \tag{3.1.3}$$

where p_j is the partial pressure of the jth mixture component and p is the total gas pressure. To describe the mixture composition, use is often made of the molar fractions $\xi_j = p_j/p$, as well as other expressions of concentrations (see Appendix E). At high temperatures the chemical composition of a mixture can change, but the ratio of concentrations of chemical elements (atoms) in all chemical reactions is retained, regardless of the constituent compounds. These relations of material balance are given in Appendix F. When reactions occur, the gas mixture composition is determined by equations of chemical kinetics in the absence of equilibrium (Sect.3.6) or by chemical balance equations in equilibrium. The form of the chemical balance equations and the procedure of determining the equilibrium composition of a reacting gas mixture are given in Appendix G. Appendices E-G take into account the presence of a condensed phase in a mixture.

The thermal properties of a gas are determined by the interdependence between pressure p, density ρ, and temperature T, that is, by the *thermal equation of a state* $p = p(\rho, T)$. It follows from the kinetic theory of an ideal gas that pressure p is proportional to the number of particles in a unit volume and temperature T [3.6],

$$p = nkT \tag{3.1.4}$$

(k is the Boltzmann constant), and the partial pressure p_j is

$$p_j = n_j kT \quad , \tag{3.1.5}$$

where n_j is the density of particles of the jth mixture component. Using (E.6, F.2), we get

$$p = \rho RT/\mu \quad , \tag{3.1.6}$$

where R is the gas constant, $R = k\mathcal{N}_A$ ($\mathcal{N}_A$ is the Avogadro number) and μ is the molecular mass of a gas mixture determined according to (E.6). For this equation there must be equilibrium over the translational degrees of freedom, regardless of the extent of deviation from equilibrium with respect to other degrees of freedom of a molecule.

Caloric properties of a gas mixture are determined by the dependence of the internal energy $\mathcal{E}$ of a gas on pressure and temperature: $\mathcal{E} = \mathcal{E}(p,T)$; this dependence is often called the *caloric equation of a state*. For an ideal gas the internal energy of each component $\mathcal{E}_j$ depends only on temperature, and the internal energy of the gas mixture is the sum of the respective contributions from each component taking into account the heat of formation $\mathcal{E}_i^0$, so that per mole we have

$$\mathcal{E} = \sum \xi_j[\mathcal{E}_j(T) + \mathcal{E}_j^0] \quad , \tag{3.1.7}$$

where ξ_j is the molar fraction of the jth component. In gasdynamic analysis it is convenient to use the specific enthalpy $h = \varepsilon + pv$; v is the specific volume and ε is the specific internal energy of a gas.[3]

Other important thermodynamic quantities are the *heat capacity* at a constant pressure $c_p = (\partial h/\partial T)_p$ and the heat capacity at a constant volume $c_V = (\partial \varepsilon/\partial T)_V$. If the heat capacity is constant and independent of temperature, such a gas is called *perfect*[4]. In this case gasdynamic equations are considerably simplified. Examples of perfect gases are gases composed of monatomic molecules or of polyatomic rotating molecules, which have no other internal degrees of freedom. Molar enthalpy of a perfect gas (bearing in mind that for a mole of an ideal gas $C_V - C_p = R$) is

$$H = C_pT \equiv \beta RT \quad , \tag{3.1.8}$$

where $\beta = \gamma/(\gamma - 1)$; $\gamma = C_p/C_V$ is the ratio of heat capacities (for a perfect gas γ = const).

Using the expressions for energy of various kinds of motion given in Appendix D, we get the respective values of enthalpy and heat capacity. For translational motion $H = 5RT/2$; this value refers to monatomic gases. Molar

---

3 Here and elsewhere small letters indicate thermodynamic quantities per unit mass, and capital letters, per mole of a substance. To pass over from molar quantities to specific ones, one should divide a molar quantity by the molecular mass of a mixture $\mu = \sum \xi_j\mu_j$ (μ_j is the molecular mass of the jth component), e.g., $\varepsilon = \mathcal{E}/\mu$, $h = H/\mu$.

4 Sometimes this model is called an "ideal" gas in contrast to a real gas with varying heat capacities. The term "perfect gas" makes it possible to distinguish this model from a more general concept of ideal gas adopted in statistical physics.

heat capacities of such gases are $C_p = 5R/2$, $C_V = 3R/2$, and their ratio $\gamma = 1.667$. The molar enthalpy of a perfect gas consisting of linear rotating molecules for $T_{rot} = T$ is $H = 7RT/2$, and the ratio of heat capacities is $\gamma = 1.4$. For a polyatomic perfect gas with nonlinear molecules $H = 4RT$ and $\gamma = 1.333$.

If we take into account the excitation of molecular vibrations, then according to (D.12) we obtain for a harmonic oscillator model the average number of vibrational quanta e_i per one mode of molecular vibrations (without degeneracy) in the form

$$e_i = 1/[\exp(\theta_i/T_i) - 1] \quad , \tag{3.1.9}$$

where θ_i is the characteristic vibrational temperature and T_i is the varying vibrational temperature for the ith mode of vibrations. If an equilibrium is maintained between rotational and translational degrees of freedom, but $T_i \neq T$, then the specific enthalpy can be represented in the form

$$h = \beta RT/\mu + \varepsilon_{vib} \quad , \quad \varepsilon_{vib} = (R/\mu) \sum \xi_i g_i \theta_i e_i \quad , \tag{3.1.10}$$

where $\beta RT/\mu$ is the specific enthalpy of a perfect gas according to (3.1.8), ε_{vib} is the specific vibrational energy, g_i is the statistical weight of the ith mode of vibrations, and ξ_i is the molar fraction of the component which includes this mode. In gas mixtures

$$\beta = 2.5 + \xi_l + 1.5\xi_{nl} \quad , \tag{3.1.11}$$

where ξ_l and ξ_{nl} are molar fractions of components consisting of linear and nonlinear molecules, respectively. The last two terms on the right-hand side of (3.1.11) account for the classical contribution of molecular rotations to the enthalpy. The values of enthalpy for more real models of molecular rotation and vibration and various electronic states can be found in reference books (see, for example, [2.8]).

For a reacting gas mixture the caloric state equation (3.1.7) can be written in the form

$$h = (1/\mu) \sum \xi_j [H_j(T) + H_j^0] \quad , \tag{3.1.12}$$

where $H_j(T)$ is the molar enthalpy of the jth component, and H_j^0 is the molar enthalpy of formation of this component. Note that the energy and enthalpy of a reacting mixture of ideal gases depend not only on temperature (as in the case of individual components) but also on pressure, which together with temperature determines the chemical composition of the mixture [in chemical equilibrium $\xi_i = \xi_i(T, p)$; see Appendix G]. The heat capacity of a reacting mixture can be found after differentiating the expressions of the types (3.1.7,12) with respect to temperature. In the differentiation, however, we must take into account the fact that the gas composition depends on temperature. Consequently, the heat capacity of a reacting mixture is not an additive function of heat capacities of its components, but depends on the nature of processes occurring in the mixture. Note that the ratio of heat capacities γ here, as in the case of a perfect gas, varies with temperature (and with pressure, if the gas composition changes).

Let us consider now the velocity of sound propagating through a gas mixture. It is known [3.7] that the sound velocity a is determined as

$$a = \sqrt{(\partial p/\partial \rho)} \quad , \tag{3.1.13}$$

where the derivative should be taken in conditions which exist in sound waves. If compression and expansion in a sound wave occur at a constant entropy (s = const), then $a^2 = (\partial p/\partial \rho)_s$. This is the case when sound propagates through a perfect gas (Sect.4.1). If gas molecules can vibrate or enter into chemical reactions, two extreme cases can be considered in which s = const. If the vibrational frequency of sound is high, then only fast energy exchange between translational and rotational degrees of freedom is possible, and all other degrees of freedom and chemical reactions do not participate in the energy exchange, that is, they are "frozen". The sound velocity computed in these assumptions is called *frozen* and is denoted by a_f. The frozen sound velocity determines the so-called high-frequency sound. When it propagates, only molecular rotations are excited, and from this view point a gas is perfect (sometimes in sound waves only the mixture composition is frozen). In contrast to this the *equilibrium* sound velocity a_{eq} corresponds to low-frequency sound; when this sound propagates, all internal degrees of freedom and chemical reactions have enough time to adjust to the change in gas parameters in a sound wave, so that equilibrium in this case also is not violated [$a_{eq}^2 = (\partial p/\partial \rho)_{eq}$]. In both cases all processes in a sound wave occur at a constant entropy.

3.2 Molecular Energy Exchange upon Collisions

The population inversion of energy levels can change either upon molecular collisions or as a result of radiation or absorption of light. Radiative transitions were discussed in Sect.2.1. Here we shall consider nonradiative transitions during collisions. In this case three results of a collision between molecules A and B exist.

1) Upon the collision of A and B no change in the internal energy of these molecules takes place. Such collisions are called *elastic*.
2) The collision of A and B is accompanied by a transfer of a part of translational energy into rotational, vibrational, and electronic energy of molecules or by redistribution of energy between these internal degrees of freedom of different molecules. These are *inelastic* collisions.
3) Sometimes inelastic collisions are accompanied by *chemical* transformations of molecules A and B (for instance, into molecules C and D). In this case the atomic composition of molecules changes.

Let v_A, v_B, ... be the velocity vectors of translational motion of molecules A, B, ...; i, j, l, m be some quantities describing internal states of molecules. Essentially each of these quantities is a set of quantum numbers which describe rotational, vibrational, and electronic states of a molecule. Then the above processes can be represented in the form

$$A(i,v_A) + B(j,v_B) \begin{cases} \rightarrow A(i,\underline{v}_A') + B(j,\underline{v}_B') & , \\ \rightarrow A(i',v_A') + B(j',\underline{v}_B') & , \\ \rightarrow C(l,\underline{v}_C) + D(m,\underline{v}_D) & , \end{cases} \qquad (3.2.1)$$

where primes denote the change of the quantity due to collision. The number of molecules A in the ith state with the velocity of translational motion within the range $\underline{v}_A$, $\underline{v}_A + d\underline{v}_A$ is $n_{A(i)}\, f_{A(i)}\, (\underline{v}_A)$, where $n_{A(i)}$ is the total number of molecules A in the ith state per unit volume [density of molecules A(i)], and similarly for other molecules. The function $f_{A(i)}\, (\underline{v}_A)$ is usually called the velocity distribution function of molecules A(i). The quantity $n_{A(i)}\, f_{A(i)}\, (\underline{v}_A)$ is determined from the *generalized Boltzmann equation*[5], which establishes the balance between molecules A (and similarly of other particles) with velocities in the range $\underline{v}_A$, $\underline{v}_A + d\underline{v}_A$, arriving at or leaving the ith state as a result of collisions,

$$\frac{d}{dt}\,[n_{A(i)} f_{A(i)}(\underline{v}_A)] = I_{el} + I_{in} + I_{ch} \quad , \qquad (3.2.2)$$

---

5 See [3.8]. The idea of Boltzmann's equation and its derivation can be found, for instance, in [3.1,3]. The generalized Boltzmann equation is considered in [3.9].

where the terms on the right-hand side denote the collision integrals corresponding to elastic, inelastic collisions and collisions resulting in chemical reactions, respectively. For example, for inelastic collisions this integral is written in the form [3.8]

$$I_{in} = \sum_{j,i',j'} \int v_{AB}\Big[\sigma_{i'i}^{j'j}(v_{AB})n_{A(i')}f'_{A(i')}(\underline{v}'_A)n_{B(j')}(\underline{v}'_B)$$

$$- \sigma_{ii'}^{jj'}(v_{AB})n_{A(i)}f_{A(i)}(\underline{v}_A)n_{B(j)}f_{B(j)}(\underline{v}_B)\Big]d\underline{v}_B \quad , \qquad (3.2.3)$$

where $\sigma_{ii'}^{jj'}(v_{AB})$, $\sigma_{i'i}^{j'j}(v_{AB})$ are the so-called inelastic collision cross sections and v_{AB} is the absolute (scalar) value of relative velocity of molecules A and B. The first term in the square brackets determines the rate of arrival of molecules A to the ith state as a result of a collision proceeding in the reverse direction to the one indicated by the arrow in (3.2.1), and the second term determines the rate at which particles A leave the ith state as a result of the collision in the forward direction [as indicated in (3.2.1)]. The Boltzmann equation assumes random and independent motion of molecules in the gas and small collision time in comparison with the mean free time between collisions. It is assumed that the distribution functions vary only as a result of collisions, and that macroscopic gas parameters do not vary over a length equal to the mean radius of molecular interaction.

By integrating both sides of (3.2.2) over all values of velocity $\underline{v}_A$, that is, by determining the variation of particle density

$$n_{A(i)} = \int n_{A(i)}f_{A(i)}(\underline{v}_A)d\underline{v}_A \quad , \qquad (3.2.4)$$

we obtain the equation for $n_{A(i)}$. Since for elastic collisions $n_{A(i)}$ is constant, the integral I_{el} is zero, and from the remaining two we shall consider now only inelastic collisions not involving chemical reactions, which will be treated in Sect.3.6. The result of integration can be considerably simplified if we consider relative motion of A and B with relative velocity v_{AB} and reduced mass $\mu_{AB} = m_A m_B/(m_A + m_B)$. Let us introduce the concept of inelastic collision *probability*

$$P = \frac{\int \sigma_{ii'}^{jj'}(v_{AB})f(v_{AB})v_{AB}dv_{AB}}{\int \sigma_{AB}(v_{AB})f(v_{AB})v_{AB}dv_{AB}} \quad , \qquad (3.2.5)$$

where σ_{AB} is the cross section of the respective elastic collision, and assume that the velocity distribution function $f(v_{AB})$ and cross section σ_{AB} are independent of the quantum state of colliding particles. Since we adopted an equilibrium distribution of molecular velocities for gas temperature T, we can use for $f(v_{AB})$ the Maxwellian distribution function

$$f(v_{AB})dv_{AB} = 4\pi(\mu_{AB}/2\pi kT)^{3/2} \exp(-\mu_{AB}v_{AB}^2/2kT)v_{AB}^2 dv_{AB} \quad . \tag{3.2.6}$$

Then the number of elastic collisions Z_{AB} of a given molecule A with all molecules B per unit time (or the collision frequency of molecule A) is

$$Z_{AB} = 4\pi(\mu_{AB}/2\pi kT)^{3/2} n_B \int \sigma_{AB}(v_{AB})\exp(-\mu_{AB}v_{AB}^2/2kT)v_{AB}^3 dv_{AB} \quad , \tag{3.2.7}$$

where n_B is the density of molecules B. When elastic collision cross section σ_{AB} does not depend on velocity v_{AB} (a rigid sphere model), we have

$$Z_{AB} = v_{AB}^{1/2}\,\sigma_{AB} n_B \quad , \tag{3.2.8}$$

where $v_{AB} = (8kT/\pi\mu_{AB})^{1/2}$ is the average thermal velocity of relative motion. For collisions of identical molecules A + A, in addition to a formal substitution $B \rightarrow A$, we must decrease Z_{AB} by half, since in each such collision two molecules A take part. Proceeding from (3.2.7), we can interpret the probability $P_{ii'}^{jj'}$ (3.2.5) as the ratio of the number of collisions between A(i) molecules, which provide for the $i \rightarrow i'$ transition in A molecules and simultaneously the $j \rightarrow j'$ transition in B molecules, and the total number of elastic collisions between A and B. Relation (3.2.5) and the assumptions made enable us to represent the results of integration of the generalized Boltzmann equation (3.2.2,3) with respect to velocities in the form of a balance equation for the number of molecules A in the ith state,

$$\frac{dn_{A(i)}}{dt} = z_{AB} \sum_{j,i',j'} \left[P_{i'i}^{j'j} n_{A(i')} n_{B(j')} - P_{ii'}^{jj'} n_{A(i)} n_{B(j)}\right] \quad , \tag{3.2.9}$$

where z_{AB} is the collision frequency per molecule of type B ($z_{AB} = Z_{AB}/n_B$). This equation shows that the rate of arrival (or leaving) of molecules A to the ith state is proportional to the number of molecules in initial states. The set of these equations for all the ith and j^{th} states of molecules A and B forms a system of balance equations that describe the variation of population of energy levels in which the population inversion can be created. These balance equations are usually called *microkinetic* equations to emphasize that we are talking about populations of individual levels in a molecule (sometimes they are called the master equations or level kinetics equations). Further, to solve the problem under consideration, we must know the probabilities of transitions $i \rightleftharpoons i'$ and $j \rightleftharpoons j'$ during collisions, that is, the dependence of the cross section of this process on the relative velocity of colliding molecules [see (3.2.7)]. The probabilities $P_{i'i}^{j'j}$ and $P_{ii'}^{jj'}$ will consequently be determined by the gas temperature and the nature of a specific inelastic process.

Of great importance is the relation between probabilities of a process in the forward and reverse directions. This relation can be obtained using the principle of detailed balancing, which has already been used in Sect.2.1 when deriving the relations between Einstein's coefficients. Indeed, in equilibrium the quantity $\bar{n}_{A(i)}$ is constant for a given temperature; thus from (3.2.9) we get

$$\frac{P_{i'i}^{j'j}}{P_{ii'}^{jj'}} = \frac{\bar{n}_{A(i)}\bar{n}_{B(j)}}{\bar{n}_{A(i')}\bar{n}_{B(j')}} \quad , \tag{3.2.10}$$

where the bar above a letter indicates that this quantity corresponds to the equilibrium state. At equilibrium the right-hand side depends only on temperature. Since $P_{i'i}^{j'j}$, $P_{ii'}^{jj'}$ are independent of concentration, ratio (3.2.10) holds for any $n_{A(i)}$, ..., that is, in the absence of equilibrium as well.

The determination of probabilities $P_{ii'}^{jj'}$ is one of the fundamental problems of physical kinetics and, in particular, dynamics of atomic and molecular collisions. This problem is dealt with in both classical mechanics and in quantum mechanics. Here also are used the combined "quasiclassical" methods, where translational motion is treated in the classical approximation, but the quantum nature of the molecules is also taken into account. In solving the problem, use is made of information on intermolecular and intramolecular interaction potentials, variation of initial conditions of collisions (such as impact parameter, mutual orientation of molecules, their initial separation), and relations between different kinds of molecular energy. Both analytical and numerical methods of solution are employed.[6] Though considerable advances have been achieved in the theory of atomic and molecular collisions, for the time being we cannot expect to obtain sufficiently precise theoretical values of probability $P_{ii'}^{jj'}$; this is mainly due to the lack of information on the nature of intermolecular interaction potentials and difficulties in obtaining an exact solution of this problem. For further discussion, however, it is important to know that the theory, as a rule, permits us to find the temperature dependence of probabilities $P_{ii'}^{jj'}$ and evaluate them. Experimental measurement of these probabilities is also not

6 Many problems of atomic and molecular collision dynamics are discussed in [3.2,10-12].

a simple problem, but can be successfully solved for a limited number of processes (for instance, some vibration-rotational transitions and electronic transitions).

More often in experiments we obtain the values that describe the variation of macroscopic parameters of the medium (the store of energy of a certain kind, some chemical component concentration, and so on). At first sight these results may seem to be hardly interesting for the analysis of processes in lasers, when information on the population of individual levels is required to determine the gain coefficient and other laser parameters. Sometimes, however, as in the analysis of processes in CO_2 gasdynamic lasers, this problem can be successfully solved assuming that there is a partial equilibrium with respect to certain degrees of freedom of molecules. In this case the population of an individual level can be found with the help of relations given in Sect.3.1 and in Appendix D, after determining the store of vibrational energy in different vibrational modes or for different molecules. Then we need equations describing the change of the store of vibrational energy with time towards equilibrium (relaxation) and similarly for other kinds of energy. These equations, called relaxation equations, can be obtained from a system of microkinetic balance equations of the type (3.2.9). This procedure is considered in Sect.3.4. Relaxation time or rate constants of energy exchange, rather than transition probabilities, appear in these equations as kinetic parameters.

In conclusion let us introduce a notation for basic collision processes occurring in infrared gasdynamic lasers. For these processes the following notation is introduced: (T) translational, (R) rotational, and (V) vibrational motion of molecules; the combination of these letters indicates the degrees of freedom of molecules which participate in a given process. Thus, T-T denotes translation-translation transition (or exchange), that is, an elastic collision. In R-T and V-T processes the translational energy is transferred into rotational and vibrational energy, respectively. The redistribution of rotational energy is designated by R-R, of vibrational energy by V-V, and the exchange between them by V-R. In the analysis of processes involving polyatomic molecules the V-V exchange refers to intramolecular transitions and intermolecular energy transfer is denoted by a prime (V-V'). During collisions often all the three kinds of energy vary simultaneously, which is designated as a V-R-T transfer. In some cases the major role in the mechanism of inversion formation is played by the transitions between electronic and vibrational degrees of freedom (E-V processes);

see, [3.13]. Here we shall pay attention mainly to the processes involving vibrational degrees of freedom of molecules.

3.3 Vibrational Transition Probabilities in Molecules

We shall give the fundamental results of the theory describing the temperature dependence of vibrational transition probabilities during collisions. Let us begin with the analysis of vibration-translation (V-T) exchange; this is possible if we neglect the contribution from rotation. Such an approximation holds better for intermolecular potentials close to spherically symmetrical potentials. The V-T exchange probability is determined most easily when vibrations are excited during a head-on collinear collision of a certain particle B (its internal state is not taken into account in the solution) with a diatomic molecule A, which is considered as a harmonic oscillator. If during collision the amplitude of vibrations of A is small, so that the vibrational state of this molecule does not affect the relative motion of A and B, and the intermolecular potential of molecule A remains harmonic and does not depend on the effect of B, then we can separate the description of the internal and external motion of the molecule. The collision is described by the effect on the oscillator of an extraneous force appearing during collision (extraneous force approximation). The magnitude of this force is determined by the "repulsion" term of the intermolecular potential, which acts at a small distance R_{AB} between molecules. Usually this potential is presented in an exponential form,

$$U(R_{AB}) = U_0 \exp(-\alpha R_{AB}) \quad , \tag{3.3.1}$$

where α is a parameter whose reciprocal determines the dimensions of the region of pronounced interaction between molecules and is called the radius of interaction (α is the reciprocal radius of interaction). The energy ΔE transferred to the oscillator can be easily found using the well-known methods of the theory of oscillations (see, [3.14]). ΔE is determined by the so-called *adiabatic factor* (or Massey parameter), which is equal to the product of cyclic frequency of vibrations and the collision time τ_c. The quantity τ_c is determined by the time during which the colliding molecules are in the region of interaction mentioned above, that is, $\tau_c \approx 1/\alpha v_{AB}$, where v_{AB} is the relative velocity of molecules A and B. The adiabatic factor is

$$\omega\tau_c = \omega/\alpha v_{AB} = 2\pi\tau_c/t_A \quad , \tag{3.3.2}$$

(t_A is the period of vibrations of molecule A); hence it is equal to the ratio of collision time and the period of vibrations. For $\omega\tau_c \gg 1$, the process of vibration excitation is adiabatic, which means that the energy exchange is very improbable. Only slow collisions are adiabatic, i.e., such that during a collision molecule A has time to "adjust" itself to external action and at the end of the collision returns almost the entire initial energy of collision to translational motion. The solution of this problem in quasiclassical and quantum approximations (see [1.8, 3.11,15])[7] has shown that the probability of vibration excitation to the first approximation of the perturbation theory linearly increases with the number n of the harmonic oscillator vibrational level,

$$P_{n-1,n}(v_{AB}) = nP_{01}(v_{AB}) \quad . \tag{3.3.3}$$

Moreover, it decreases with the increase of the adiabatic factor $\omega\tau_c$,

$$P_{01}(v_{AB}) \propto \exp(-2\pi\omega\tau_c) \quad . \tag{3.3.4}$$

Finally, the excitation probability is zero for all the transitions for which $\Delta n \geqq 2$, so that transitions occur only between neighboring levels. Then, putting $\sigma_{01} = \sigma_{AB}P_{01}(v_{AB})$, where σ_{AB} is the elastic collision cross section for the rigid sphere model, we get, according to (3.2.5,6),

$$P_{01}(T) \sim \int_0^\infty \exp[-(2\pi\omega/\alpha v_{AB}) - (\mu_{AB}v_{AB}^2/2kT)]v_{AB}^3 dv_{AB} \quad . \tag{3.3.5}$$

The integrand in (3.3.5) has a sharp maximum determined by the minimum of the exponent. By determining this minimum, we obtain the relative velocity v_* that is most effective in excitation of gas molecules at a given temperature (the subscript AB will be omitted below),

$$-\frac{2\pi\omega}{\alpha v_*^2} + \frac{\mu v_*}{kT} = 0 \quad \text{or} \quad v_* = \left(\frac{2\pi kT\omega}{\mu\alpha}\right)^{1/3} \quad . \tag{3.3.6}$$

It should be noted that v_* noticeably exceeds the average thermal velocity of relative motion of molecules. Performing in (3.3.5) integration by the steepest descent method near $v = v_*$, we arrive at

---

7 The theory presented in [3.15,16], which accounts for short-range forces of repulsion, is sometimes called the Schwartz-Slawsky-Herzfeld theory (SSH theory).

$$\left.\begin{aligned} P_{01}(T) &\propto \exp\left(-\frac{2\pi\omega}{\alpha v_*} - \frac{\mu v_*^2}{2kT}\right) , \quad \text{or} \\ P_{01}(T) &\propto \exp(-3\chi) \equiv \exp\left(-\frac{3}{2}\frac{\theta'}{T}\right)^{1/3} \end{aligned}\right\} , \tag{3.3.7}$$

where χ, θ' are the parameters often used in the theory of vibrational relaxation,

$$\chi = \left(\frac{\pi^2\mu\omega^2}{2\alpha^2kT}\right)^{1/3} , \quad \theta' = \frac{4\pi^2\mu\omega^2}{\alpha^2k} . \tag{3.3.8}$$

Thus, the logarithm of the vibrational transition probability in a harmonic oscillator during adiabatic collisions is proportional to $T^{-1/3}$. This fundamental result, obtained by LANDAU and TELLER in 1936 [3.17], was confirmed by many experiments, despite considerable simplifications adopted in the statement of the problem. Expressions of the type (3.3.7) are usually called the *Landau-Teller relations*.

When deriving (3.3.5,7), we assumed that the nature of the relative motion A-B did not depend on the variation of the vibrational state of A. In this approximation the excitation and deactivation probabilities are equal: $P_{nm}(T) = P_{mn}(T)$. However, this contradicts the principle of detailed balancing (3.2.10). The contradiction can be removed if we take into account the change in kinetic energy of relative motion as a result of excitation or deactivation (see [3.11]). It follows from the solution of the quantum-mechanical problem that (3.3.5,7) must be related to the deactivation process, so that

$$\log P_{10}(T) \propto T^{-1/3} . \tag{3.3.9}$$

Note that according to (3.3.3) and (3.2.10), for a harmonic oscillator we have

$$P_{n+1,n}(T) = (n + 1)P_{10}(T) . \tag{3.3.10}$$

The dynamic problem of the vibrational excitation of harmonic oscillators has also been solved in a more complicated treatment with the account of forces of intermolecular attraction and the effect of molecular orientation by treating collisions in a three-dimensional representation. The transfer of not only translational but also rotational energy into vibrational energy has been taken into account. The results have been obtained taking account of initial vibrations of molecules. Some methods and results of the solution of such problems can be found in [1.8, 3.11], and reviews with numerous references in [3.12,18]. It should be noted, however, that within the adiabatic approximation, despite these refinements, the temperature dependence of P_{10} in general remains of the form (3.3.9). For nonadiabatic (fast) collisions this dependence is not observed (usually $P_{10} \propto T$). Considerable deviations from (3.3.9) take place due to the contribution to the V-T exchange of strong intermolecular attraction and dipole interactions between colliding particles (e.g., for H_2O molecules and radicals) or low excited electronic states (e.g., in the NO molecule), and also the effect of rotation on the V-T exchange probability. This is especially important for molecules composed of atoms with considerably differing masses (e.g., halogen hydride), and in a number of other cases, examples of which are given in Sect.3.5.

The problem becomes somewhat more complicated when an oscillator cannot be considered harmonic. In the simplest case account of anharmonicity can be taken by introducing into the exponent of (3.3.7) a correction corresponding to the energy transferred to the anharmonic oscillator during collision. For the anharmonic Morse oscillator (2.1.4) the frequency difference between the levels decreases with the level number n, so that to a good approximation we have

$$\omega_{n+1,n} = \omega_{10}(1 - 2x_e n) \quad , \tag{3.3.11}$$

where ω_{10} is the cyclic frequency of the ground transition $1 \to 0$ [$\omega_{10} = 2\pi\nu_0(1 - 2x_e)$, ν_0 is the frequency of the corresponding harmonic oscillator] and $x_e = \theta/4D$ is the anharmonicity parameter [θ is the characteristic vibrational temperature, D is the dissociation energy in degrees, which, according to (2.1.4), is close to D_e]. Using the approximation

$$\omega_{n+1,n}^{2/3} = \omega_{10}^{2/3}(1 - 2x_e n)^{2/3} \approx \omega_{10}^{2/3}\left[1 - (4/3)x_e n\right] \tag{3.3.12}$$

and substituting (3.3.12) into (3.3.10) and (3.3.7), we obtain

$$P_{n+1,n} = (n + 1)P_{10}\gamma^n \quad , \tag{3.3.13}$$

(see [3.11]), where

$$\gamma = \exp[(\theta'/T)^{1/3}\theta/4D] \equiv \exp(\chi\theta/2D) \quad , \tag{3.3.14}$$

and θ' and χ are determined according to (3.3.8). Note that $\log \gamma \propto T^{-1/3}$ [for N_2, O_2, and CO molecules $\gamma \approx \exp(1.87\ \theta^{2/3}T^{-1/3}x_e)$]. For a harmonic oscillator $D \to \infty$, $\gamma \to 1$. The expressions for the probabilities of the V-T exchange for anharmonic oscillators in the more general case can be found in [3.11,18]. During excitation of an anharmonic oscillator multiquantum transitions are possible even in first order perturbation theory. If a collision occurs adiabatically, the probability of multiquantum excitation (or deactivation) is much lower than the probability of a single-quantum process.

Now we shall give the results of the solution for the vibrational energy exchange between two colliding molecules. This is important for understanding the nature of phenomena occurring in infrared gasdynamic lasers. During such collisions r vibrational quanta of one molecule can be transformed, with a certain probability, into s quanta of the other molecule. Let us denote the probability of this process per one collision by $Q_{n,n-s}^{m,m+r}$, where n and m are the numbers ascribed to vibrational levels of the colliding molecules. As before, we shall be interested in the temperature dependence of the energy transition probability, and also its dependence on the level number (i.e., numbers n and m). If we generalize relations (3.3.3) for the case of multiquantum energy transfer in a system of harmonic oscillators of two types, we arrive at the expression [3.11,19]

$$Q_{n,n-s}^{m,m+r} = \frac{1}{r!s!}\frac{n!(m+r)!}{(n-s)!m!}Q_{s0}^{0r} \quad , \qquad (3.3.15)$$

so that for a single-quantum exchange we have

$$Q_{n,n-1}^{m,m+1} = n(m+1)Q_{10}^{01} \quad . \qquad (3.3.16)$$

In the case of anharmonic oscillators the single-quantum energy exchange is of greatest interest. Suppose the energy defect in this process is $\Delta E_n^{m+1} = |E(m+1) - E(m) - E(n) + E(n-1)|$; the value of ΔE_n^{m+1} is usually small compared with the vibrational quantum, so the exchange takes place at the near-resonance conditions. Then the most important contribution to the energy transfer (especially at low temperatures) is that from the long-range part of the intermolecular potential, due largely to the interaction between electrostatic, magnetic and induced dipoles, and multipoles [3.1]. The short-range part of the potential, determining repulsion forces according to (3.3.1), causes quantum transitions and curvature of the paths of colliding molecules, while the long-range part (attraction forces) results only in the change of vibrational states and almost does not affect the relative motion. With the account of the short-range part of the potential near the resonance, the exchange probability increases with the increase of the relative energy of collision, so that [3.11,20,21]

$$\left[Q_{sh.r}\right]_{n,n-1}^{m,m+1} \propto n(m+1)Tf_{sh.r}\left(\Delta E_n^{m+1}/T^{1/2}\right) \quad , \tag{3.3.17}$$

where $f_{sh.r}(\Delta E_n^{m+1}/T^{1/2})$ is the function of the energy defect. This function decreases monotonically with increase of the argument. For the long-range part of the potential, the probability is mainly determined by the collision time, that is, it is inversely proportional to the relative energy of collision and hence to temperature [3.2,12,22,23],

$$\left[Q_{l.r}\right]_{n,n-1}^{m,m+1} \propto n(m+1)T^{-1}f_{l.r}\left(\Delta E_n^{m+1}/T^{1/2}\right) \quad , \tag{3.3.18}$$

where $f_{l.r}$ is close to $f_{sh.r}$ in (3.3.17). At low temperatures and small energy defect usually $Q_{sh.r} << Q_{l.r}$. Far from the resonance the contribution from the repulsion forces is very small. The account of the actual decrease of the energy defect as a result of simultaneous multiquantum transitions permits us to obtain a good approximation to experimental results in a wide range of values ΔE_n^{m+1}, using in the solution only the long-range part of the potential [3.24]. In any case at high temperatures the effect of short-range repulsion forces increases. Thus, the temperature dependence of the V-V exchange probability is obtained by combining (3.3.17) and (3.3.18) and consequently can be nonmonotonic. The concrete examples of these dependencies for the processes involving CO_2 and CO molecules are given in Sects.3.5, 5.1, and 6.1.

3.4 Vibrational Relaxation Kinetics

Let us go over from the dynamic problem for separate collisions to the statistical description of the vibrational state population of molecules in a gas, using microkinetic balance equations for a certain vibrational level (3.2.9). Let us begin with the analysis of the V-T exchange in the assumption that levels are equidistant, transitions are single-quantum, transition probabilities are of the forms (3.3.10) and (3.3.13)[8], and internal degrees of freedom of the partners in collision with the given molecule do not participate in the V-T exchange. Then in (3.2.9) we can carry out the independent summation over j, j' and take outside the brackets the resulting value of the

8 The application of (3.3.13), obtained for an anharmonic oscillator with nonequidistant levels, to this model is inconsistent, but it allows us, however, to take account of anharmonicity in the simplest way.

total concentration of molecules n_B. For convenience let us introduce here somewhat different notation for levels and populations: x_n is the population of the nth level of the molecule under consideration ($x_n \equiv n_{A(i)}$ of Sect. 3.2). The variation of x_n is determined by populating this level at the expense of depopulation of levels n + 1 and n - 1, and vice versa. Then the following system of microkinetic balance equations is obtained:

$$\frac{dx_0}{dt} = Z(P_{10}x_1 - P_{01}x_0) \quad ,$$

$$\frac{dx_1}{dt} = Z(P_{21}x_2 - P_{10}x_1 - P_{12}x_1 + P_{01}x_0) \quad ,$$

$$\frac{dx_n}{dt} = Z(P_{n+1,n}x_{n+1} - P_{n,n-1}x_n - P_{n,n+1}x_n + P_{n-1,n}x_{n-1}) \quad . \tag{3.4.1}$$

Here, as before, we assume that the collision frequency Z of the molecule under consideration with all the other molecules does not depend on its vibrational state (the subscript AB is omitted here). Let us determine macroscopic variables — the total vibrational energy ε and the total number of molecules N — in terms of population x_n,

$$\varepsilon = h\nu \sum nx_n \quad , \quad N = \sum x_n \quad . \tag{3.4.2}$$

At equilibrium, when the population distribution over the levels is of the Boltzmann type, the overall vibrational energy, according to (D.12), for $T_{vib} = T$ is[9]

$$\bar{\varepsilon} = \frac{Nh\nu\exp(-\theta/T)}{1-\exp(-\theta/T)} \equiv \frac{Nh\nu}{\exp(\theta/T) - 1} \quad . \tag{3.4.3}$$

Using the principle of detailed balancing (3.2.10), we obtain from the first equation of (3.4.1)

$$P_{01} = P_{10}\exp(-\theta/T) \quad . \tag{3.4.4}$$

To derive the relaxation equation for energy ε, we multiply each of equations (3.4.1) by $h\nu$ and the corresponding level number, and then sum up the left-hand and the right-hand sides of these equations. Then, after transformation of the series obtained on the right-hand side, we obtain

9 Here, unlike (D.12), the value of energy is computed not for one molecule, but for the total number of molecules N.

$$\frac{d\varepsilon}{dt} = Zh\nu\left(\sum_{n=0}^{\infty} P_{n,n+1}x_n - \sum_{n=1}^{\infty} P_{n,n-1}x_n\right) \quad . \qquad (3.4.5)$$

For a harmonic oscillator, taking into account (3.3.10) and (3.4.4), we get

$$\frac{d\varepsilon}{dt} = Zh\nu P_{10}\left[\exp(-\theta/T) \sum (n+1)x_n - \sum nx_n\right]$$

$$= ZP_{10}\{[\exp(-\theta/T) - 1]\varepsilon + Nh\nu \exp(-\theta/T)\} \ ,$$

and as a result, with (3.4.3), we arrive at the relaxation equation for vibrational energy,

$$\frac{d\varepsilon}{dt} = \frac{\bar{\varepsilon} - \varepsilon}{\tau} \quad , \qquad (3.4.6)$$

where the quantity

$$\tau = \{ZP_{10}[1 - \exp(-\theta/T)]\}^{-1} \qquad (3.4.7)$$

is the so-called *vibrational relaxation time* for a system of harmonic oscillators.

The relaxation equation (3.4.6) is valid for any stage of deviation from equilibrium only for a system of harmonic oscillators (or for any oscillator near equilibrium). For harmonic oscillators the relaxation time τ is independent of the degree of deviation from equilibrium and is determined only by the temperature and concentration of the partners in collision.

To verify the validity of (3.4.6), let us turn to experiments. Among experimental methods of studying energy exchange kinetics the most widespread are ultrasonic acoustics methods, shock waves, and optical excitation (in particular the laser fluorescence method) (see [1.8, 3.2,16]). The measurement of vibrational energy when the system returns to equilibrium and the application of relaxation equation (3.4.6) make it possible to determine the value of τ. The results of measurements of τ in the widest temperature range were obtained using shock waves in shock tubes [1.8]. Systematic data on the results of many investigations are presented in [3.2,16,25,26]. One of the latest reviews on the study of diatomic molecules is given in [3.12]. The results of analysis of vibrational energy relaxation in carbon dioxide will be given in Sect.3.5. All these investigations basically confirm the possibility of describing the V-T exchange with the help of (3.4.6). However, large errors in determining τ are obtained. These errors are not only due to experimental difficulties but also due to the effect of deviation from equilibrium on τ when the relaxation equation in the form (3.4.6) is used [3.27].

The dependence of τ on the concentration of partners in collision is manifested in the fact that at the same temperature τ is inversely proportional to the partial pressure of the gas — the partner in collision with the molecules under consideration. This is completely confirmed by experiments and indicates that the vibrational energy exchange occurs during double collisions. Excluding this trivial dependence of τ on pressure, we shall relate the vibrational relaxation time τ to a certain pressure, e.g., atmospheric pressure, and rewrite (3.4.6) in the form

$$\frac{d\varepsilon}{dt} = \frac{p}{\tau}(\bar{\varepsilon} - \varepsilon) \quad , \tag{3.4.8}$$

where τ has the dimension [time]$\cdot$[pressure][s $\cdot$ atm] and depends only on temperature. In a mixture the effective value of the relaxation time is determined by the relation

$$1/\varepsilon = \sum \xi_s/\tau(s) \quad , \tag{3.4.9}$$

where $\tau(s)$ is the relaxation time for collisions of molecules under consideration with molecules of type s, whose molar fraction is ξ_s. According to (3.4.7), the temperature dependence of τ is determined by the dependence $P_{10}(T)$, that is by the Landau-Teller relations (3.3.7,9). For this reason the results of measurement of τ are represented in the coordinates $(\log\tau,\ T^{-1/3})$, i.e., in the Landau-Teller coordinates.

From the analysis of many experimental results, MILLIKAN and WHITE [3.25] suggested for τ a semiempirical formula

$$\ln(p\tau) = 1.16\times 10^{-3}\sqrt{\mu}\theta^{4/3}(T^{-1/3} - 0.015\mu^{1/4}) - 18.42 \quad , \tag{3.4.10}$$

where τ is expressed in seconds, p in atmospheres, μ is the reduced mass of colliding pairs in atomic units, and θ is the characteristic vibrational temperature in K. In practical calculations it is expedient to use analytical approximations of available experimental data for certain molecules (see Sect.5.1), or, if they are unknown, to use formula (3.4.10) for estimates.

At vibrational relaxation, in a system of harmonic oscillators there is always the Boltzmann distribution of populations x_n of the levels with the current value of the vibrational temperature T_{vib}, (D.9),

$$x_n = N[1 - \exp(-\theta/T_{vib})]\exp(-n\theta/T_{vib}) \quad . \tag{3.4.11}$$

This is shown for relaxation in isothermal conditions in [3.28]. The more general case of nonisothermal relaxation in a system of harmonic oscillators taking account of the V-V exchange is considered in Sect.4.2 and Appendix M. After determining the value of ε by solving (3.4.6) and computing

T_{vib} according to (D.12), we can, using (3.4.11), find the population of any level, which is important for studying gasdynamic lasers. This set of computations has been used in the analysis of processes in a CO_2 gasdynamic laser, for which the harmonic oscillator model proved to be sufficient.

The simplest account of the effect of anharmonicity on the values of transition probabilities of the form (3.3.13) permits us to obtain the relaxation equation which accounts for the anharmonicity of molecules. In this case it is expedient to neglect the deviations from the Boltzmann distribution of populations, i.e., to adopt (3.4.11) and use as a macroscopic parameter the vibrational temperature T_{vib}. Then, proceeding from (3.3.13), (3.4.4), and (3.4.11), we can obtain from (3.4.5) for $T_{vib} < \theta/\ln\gamma$

$$\frac{d\varepsilon}{dt} = ZP_{10}\,\frac{[1 - \exp(-\theta/T)][1 - \exp(-\theta/T_{vib})]^2}{[1 - \gamma\exp(-\theta/T_{vib})]^2}\,(\bar{\varepsilon} - \varepsilon) \quad . \tag{3.4.12}$$

Retaining the general form of the relaxation equation in the simplest form (3.4.6), we introduce a correction into the relaxation time τ_h obtained for a system of harmonic oscillators (3.4.7), so that

$$\tau_{an} = \tau_h\left[\frac{1 - \gamma\exp(-\theta/T_{vib})}{1 - \exp(-\theta/T_{vib})}\right]^2 \quad , \tag{3.4.13}$$

where τ_{an} is the vibrational relaxation time for a system of anharmonic oscillators in conditions where anharmonicity affects only the probability of transitions according to (3.3.13). It follows from (3.4.13) that τ depends not only on the gas temperature T but also on the vibrational temperature T_{vib}, i.e., on the extent of deviation from vibrational equilibrium. Relation (3.4.13) was confirmed in experiments investigating vibrational relaxation behind the front of a strong shock wave in oxygen, nitrogen, and other gases[10]. The explicit dependence of τ on T_{vib} disappears in two extreme cases: for a very low degree of excitation (in this case $\tau_{an} \approx \tau_h$) and at equilibrium.

Let us go over to the V-V exchange kinetics and derivation of the respective relaxation equations. As before, we shall begin with the analysis of processes in molecular mixtures, which are approximated by the harmonic os-

10 Thus, in the O_2-Ar mixture, for T = 6000 K the value of τ in the process of vibrational excitation of O_2 decreases by a factor of 2 to 3 with an increase of T_{vib} [3.29]. In nitrogen at a higher temperature τ decreases by an order of magnitude.

cillator model. The sequence of operations in this case remains the same as at the beginning of the section, the only difference being that both colliding molecules participate in the energy exchange (or the two types of vibrations which can exist in one molecule, as in the case of the CO_2 molecule). In the simplest case of nondegenerate harmonic oscillators of two types, i and j (level populations x_n and y_m, respectively), for a single-quantum V-V exchange, the balance equation, according to (3.2.9), should be written as

$$\left(\frac{dx_n}{dt}\right)_{VV} = z\sum_m \left(Q^{m-1,m}_{n+1,n}\, x_{n+1}y_{m-1} - Q^{m,m-1}_{n,n+1}x_n y_m - Q^{m,m+1}_{n,n-1}x_n y_m + Q^{m+1,m}_{n-1,n}x_{n-1}y_{m+1}\right) \quad , \qquad (3.4.14)$$

where z is the collision frequency for these molecules per unit concentration. Here the first term in the parentheses corresponds to the molecular transition from the upper level to the level under consideration, the second, to the transfer of molecules to the upper level, the third, to the transition of molecules to the lower level, and the last, to the arrival of molecules from the lower level. The rate of each process is proportional to the initial population of the levels, and the probability plays the role of proportionality factor. The principle of detailed balancing enables us to obtain in this case too the relation between the probabilities of forward and reverse processes of energy transfer,

$$Q^{m,m+1}_{n,n-1} = Q^{m+1,m}_{n-1,n}\exp[-(\theta_j - \theta_i)/T] \quad , \qquad (3.4.15)$$

where θ_i and θ_j are the characteristic vibrational temperatures of the respective vibrational modes. The mathematical treatment, as before, consists in the multiplication of each of the equations (3.4.14) by $nh\nu$ and summation over all n (see [1.8]). Thus we arrive at the following rate equation for the total vibrational energy ε_i for the single-quantum V-V transfer [3.30]:

$$\left(\frac{d\varepsilon_i}{dt}\right)_{VV} = zQ^{01}_{10}\left[\frac{\varepsilon_j}{k\theta_j}(\varepsilon_i + k\theta_i N_i)\exp\left(\frac{\theta_j - \theta_i}{T}\right) - \varepsilon_i\left(\frac{\varepsilon_j}{k\theta_j} + N_j\right)\right] \quad , \qquad (3.4.16)$$

where N_i and N_j are the numbers of the kth and j^{th} types of oscillations, respectively, in the volume under consideration and k is the Boltzmann constant. It is convenient to represent the product $zQ^{01}_{10}N_j$ as the rate constant k_{ij} of the V-V transfer, and express the vibrational energy in terms of the number of vibrational quanta e_i taken on the average per given mode[11]. (Footnote see next page.) In the harmonic oscillator approximation, according to (3.1.9), $e_i = [\exp(\theta_i/T_i) - 1]^{-1}$, where T_i is the vibrational temperature, so that $\varepsilon_i = k\theta_i N_i e_i$.

The rate constant k_{ij} of the V-V exchange is proportional to the concentration of the partners of the given molecule in collisions, i.e., it is proportional to the gas pressure. Then, separating out the pressure dependence of k_{ij} and passing over in (3.4.16) to e_i, we obtain

$$\left(\frac{de_i}{dt}\right)_{VV} = pk_{ij}\left[e_j(e_i + 1)\exp\left(\frac{\theta_j - \theta_i}{T}\right) - e_i(e_j + 1)\right] \quad , \tag{3.4.17}$$

where k_{ij} has the dimensions reciprocal in time and pressure, e.g., $s^{-1} \cdot atm^{-1}$ and depends only on temperature. In (3.4.17) the effect of the deviation from equilibrium on the value of k_{ij} is not taken into account, and all the resulting inaccuracies are included into the error in determining k_{ij}. The absolute value and the temperature dependence of k_{ij} are basically determined by the same experimental technique as used in measuring τ_{V-T}. Concrete examples of the results of measurement of k_{ij} will be given in the next section. Note that in gas mixtures the rate constant is

$$k_{ij} = \sum \xi_s k_{ij}(s) \quad , \tag{3.4.18}$$

where $k_{ij}(s)$ is the rate constant of the V-V exchange for molecules under consideration with the molecules of the sth type.

The passage from the vibrational energy to the average number of vibrational quanta at the given mode does not change the form of the relaxation equation for the V-T exchange; thus for both processes of V-T and V-V exchange we have

$$\frac{de_i}{dt} = \frac{p}{\tau_{VT}}(\bar{e}_i - e_i) + pk_{ij}\left[e_j(e_i + 1)\exp\left(\frac{\theta_j - \theta_i}{T}\right) - e_i(e_j + 1)\right] \quad , \tag{3.4.19}$$

and similarly for e_j. Here $\bar{e}_i$ is the value of e_i at vibrational equilibrium. It is important that the account of the V-V exchange makes the relaxation equations nonlinear; the linear form of equations is retained only when the V-V exchange is a resonance exchange, i.e., when $\theta_i = \theta_j$. In this case

---

11 Sometimes, for clarity, the quantity $\tau_{VV} = 1/k_{ij}$ is introduced, which conventionally describes the relaxation time of the V-V exchange, and conversely, V-T transfer is described in terms of the rate, putting $k_{VT} = 1/\tau_{VT}$.

$$\left(\frac{de_i}{dt}\right)_{VV} = pk_{ij}(e_j - e_i) \quad . \qquad (3.4.20)$$

The analysis of the processes occurring in infrared gasdynamic lasers on triatomic molecules deals with the multiquantum V-V exchange involving degenerate harmonic oscillators. Let n_i and ℓ_i be the number of the initial level and the number of the ith quanta being exchanged, respectively, in such a multiquantum transition (and similarly for j), and during molecular collisions the transition $n_i \pm \ell_i \rightleftharpoons n_j \pm \ell_j$ is assumed to occur. Then as a result of the transformation of the microkinetic equation, in the square brackets of (3.4.17,19) there appear ℓ_i products $e_i(e_i + 1)$(and similarly for j) with the respective account of the transition energy at the exponent. The rate of variation of the average number of vibrational quanta e_i in this case is proportional to the number of lost (or gained) quanta and should be divided by the statistical weight g_i of this mode of vibrations,

$$\left(\frac{de_i}{dt}\right)_{VV} = pk_{ij}\frac{\ell_i}{g_i}\left[e_j^{\ell_j}(e_i + 1)^{\ell_i}\exp\left(\frac{\ell_j\theta_j - \ell_i\theta_i}{T}\right) - e_i^{\ell_i}(e_j + 1)^{\ell_j}\right] \quad . \qquad (3.4.21)$$

Moreover, according to the principle of detailed balancing, we have

$$k_{ji} = k_{ij}\exp[(\ell_j\theta_j - \ell_i\theta_i)/T] \quad . \qquad (3.4.22)$$

The equations are of a similar form when three or more modes of vibrations simultaneously take part in the V-T transfer. In the most general form these equations have been obtained, for example, in [3.19,31]. Examples of these relaxation equations are given in the next section. The form of (3.4.21) depends on the specific nature of processes of energy transfer, so for different channels of the V-V exchange the relaxation equations, and hence the rate constants, are different. In the case of polyatomic molecules these channels are usually unknown, which not only complicates the form of relaxation equations but also is a source of systematic errors.

Let us discuss an important particular case of vibrational relaxation in a gas mixture, when the rate of the vibrational exchange between different levels and even between different modes (i and j) considerably exceeds the rate of the V-T energy exchange, so that $\tau_{VV} \ll \tau_{VT}(Q_{10}^{01} \gg P_{10})$. Then at times $\tau_{VV} \ll t \ll \tau_{VT}$ a situation can develop in a relaxing system where the V-T exchange has not yet had time to occur and the fast V-V exchange provides for a certain quasi-stationary distribution of vibrational energy over vibrational modes or over molecular components of the j, j mixture. In this case $(de/dt)_{VT}$ is still close to zero because of the large value of τ_{VT}, and $(de/dt)_{VV}$ is already close to zero due to completion of the relaxation process of the V-V exchange (due to smallness of τ_{VV}). Then, equating to zero the right-hand side of (3.4.17) and taking into account the Boltzmann distribution over the levels within each component with current values of vibrational temperature T_i and T_j, we get after simple transformations

$$(\theta_j/T_j) - (\theta_i/T_i) = (\theta_j - \theta_i)/T \quad . \tag{3.4.23}$$

This expression can be easily generalized for the case of a fast multiquantum exchange [3.32,33] ,

$$(r\theta_j/T_j) - (s\theta_i/T_i) = (r\theta_j - s\theta_i)/T \quad , \tag{3.4.24}$$

where r and s are the numbers of exchanged quanta. Equation (3.4.23) was confirmed by experiments on nonequilibrium cooling of the N_2-CO_2 mixture in a nozzle [3.34]. In [3.35] these relations were used for the analysis of quasi-stationary equilibrium in mixtures of diatomic and triatomic molecules used for development of gasdynamic lasers.

Let us consider now the vibrational relaxation kinetics, when the nonresonant V-V exchange occurs between anharmonic oscillators. Formally, for single-quantum transitions the microkinetic balance equations for populations in this case are similar to (3.4.14) for a harmonic oscillator. Using the principle of detailed balancing, we can write

$$Q_{n-1,n}^{m+1,m} = Q_{n,n-1}^{m,m+1} \exp\{[E(m+1) - E(m) - E(n) + E(n-1)]/kT\} \quad ,$$

so that, using (2.1.5) and introducing $x_e = \theta/4D$, see (3.3.11), we get

$$Q_{n-1,n}^{m+1,m} = Q_{n,n-1}^{m,m+1} \exp[-2\theta x_e(m - n + 1)/T] \quad . \tag{3.4.25}$$

Then, proceeding from (3.4.14), we obtain the system of balance equations, which take account of the vibrational exchange,

$$\left(\frac{dx_n}{dt}\right)_{VV} = z\left(\sum_m Q_{n+1,n}^{m,m+1}\{x_m x_{n+1} - x_{m+1}x_n \exp[-2\theta x_e(m-n)/T]\}\right.$$
$$\left. - \sum_m Q_{n,n-1}^{m,m+1}\{x_m x_n - x_{m+1}x_{n-1} \exp[-2\theta x_e(m-n+1)/T]\}\right) \quad . \tag{3.4.26}$$

If in a system of anharmonic oscillators the V-V exchange probability is much higher than that for V-T transitions, then, as a result of fast intermolecular energy transfer, a quasi-stationary state sets in that is analogous to the one described above. In this case $(dx_n/dt)_{VV} \to 0$, and from (3.4.26) it follows that

$$(x_{m+1}/x_m)\exp(-2\theta x_e m/T) = (x_{n+1}/x_n)\exp(-2\theta x_e n/T) \quad ,$$

from which it is clear that the combination of values presented here does not depend on the level number and is equal, for example, to $\zeta = x_1/x_0$ (for $n = 0$). The quantity ζ can be conveniently expressed in terms of the effective "temperature" of the first level as a measure of relative population x_1/x_0

$$T_1 = \frac{\theta}{\ln(x_0/x_1)} \quad , \tag{3.4.27}$$

where θ is the value of the ground vibrational quantum in degrees. Then $\zeta = (\theta/T_1) - (\theta/T)$ and

$$x_n = x_0 \exp\{-n[\theta/T_1 - (n - 1)\theta x_e/T]\} \quad . \tag{3.4.28}$$

The molecular distribution over the levels presented in the form (3.4.28) is often called the *Treanor distribution* after the first author of [3.36]. The Treanor distribution describes a quasi-stationary state of a system of anharmonic oscillators in the absence of the V-T exchange. For harmonic oscillators ($x_e = 0$), (3.4.28) becomes the Boltzmann distribution, and T_1 becomes the vibrational temperature T_{vib}. From (3.4.28) it is clear that for $T_1 > T$ the upper levels of anharmonic oscillators are populated to a much greater extent than in the case of the Boltzmann distribution. For levels with $n > n^*$, where n^* is the integer closest to

$$n^* = (T/2x_eT_1) + 1/2 \quad , \tag{3.4.29}$$

the Treanor distribution corresponds to the population inversion of the levels, for which $x_{n+1} > x_n$ (Fig.3.1). The creation of such conditions is of great interest for the development of laser systems. According to (3.4.29), it is expedient to decrease the gas temperature T and to increase the initial store of vibrational energy (by increasing T_1), simultaneously taking care of the increasing rate of the V-V exchange and cessation of the V-T exchange. Such a distribution is formed due to the decrease in the exchange probability for higher vibrational levels, as a result of which their population retains a relatively high value.

Actually, however, the dependence of the probabilities $P_{n,n-1}$ and $Q^{m,m+1}_{n,n-1}$ on level numbers is such (see Sect.3.3) that already for $n \approx n^*$ the V-T exchange probability approaches the V-V transfer probability even if $Q^{m,m+1}_{n,n-1} \gg P_{n,n-1}$ for the lower levels. For this reason the quasi-stationary Treanor distribution is not realized for the upper levels, because a part

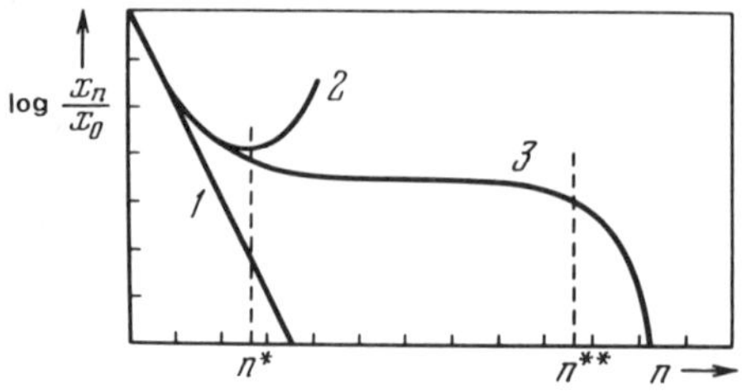

Fig.3.1. Population distribution for the nonresonant vibrational exchange in a quasi-stationary state. (*1*) Boltzmann distribution; (*2*) Treanor distribution; (*3*) Gordiets distribution

of the energy is transferred into translational motion of molecules. The distribution of anharmonic molecules over the levels with simultaneous V-V and V-T exchanges is obtained either as a result of a numerical solution of the system of microkinetic balance equations (3.4.1) and (3.4.26) [3.37] or, in the approximate treatment, analytically [3.32,38,39]. The shape of this distribution (*3*) is shown in Fig.3.1. Three regions can be distinguished: lower levels ($0 \leqq n \leqq n^*$), for which the distribution is close to the Treanor distribution; middle levels ($n^* \leqq n \leqq n^{**}$), where the number of molecules changes little with the level number n; and upper levels ($n > n^{**}$), where the number of molecules sharply decreases. The existence of a plateau for middle levels is due to the competition between the nonresonance V-V exchange, which tends to increase the population of this region, and V-T transitions, which result in depopulation of excited levels with $n > n^*$; eventually, for $n > n^{**}$ the V-T exchange prevails. The value of n^{**} is obtained in [3.38] and is

$$n^{**} = \frac{1}{\ln\gamma} \ln\left(\gamma^{n^*} + \frac{12\theta x_e[1 - \exp(-\theta/T_1)]}{T(\ln\gamma)^2} \times \frac{Q_{10}^{01}}{P_{10}} (n^* + 1)\exp\left\{[-(n^*)^2\theta x_e/T] + \frac{1}{2}\right\}\right) . \tag{3.4.30}$$

The value of γ is determined from (3.3.14).

In the general case the derivation of relaxation equations, which describe the simultaneous V-T and V-V exchange in a system of anharmonic oscillators, is difficult. If the majority of molecules is concentrated at lower levels, $n \ll n^*$ (these are exactly the typical conditions in gasdynamic lasers), then approximate relations can be considered [3.38], which provide for the phenomenological description of these processes for $T \ll \theta$ and low resonance defect ($2\theta x_e|m - n| \ll T$). These relations connect the vibrational energy store with the gas temperature T and the effective "temperature" of the first level T_1 (which in this case plays the role of a macroscopic parameter), and also establish the value of the vibrational relaxation time τ_{an},

$$\varepsilon = k\theta\left[\frac{\exp(-\theta/T_1)}{1 - \exp(-\theta/T_1)} + f_{Tr}(n^*)\,\frac{n^* + 1}{\sqrt{e}}\,(n^{**} - n^*)\right] \quad , \qquad (3.4.31)$$

$$\tau_{an} = \frac{\varepsilon}{k\theta P_{10}}\left\{\frac{\exp(-\theta/T_1)[1 - \exp(-\theta/T_1)]}{[1 - \gamma\exp(-\theta/T_1)]^2}\right.$$

$$\left. + f_{Tr}(n^*)\,\frac{n^* + 1}{\sqrt{e}\,\ln\gamma}\,(\gamma^{n^{**}} - \gamma^{n^*})\right\}^{-1} \quad . \qquad (3.4.32)$$

Here f_{Tr} is the relative value of the nth level population according to the Treanor distribution,

$$f_{Tr}(n) = [1 - \exp(-\theta/T_1)]\exp\{-n[\theta/T_1 - (n - 1)\theta x_e/T]\} \quad . \qquad (3.4.33)$$

When the importance of the nonresonant vibrational exchange decreases, $n^{**} \to n^*$ and the molecular distribution is close to the Boltzmann distribution, then the ratio τ_{an}/τ_h is obtained from (3.4.32) in the form of (3.4.13).

3.5 Vibrational Exchange Kinetics in CO_2

The progress in the study of vibrational relaxation kinetics can be demonstrated by using as an example the energy transfer during collisions of CO_2 molecules with each other and with other molecules. We are especially interested here in the experimental results. At an early stage of investigations acoustical methods were widely used (the measurement of ultrasonic absorption and dispersion in a gas mixture)[3.16]. In recent years much information was obtained using the laser fluorescent method, where the change in the population of vibrational levels was studied in a gas excited by the CO_2 laser radiation [3.85-87]. This method was supplemented by the analysis of the phase shift of the fluorescence damping signal with periodic excitation [3.77]. Considerable progress was achieved in the analysis of absorption and amplification of laser radiation in the mixtures of CO_2 with other gases, and also in the study of laser operating conditions in these mixtures and damping of oscillations after the pumping has been switched off [3.75,106]. Gasdynamic methods, in particular the analysis of vibrational kinetics of CO_2

molecules excited by shock waves, proved to be useful [1.8, 3.57]. Most of these works have, however, the same drawback: in each experiment as a rule the population of only one vibrational state of CO_2 molecules was investigated or the energy of only one vibrational mode was determined. This hampered the separation of the energy transfer channels in experiments, which resulted in considerable systematic errors, due to which the data presented by different authors sometimes do not agree, especially when different techniques were used. Apparently, some processes are in fact more complicated than it was assumed when treating primary experimental results. Typically random errors in experiments are smaller than systematic ones and are (in the standard deviation representation) from 20 to 50%.

The vibrational exchange parameters are the vibrational relaxation time τ for V-T transitions and the rate constant k_{ij} for the V-V energy transfer between indicated modes through certain channels[12]. Let us introduce the notation for vibrations participating in the exchange as it was done in Sect.2.1 for CO_2 molecules; further, let the subscripts 1, 2, 3 correspond to the first [(ν_1) symmetrical valence], the second [(ν_2) antisymmetrical deformation], and the third [(ν_3) antisymmetrical valence] modes of the CO_2 vibrations, respectively, and the subscript M (or 4) be related to other molecules. It should be noted that the gas-kinetic collision cross section in CO_2 for T = 300 K corresponds to the T-T exchange rate of about $6 \times 10^9\ s^{-1} \cdot atm^{-1}$.

The transitions between the symmetrical valence ν_1 and deformation ν_2 vibrational modes are the fastest vibrational exchange processes in CO_2 molecules. High rates of these processes are due to the Fermi resonance between the levels of these modes. The results of experiments in which the rate of the $10^00 \rightleftharpoons 02^00$ exchange was measured (notation for the levels of the CO_2 molecule is given in Sect.2.1) differed by two orders of magnitude.

Year	1968	1969	1972	1972	1973	1975	1975
Rate constant k_{12} ($s^{-1} \cdot atm^{-1}$)	3×10^8	7×10^6	1×10^7	8×10^5	1.1×10^8	1.7×10^6	$2.3 \times 10^{[illegible]}$
Reference	[3.42]	[3.43]	[3.44]	[3.45]	[3.46]	[3.47]	[3.48]

This difference does not allow us to make an unambiguous conclusion. The analysis [3.49] of the $10^0 \rightleftharpoons 02^00$ exchange rate, taking account of the

---

12 See [3.26]. The later works are ones by KERBER et al. [3.40] and BLAUER and NICKERSON [3.41].

short-range repulsion forces between the molecules in the Schwartz-Slawsky-Herzfeld theory (SSH, see Sect.3.3) shows that for T = 300 K this rate must be about 5×10^7 $s^{-1} \cdot atm^{-1}$. The effect of the long-range repulsive forces between the molecules upon the multipole interaction and the possbility of the V-R transfer according to [3.50] increases the exchange rate by more than the order of magnitude (up to 8×10^8 $s^{-1} \cdot atm^{-1}$), which exceeds the experimental results. Thus, the theory in any case predicts higher values of the probability of the $\nu_1 \rightleftharpoons \nu_2$ exchange. Hence in many problems we can assume that the symmetrical modes of the CO_2 molecules vibrations are at mutual equilibrium.

If the distribution of molecules over the levels of the ith mode is of the Boltzmann type, we can introduce, according to (3.1.1), the vibrational temperature T_i, which determines the vibrational energy of this mode. If there is an equilibrium in $\nu_1 \rightleftharpoons \nu_2$, then we have $T_1 = T_2$. It should be noted, however, that if we take into account the Fermi resonance in the interaction of these twin modes of the CO_2 vibrations, then it is meaningless to introduce different temperatures T_1 and T_2: if these temperatures exist, they are equal [3.51]. Unfortunately, in some works on CO_2 gasdynamic lasers we can find unreasonable assumptions about different values of T_1 and T_2. If so, it would be correct to give up the assumption about the Boltzmann distribution of molecules in these modes and consider a kinetic model which would take into account each vibrational level and transitions between them separately (level kinetics). For example, such an approach is required in the analysis of processes in mixtures with a low CO_2 content [3.52], where the Boltzmann distribution has no time to set in. The required temperature-dependent values of transition probabilities between many levels of the CO_2 molecule were obtained on the basis of the SSH theory including the Fermi resonance [3.41,53].

When the V-T transfer can be ignored, the high probability of the V-V transfer between lower vibrational levels of the twin symmetrical modes of the CO_2 molecule promotes the universal (independent of the rate constants) quasi-stationary distribution of molecules. As in the preceding section, the form of this distribution for $\tau_{VV} \ll \tau_{VT}$ can be obtained by equating the number of forward and reverse V-V transitions per unit time between all the levels of the twin modes in CO_2. Of major importance here is the anharmonicity (the levels are not equidistant), which is the result of the energy levels splitting owing to the Fermi resonance (see Sect.2.1). According to [3.51,54] such a distribution has the form

$$x_{v_1v_2^{\ell}v_3} = x_{00^00}g_{\ell}\exp[-\zeta_s(2v_1 + v_2) - \zeta_a v_3 - (E_{v_1v_2^{\ell}v_3}/kT)] \quad , \quad (3.5.1)$$

where $E_{v_1v_2^{\ell}v_3}$ is the energy of the respective level, $g_{\ell} = 1$ for $\ell = 0$, $g_{\ell} = 2$ for $\ell \neq 0$. ζ_s and ζ_a are the parameters for the symmetrical (twin) and antisymmetrical CO_2 modes, respectively, which, as in Sect.3.4 [(3.4.27) and (3.4.28)], determine the vibrational energy: if the levels are equidistant, $\zeta = (\theta/T_{vib}) - (\theta/T)$, where T_{vib} is the vibrational temperature and T is the translational gas temperature. For $v_3 = 0$, (3.5.1) can be represented in the form

$$x_{v_1v_2^{\ell}0} = x_{00^00}g_{\ell}\alpha_s^{v}\exp(-\Delta_{v_1v_2^{\ell}0}/T) \quad , \quad (3.5.2)$$

where $\alpha_s = \exp(-\theta/T_2)$, T_2 is the vibrational temperature of the twin modes, and $\Delta_{v_1v_2^{\ell}0} = E_{v_1v_2^{\ell}0}/k) - \theta_2 v$ is the shift of the level under consideration with respect to the corresponding harmonic level $2v_1\theta_2 + \theta_2 v_2$. For the high vibrational and low translational temperatures the distributions (3.5.1,2) have a sawtooth shape as is shown in Fig.3.2. In this case the population inversion for a number of levels can be created [3.54]. The dependence of the transition probability on the vibrational number is required for computations in the absence of a quasi-stationary distribution as established in [3.51]. In this case the relaxation equation describing the vibrational energy of twin modes as a whole is obtained in the general form (3.4.6).

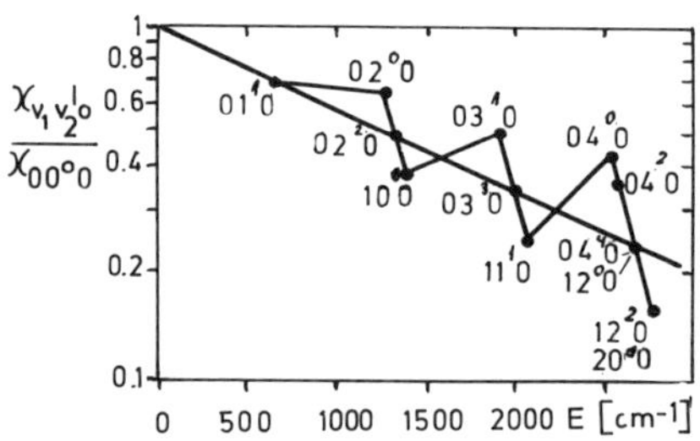

Fig.3.2. Relative population distribution of the CO_2 twin symmetrical modes for T = 300 K, $\alpha_s = 0.7$, depending on energy levels [3.54]

The main channels of the V-T exchange in CO_2 are the transitions between the bending vibrational and translational degrees of freedom. The probability of the V-T exchange with other vibrational modes is much lower. For $10^00 \rightleftharpoons 00^00$ transitions and their multiples, lower probability follows from the theory (which includes Fermi resonance) since here we have two-quantum transitions. The probability of the collisional $00^01 \rightleftharpoons 00^00$ transitions is

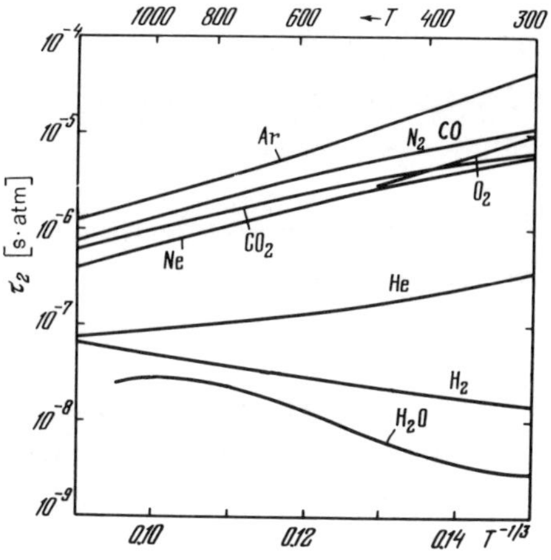

Fig.3.3. Averaged τ_2 values for collisions of the CO_2 molecule with atoms of He, Ne, Ar [3.57,58]; He [3.59]; with molecules N_2 [3.57,60,61]; O_2 [3.60,62]; H_2 [3.57,63-65]; CO [3.66,67]; CO_2 [3.56,58,60,61,68]; H_2O [3.69-71]

still lower. Thus, in the analysis of the V-T exchange in CO_2 we obtain relaxation equation (3.4.8) in the form

$$\frac{de_2}{dt} = \frac{p}{\tau_2}(\bar{e}_2 - e_2) \quad , \tag{3.5.3}$$

where τ_2 is the relaxation time for the energy of the bending vibrations. The measurement of τ_2 is complicated by the fact that the V-T transitions in CO_2 are accompanied by the fast V-V exchange with other vibrational modes, so that in experiments a certain effective total time τ_Σ of the V-T relaxation for the whole molecule is registered. Then τ_2 is computed using the measured value of τ_Σ by the relation $\tau_2 = \tau_\Sigma(c_2/\sum c_i)$, where c_i is the contribution of the ith mode to heat capacity, and the summation is carried out over all modes (see [3.2,55]); this relation was confirmed experimentally [3.56].

Figure 3.3 shows, in the Landau-Teller coordinates, the results of experiments (mainly of the last years) on the V-T exchange in CO_2 during collisions with different atoms and molecules. As has been shown in Sect.3.3, the adiabatic approximation with the account of the short-range repulsion forces provides the explanation for the general regularities of the V-T exchange; in this case $\log\tau_2$ increases with $T^{-1/3}$ (at the first approximation, linearly). Just this dependence is observed during collisions of CO_2 molecules with each other, with the atoms of inert gases and some simple homonuclear diatomic molecules. For other polyatomic molecules the probability of the V-T exchange is much higher (the relaxation time τ_2 is small). For the collisions of $CO_2 + H_2O$, $CO_2 + H_2$ the inverse dependence of the relaxation time on $T^{-1/3}$ was observed. This dependence was satisfactorily explained by the Sharma theory [3.72,73]. The fast V-T exchange in $CO_2 + H_2O$ was explained by the quasi-resonance energy transfer between these molecules and the following V-T deactivation of H_2O. For $CO_2 + H_2$, the quasi-classical problem about collisions between a harmonic oscillator and a rigid rotator, taking account of the CO_2 vibrational energy transfer into the H_2 rotation, was solved. The transitions of the V-R-T type are essential also for the collisions of $CO_2 + H_2O$ [3.70].

During dissociation of CO_2 and in some other cases, CO molecules and O atoms appear in a gas. The values of τ_2 for the $CO_2 + O$ collisions at high temperatures were obtained in shock tube experiments [3.74]. It turned out that oxygen atoms are more effective than argon atoms by an order of magnitude. The results for $CO_2 + CO$ are contradictory [3.66,67,75], which is evidently connected with a strong influence of impurities on the deactivation of CO molecules. On the basis of experiments [3.66,67,76] we can conclude that the effectiveness of CO in the V-T exchange is close to the effectiveness of N_2.

Let us go over to the analysis of energy exchange between the symmetrical and antisymmetrical vibrational modes in CO_2. The laser fluorescence method is the principal technique for obtaining almost all the data on the deactivation probability of the first excited level 00^01 of the antisymmetrical mode of CO_2 vibrations. The interpretation of these results is usually complicated by the lack of information on the state of the products. When the deactivating partner of the CO_2 molecule is a diatomic or, especially, a polyatomic molecule with a small resonance defect with respect to the 00^01 level, then the intermolecular transfer of vibrational energy, V-V' exchange, becomes quite probable. The separation of the V-V and V-V' exchange in this case can be ambiguous, and we simply speak of the deactivation rate of the 00^01 state of CO_2. Nevertheless, for some applications it is important to separate a component of this process, which is related to the exchange between the ν_3 and ν_1, ν_2 modes of the CO_2 vibrations, i.e., the intramolecular V-V exchange. This can be done by varying the conditions of the experiment (first of all, the composition of mixtures). The estimation of the role of an energy exchange channel between the ν_3 and ν_1, ν_2 modes can be made on the basis of the SSH theory [3.41,53]. Theoretical values of transition probabilities promoting depopulation of the 00^01 level are shown in Fig.3.4. These values are related to the probability of the V-T transition $01^10 \rightleftharpoons 00^00$. It is clear from the figure that the preferable channels of

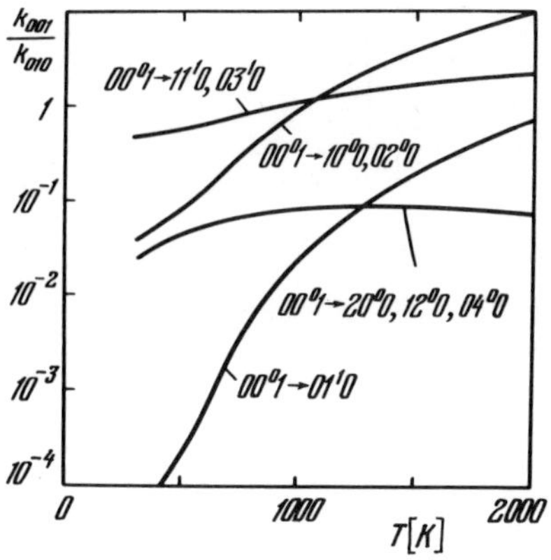

Fig.3.4. Effect of the gas temperature on the relative values of deactivation probability of the 00^01 state of CO_2 during the V-V exchange

the V-V exchange are the transitions $00^01 \rightleftharpoons (11^10, 03^10)$ and $00^01 \rightleftharpoons (10^00, 02^00)$. This is confirmed by experiments [3.77] in which the effects of the nonisothermal deactivation of the 00^01 state through different channels was taken into account using the phase shift method of laser fluorescence. For simplicity let us limit ourselves to one of these channels, namely, owing to relatively low gas temperature, to the transition $00^01 \rightleftharpoons (11^10, 03^10)$. In [3.19,78-82] it is considered to be the principal channel in the kinetics of the V-V exchange in CO_2. Then, according to (3.4.21), (see also [3.19,31]), the rate of the change of the average number of quanta e_3 through this channel is

$$\frac{de_3}{dt} = pk_{3(1,2)}\left[e_1e_2(e_3+1)\exp\left(-\frac{\theta_3-\theta_2-\theta_1}{T}\right) - (e_1+1)(e_2+1)e_3\right] + pk_{3(2)}\left[e_2^3(e_3+1)\exp\left(-\frac{\theta_3-3\theta_2}{T}\right) - (e_2+1)^3e_3\right] \quad , \tag{3.5.4}$$

where the first term corresponds to the $00^01 \rightleftharpoons 11^10$ transition and the second, to the $00^01 \rightleftharpoons 03^10$ transition; $k_{3(1,2)}$ and $k_{3(2)}$ are the exchange rate constants for these processes. Putting $T_1 = T_2$ and taking into account that $\theta_1 \approx 2\theta_2$, we obtain

$$e_1 = e_2^2/(2e_2+1) \quad . \tag{3.5.5}$$

Then, eliminating e_1 from (3.5.4), we get

$$\frac{de_3}{dt} = pk_{32}\left[e_2^3(e_3 + 1)\exp\left(-\frac{\theta_3 - 3\theta_2}{T}\right) - (e_2 + 1)^3 e_3\right] \quad , \tag{3.5.6}$$

where k_{32} is the overall rate constant of the exchange

$$k_{32} = k_{3(1,2)}/(2e_2 + 1) + k_{3(2)} \quad . \tag{3.5.7}$$

If $(2\bar{e}_2 + 1)/(2e_2 + 1) \approx 1$, then k_{32} depends only on the translational temperature T. Let us assume that this ratio is in fact close to unity; the error due to this assumption can be counted as the inaccuracy in determining k_{32}. For $T_1 = T_2 = T$ (below we shall write in this case that $T_2 = T$) (3.5.6) can be linearized,

$$\frac{de_3}{dt} = pk_{32}^0(\bar{e}_3 - e_3) \quad , \tag{3.5.8}$$

so that

$$k_{32} = k_{32}^0 \frac{[1 - \exp(-\theta_2/T)]^3}{1 - \exp(-\theta_3/T)} \quad . \tag{3.5.9}$$

The characteristic feature of the laser fluorescence method, which considerably simplifies the representation of the results of measurement, is that these experiments are carried out in such conditions that the vibrational state of the ν_1 and ν_2 modes of the CO_2 is practically at equilibrium with the translational motion ($T_2 = T$), and the linearized equation (3.5.8) can be used. A summary of some results of measurements of k_{32}^0 is given in Fig. 3.5. The graph shows the dependence of log k_{32}^0 on $T^{-1/3}$; as before, we assume that the adiabatic mechanism of exchange plays the major role, when we account for the short-range repulsion forces. In fact, for most of the partners in collision with the CO_2 molecule, including such molecules as H_2 and CO, there is a qualitative agreement with the increase in the probability of the $\nu_3 \rightleftharpoons \nu_1, \nu_2$ exchange with the temperature predicted by the SSH theory. However, the quantitative agreement between the theory and experimental data can be achieved only by introducing into the theory the corrections accounting for the orientation of molecules during the collision, the effect of anharmonicity, Coriolis forces, etc. This agreement can also be achieved by choosing other parameters (for example, by varing the intermolecular interaction potential) and considering molecular rotation. The dependence of log k_{32}^0 vs $T^{-1/3}$ for such partners as H_2O and NO changes sign, which indicates a change in the mechanism of the V-V exchange[13]. The inverse

---

13 For the V-V exchange during the $CO_2 + H_2O$ collision the results obtained in [3.92] and the discussion of previous works carried out in this paper were used.

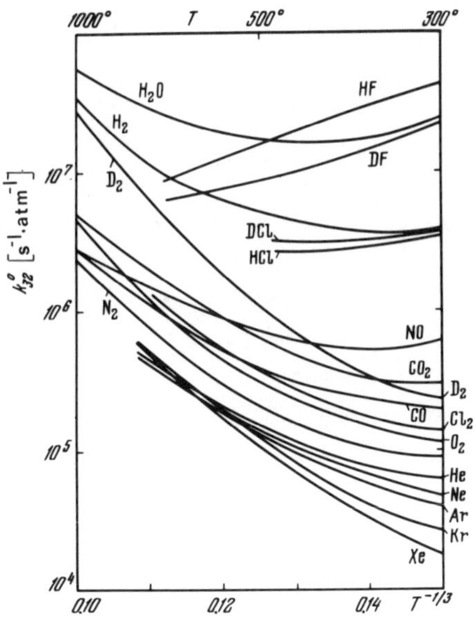

Fig.3.5. Averaged values of the V-V exchange rate constant in CO_2 (in the linearized relaxation equation) during collisions with the atoms of He [3.83-86], Ne, Ar, Kr, Xe [3.86] and the molecules of N_2 [3.83,84,87], O_2 [3.85,88], H_2 [3.86,89], D_2 [3.86,90], Cl_2 [3.89], CO [3.88,90, 91], NO [3.89], H_2O [3.92], CO_2 [3.84,89,92], HCl, DCl, HF, DF [3.93]

temperature dependence of the exchange probability at low temperatures can be explained by account of the long-range attraction forces due to multiple intermolecular interaction [3.94,95]. The role of the short-range repulsion forces increases as the relative velocity of collisions rises with temperature. Thus the exchange probability grows (Sect.3.3). For collisions of CO_2 with halogen hydride molecules the inverse temperature dependence of the exchange rate constant was established (or a weak dependence of k^0_{32} on T in the case of HCl and DCl). On the basis of the solution of the dynamic collision problem in the frames of classical mechanics [3.96] it is possible to explain this dependence by the effect of the V-R exchange and attraction forces. If the energy exchange between the ν_3 and ν_1, ν_2 modes of CO_2 vibrations is investigated in the absence of equilibrium between the symmetrical vibrations and translational motion and if $T_2 \neq T$ (for instance, in shock waves [3.97,98]), then the interpretation of the experiment is considerably complicated and becomes ambiguous.

In gas mixtures at high temperatures chemically active atoms can be present — the products of reactions occurring in a gas. The measurement of the rate constant k^0_{32} of the deactivation of the 00^01 CO_2 level during collisions with these atoms [3.99-101] led to the following results (T = 298 K):

A	O	H	N	Cl	F	Na
k^0_{32}, $s^{-1}\cdot atm^{-1}$	4.6×10^6	$<2.5\times10^6$	$<7.3\times10^5$	3.7×10^8	7.9×10^7	6.1×10^6 .

Let us consider now the vibrational energy exchange during collisions of CO_2 with other molecules. To do this we must know precisely the degrees of freedom and the states which take part in these exchange. The principal role is played here by the states of the colliding molecules which are near resonance. For this reason during the collisions of $CO_2 + N_2$, $CO_2 + CO$, $CO_2 + HCl$ the exchange involving the ν_3 mode of CO_2 vibrations is the most probable one. Thus, for the collisions of $CO_2 + CO$ the rate constant k_{2M} of the V-V' exchange is two orders of magnitude smaller than k_{3M} [3.102]. Owing to the very fast V-T deactivation of H_2O the exchange of CO_2 (ν_2) with the H_2O molecule is in fact reduced to the V-T exchange in CO_2 under the effect of the H_2O molecules. For the exchange of CO_2 (ν_2) $\rightleftharpoons O_2$ during absorption of ultrasound it was found that k_{2M} is equal, in $[s^{-1} \cdot atm^{-1}]$, to 1.6×10^5 for T = 300 K, 2.1×10^5 for T = 450 K, and 2.5×10^5 for T = 600 K [3.62].

Many other results were obtained for the V-V' exchange with the 00^01 state of the CO_2 molecule; the summary of these data is given in Fig.3.6. Experiments indicate (at least for not too high temperatures) the existence of the inverse dependence of the exchange probability on the temperature for all the partners in collisions. This temperature dependence of k_{3M} can be explained in the frames of the theory leading to relations (3.3.17) and (3.3.18) (see [3.12,95,105,112]). According to [3.108], for the collisions of $CO_2 + CO$ the exchange probability first drops with the increase in temperature, then begins to grow. This increase is almost the same in [3.103] and [3.91]. The difference in values of k_{3M} obtained by different authors for the collisions of $CO_2 + N_2$ is more pronounced; this difference appears only for high temperatures.

Among these data the results of [3.103] are preferable, since they have been obtained using the most precise phase method, whereas almost all the other results were obtained in experiments with shock waves, where a more complicated and less reliable interpretation of the results is required. It should be noted that the results of [3.102] for $CO_2 + CO$ are apparently unreliable because of low sensitivity of the registered signal to the variation of the required constant (see discussion in [3.66]).

The electronic energy of atoms can also be effectively transfer into vibrational energy. In [3.13] the induced infrared radiation as a result of E-V transfer of energy from the Br atoms to the vibrational states (101) and (021) of the CO_2 molecules was observed. The theory of these processes for the N_2 + Na collisions is given in [3.11].

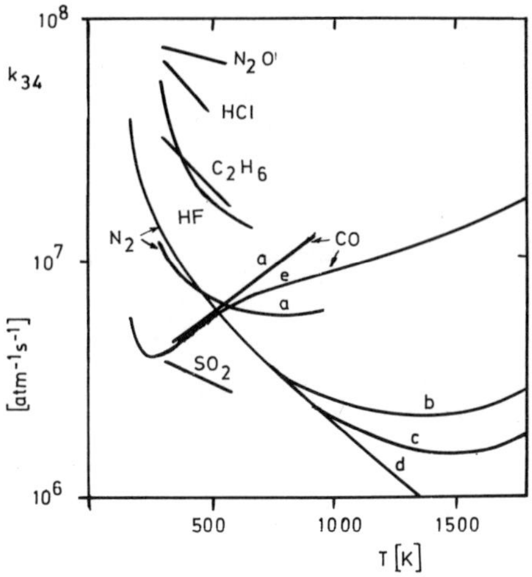

Fig.3.6. Averaged values of the V-V exchange rate constant of CO_2 (ν_3) during collisions with molecules of N_2 [3.87,90,103-107], CO [3.66,90,91, 103,107,108], HCl, HF [3.93,109,110], N_2O [3.90], SO_2 and C_2H_6 [3.111]. (*a*) [3.103]; (*b*)[3.105]; (*c*) [3.66]; (*d*) [3.104]; (*e*) [3.66,91]

3.6 Chemical Reactions in Gases

The theoretical solution of the problem of kinetics of chemical reactions in a gas, just as in the case of vibrational energy exchange, consists of three stages: 1) the analysis of dynamics of an elementary act of chemical transformation, 2) the statistical analysis of a large number of these processes in a gas, and 3) the derivation of phenomenological equations of chemical kinetics. Suppose that during a molecular collision the following process occurs:

$$A(i,\underline{v}_A) + B(j,\underline{v}_B) \rightarrow C(\ell,\underline{v}_C) + D(m,\underline{v}_D) \quad , \qquad (3.6.1)$$

where, as in Sect.3.2, the quantities i, j, ℓ, m describe the internal state of the molecules (we shall denote the set of these quantities by ζ). Each collision of A + B is characterized by an initial value of the velocity vector of the relative motion of colliding molecules $\underline{v} = \underline{v}_A - \underline{v}_B$ and the impact parameter b, which is equal to the distance between the vector $\underline{v}_A$, passing through the center of molecule A, and the center of molecule B. As a result of the solution of the dynamic problem we can obtain the so-called reaction cross section, which is equal to

$$\sigma(\zeta,v) = 2\pi \int p(\zeta,v,b)bdb \quad , \qquad (3.6.2)$$

where $p(\zeta,v,b)$ is the reaction probability for given values of i, j, ℓ, m, v, b. The total cross section of a chemical reaction is obtained by summation of (3.6.2) over all internal states of the reagents taking account of the relative density of particles in each state of colliding molecules,

$$(v) = \sum_{i,j,\ell,m} \frac{n_{A(i)}}{n_A} \cdot \frac{n_{B(j)}}{n_B} \sigma(\zeta,v) \quad , \tag{3.6.3}$$

where the same notation as in Sect.3.2 is used for concentrations, and $n_A = \sum_i n_{A(i)}$, $n_B = \sum_j n_{B(j)}$. Suppose that a chemical reaction occurs only when the relative energy of collisions exceeds a certain activation energy Q, and in other respects the molecules can be treated as solid spheres. In this case the reaction cross section $\sigma(v)$ is obtained in the simplest form (see Appendix H). Some generalizations of this model are given in [3.8,113]; the review of the elementary chemical transformation theory for gases can be found in [3.10,11].

In the solution of the statistical problem the generalized Boltzmann equation (3.2.2) is used, where the last term on the right-hand side, determining the rate of a chemical process (3.6.1), is [3.8]

$$I_{ch} = \sum_{j,\ell,m} \int v\sigma(\zeta,v)[n_{C(\ell)} f_{C(\ell)}(\underline{v}_C) n_{D(m)} f_{D(m)}(\underline{v}_D)$$

$$- n_{A(i)} f_{A(i)}(\underline{v}_A) n_{B(j)} f_{B(j)}(\underline{v}_B)] d\underline{v}_B \quad . \tag{3.6.4}$$

The meaning of every quantity in (3.6.4) is clear from Sect.3.2. By integrating (3.2.2) with respect to velocity $\underline{v}_A$ and summing over all the states i, j, ℓ, m (that is, taking into account only chemical transformations of molecules) and using (3.6.4), we obtain

$$-\frac{dn_A}{dt} = k_f n_A n_B - k_r n_C n_D \quad , \tag{3.6.5}$$

where

$$k_f = \sum_{i,j,\ell,m} \frac{n_{A(i)}}{n_A} \cdot \frac{n_{B(j)}}{n_B} \iint v\sigma(\zeta,v)$$

$$\times f_{A(i)}(\underline{v}_A) f_{B(j)}(\underline{v}_B) d\underline{v}_A \, d\underline{v}_B \quad . \tag{3.6.6}$$

The expression for k_r is written in a similar form.

Essentially, (3.6.5) is the required phenomenological equation, which describes the variation of particle concentration as a result of chemical reactions (in the given case the reaction $A + B \rightleftharpoons C + D$), if the coefficients k_f and k_r depend only on macroscopic parameters. It is clear from (3.6.6) that these coefficients are the result of integration of complicated relations for microscopic quantities. For this reason, obtaining phenomenological equations for concentrations reduces to a further transformation of (3.6.6) for k_f, k_r. We shall consider that the Maxwell-Boltzmann equilibrium distribution of molecules over velocities and internal energy is realized in a gas, that is, we admit the so-called equilibrium hypothesis; moreover, the velocity distribution functions are independent of internal states of molecules. In these assumptions the required coefficients for the simplest models of intermolecular interaction depend only on the gas temperature. This dependence is of the form of the *Arrhenius formula* (see Appendix H),

$$\left.\begin{aligned} k^0(T) &= A\sqrt{T}\exp(-Q/kT) \quad \text{or} \\ k^0(T) &= A'T^q\exp(-Q/kT) \end{aligned}\right\} , \tag{3.6.7}$$

where A, A', q, Q are constants, and the subscript 0 indicates that the expression is obtained from the assumptions of the equilibrium hypothesis. Since $k^0(T)$ is independent of concentration, it is usually called the *reaction rate constant*. Many experiments carried out using various techniques confirmed the validity of the Arrhenius formula (3.6.7), which is the fundamental equation of chemical kinetics. Usually the values of the parameters A', q, Q for each reaction are determined experimentally. Reference [3.114] gives a comprehensive list of these values. It should be noted that errors in the values of these rate constants and systematic differences in the results obtained by different authors are usually rather large (hundreds of percent and sometimes even orders of magnitude). As has been noted in Sect. 3.4, this is due to, apart from the drawbacks of the measuring techniques, the deviation of the systems from equilibrium not being taken into account, which can differ in different experiments. Neither is it taken into consideration by the Arrhenius formula (3.6.7). The great advantage of the application of (3.6.7) is that the reaction rate constant depends on only one macroscopic parameter—temperature.

When the accuracy requirements are raised, we should reject the equilibrium hypothesis, which inevitably leads to the increase of the number of parameters required for determining k. The rejection of the equilibrium hypothesis signifies the transition from the so-called "equilibrium" (or Arrhenius) chemical kinetics to "nonequilibrium" kinetics [3.2,115]. In nonequilibrium kinetics k is dependent on the extent of deviation from equilibrium. In the simplest case k is dependent on the vibrational temperature T_{vib} of molecules entering into a reaction (when $T_{vib} \neq T$)[14]. (Footnote see next page.) The use of the vibrational temperature implies the existence of a Boltzmann distribution of molecules over vibrational levels regardless of the chemical reaction (when the vibrational exchange rate is higher than the rate of the chemical reaction).

Suppose that only the molecules having the vibrational energy E^* can enter into a chemical reaction; in this case only the vibrationally excited molecules take part in the chemical reaction (an example is the dissociation of diatomic molecules in a shock wave [1.8]). Let the vibrational level m^* correspond to the energy E^*. Then the reaction rate constant is proportional to the product of the reaction probability P_{m^*} (for the molecules n_{m^*} on the level m^*) and the concentration of these molecules, that is, $k \propto P_{m^*}n_{m^*}/n$. Since P_{m^*} is independent of the population of the m^{*th} level, the ratio of the "nonequilibrium" value of this constant $k(T, T_{vib})$ to the equilibrium one $k^0(T)$ is $n_{m^*}/n^0_{m^*}$, where $n^0_{m^*} = n_{m^*}$ for $T_{vib} = T$. Assuming in this approximate model the existence for n_{m^*} and $n^0_{m^*}$ of the Boltzmann distribution over the levels with temperature T_{vib} and T, respectively, we arrive at the ratio

$$\frac{k(T, T_{vib})}{k^0(T)} = \frac{Q(T)}{Q(T_{vib})} \exp\left[-\frac{E^*}{k}\left(\frac{1}{T_{vib}} - \frac{1}{T}\right)\right] \quad , \qquad (3.6.8)$$

where $Q(T_{vib})$ is the vibrational partition function [3.116-119]. For a harmonic oscillator $Q(T_{vib}) = [1 - \exp(-\theta/T_{vib})]^{-1}$ (D.10), so that for not very high temperatures

$$k(T, T_{vib}) = k^0(T) \frac{T}{T_{vib}} \exp\left[-\frac{E^*}{k}\left(\frac{1}{T_{vib}} - \frac{1}{T}\right)\right] \quad , \qquad (3.6.9)$$

where $k^0(T)$ has the form of the Arrhenius dependence (3.6.7). Actually, as a result of a chemical reaction, the population differs considerably from the Boltzmann one [1.8,3,11,120]. In quasi-stationary states with respect to the vibrational exchange, when anharmonicity plays a considerable role, the nonresonance V-V exchange must be taken into account; see Sect.3.4 [3.121-123]. It is clear that the chemical reaction probability differs from zero not only for m^* but also for other vibrational levels (for example, it was found in [3.124] that the probability of dissociation of molecules increases exponentially with the level number). Nevertheless, in

---

14 In this case the notion of the rate constant losses its meaning; for this reason some authors replace it by a more general concept of the reaction rate coefficient. Here we retain the historically formed concept of the rate constant in the nonequilibrium kinetics as well.

experiments with shock waves (3.6.8) and (3.6.9) were confirmed for a large number of diatomic molecules (see [3.27,125]).

In the development of infrared gasdynamic lasers of great interest is the reverse process, i.e., the formation of vibrationally excited molecules is a result of a chemical reaction. It can be described in the simplest form when deviations from the Boltzmann distribution of molecules over the levels are negligibly small. For a harmonic oscillator of the type A the relaxation equation for the balance of vibrational energy e_A per oscillator is then obtained as the sum of the energy fluxes due to the V-T relaxation and as a result of a chemical reaction

$$\frac{de_A}{dt} = \frac{p}{\tau}(\bar{e}_A - e_A) + (E_A^* - e_A)\frac{1}{n_A}\frac{dn_A}{dt} \quad , \tag{3.6.10}$$

where n_A is the density of particles A and E_A^* is the mean vibrational energy lost (or acquired) at each act of the chemical reaction. In a more general case the change of the vibrational energy e_i of the ith mode in the jth molecule is described by an equation which generalizes (3.6.10) and takes account of the results of Sect.3.4,

$$\frac{de_i}{dt} = \left(\frac{de_i}{dt}\right)_{VT} + \left(\frac{de_i}{dt}\right)_{VV} + \frac{1}{n_j}\sum_{\ell}(E_{i\ell}^* - e_i)(a'_{j\ell} - a_{j\ell})S_\ell \quad , \tag{3.6.11}$$

where the summation is performed over all ℓ reactions in which the jth molecule participates. Here S_ℓ is the rate of the ℓth reaction, $a_{j\ell}$, $a'_{j\ell}$ are the stoichiometric coefficients for the substance j in this reaction (see Appendix I) and $E_{i\ell}^*$ is the value of E^* for the ℓth reaction and the ith mode. The determination of $E_{i\ell}^*$ is of great importance for the development of chemical lasers. For reactions of dissociation and recombination of diatomic molecules $E_{i\ell}^*$ is either close to the dissociation energy D or differs from it by the order of the average thermal energy of a gas ($E^* \approx D - kT$). When polyatomic molecules, e.g., CO_2 are formed, it is not always clear how the vibrational energy is distributed over different vibrational modes. Sometimes it is assumed that in this case E^* is distributed uniformly over all the modes (see Sect.5.8).

If the average time for the redistribution of molecules over energy levels differs very little from the characteristic time of a chemical reaction, considerable deviations from the Boltzmann distribution of molecules over the levels are possible. To compute populations of the levels in such conditions we must return to the generalized Boltzmann equation (3.2.2). Assuming that energy relaxation occurs sufficiently fast and simplifying the respective collision integrals (3.2.3) and (3.6.4), we obtain, after integration with respect to velocity of the relative motion of molecules, the balance equation for populations of the levels. Thus, generalizing the form of the microkinetic balance equations (3.4.1,14) and taking into account the possibility of transitions between any levels simultaneously with the chemical reaction $A + B \rightleftharpoons C + D$, we can write for the population of the ith levels of the molecule $[n_{A(i)} \equiv x_n]$

$$\frac{dn_{A(i)}}{dt} = \sum_M Z_{AM}\left[\sum_a Q_{si}(M)n_{A(s)} - \sum_s Q_{is}(M)n_{A(i)}\right]$$

$$+ k_{ri}n_C n_D - k_{fi}n_A n_B \quad , \qquad (3.6.12)$$

where Z_{AM} is the frequency of collisions of molecules A with a molecule of the type M, $Q_{si}(M)$, $Q_{is}(M)$ are the probabilities of the transitions $s \rightarrow i$, $i \rightarrow s$ in the molecule A during the collisions of A + M, and k_{ri} and k_{fi} are the rate constants for the formation and disappearance of molecules A in the ith state. To solve (3.6.12) we must know the probabilities $Q_{si}(M)$ and the rate constants k_{ri} (the quantities with reverse directions are easily determined from the principle of detailed balancing). Equations (3.6.12) can be used to describe the simultaneous course of chemical reactions and vibrational energy exchange (in gasdynamic and chemical lasers) as well as the redistribution of molecules and atoms over electronic energy levels in the presence of simultaneous ionization and other gas transformations (in recombination and plasmadynamic lasers). Examples of the solution of this problem are given in the last chapters of this book.

Until now we have considerd bimolecular exchange reactions of the type $A + B \rightleftharpoons C + D$. Apart from this reaction, other reactions can occur in a gas, such as unimolecular dissociation $XY \rightarrow X + Y$; bimolecular recombination of atoms $X + Y \rightarrow XY + h\nu$, accompanied by emission of quantum $h\nu$; dissociation of XY upon the impact of particle M, $XY + M \rightarrow X + Y + M$; and the reverse process of trimolecular recombination $X + Y + M \rightarrow XY + M$[15]. In a cooling gas or plasma the most important are the recombinations leading to the formation of molecules (during chemical recombination) and neutral particles (during electron-ion recombination). In the reaction

---

15 Symbols X, Y, Z, ... designate atoms or radicals composing molecules A, B, etc. It should be noted that reactions of the type $A + B \rightarrow C + D$ can be simple exchange reactions ($X + YZ \rightarrow XY + Z$), when three centers X, Y, and Z participate in the reaction, or double exchange ($XY + ZW \rightarrow XW + YZ$) with four centers.

$$X + Y \rightarrow XY + h\nu \tag{3.6.13}$$

the excess energy of relative motion of recombining particles X and Y is removed from XY in the form of radiation (optical stabilization of the XY molecule), and in the process

$$X + Y + M \rightarrow XY + M \tag{3.6.14}$$

a stable molecule is formed as a result of triple collision and transfer of the excess energy to the third particle M (mechanical stabilization). The probability of optical stabilization (3.6.13) is determined by the ratio of the lifetime, τ_c, of a complex of colliding atoms (quasimolecule XY') to the time interval from the beginning of collision to the moment when this quasimolecule emits energy, i.e., to the lifetime for the respective spontaneous optical transition. τ_c is close to 10^{-13} to 10^{-14} s, and the mean lifetime of spontaneous emission is 10^{-7} to 10^{-8} s for allowed transitions and is considerably larger for optical forbidden transitions. Thus even for quasimolecules possessing the appropriate allowed optical transitions the indicated ratio is close to 10^{-5}, which corresponds to an extremely low probability of optical stabilization (3.6.13). Consequently, the principal mechanism of molecular formation is mechanical stabilization during triple collision (3.6.14), except for very low pressures (about 1 torr and lower), when the number of triple collisions in a gas is small.

The phenomenological equation of chemical kinetics, which describes the time variation of concentration of formed molecules, n_{XY}, for the process (3.6.14) with the account of the reverse process of dissociation can be written in the form similar to (3.6.5),

$$\frac{dn_{XY}}{dt} = k_r n_X n_Y n_M - k_d n_{XY} n_M \quad , \tag{3.6.15}$$

where k_r, k_d are the rate constants of recombination and dissociation, respectively. There is no sufficiently rigorous substantiation of this equation in the framework of the kinetic theory of gases so far, since it is rather difficult to formulate the generalized Boltzmann equation for triple collisions [3.8,9]. Nevertheless, this equation is widely used in practice, where it has received good confirmation. In the theory sometimes various approximate methods are employed, which allow one to reveal the mechanism of triple recombination and to achieve sufficient agreement with experimental results. Thus, this problem is often reduced to the two-body problem by considering the binary collision of a quasimolecule with a third particle [3.2].

The description of chemical kinetics equations for any reaction in the general form is given in Appendix I.

3.7 Processes in Heterogeneous Mixtures

So far we have only considered processes in a gas free of condensed admixtures. But as was mentioned in Sect.2.5, in a gasdynamic laser flow liquid and solid condensate particles may be present. We shall discuss here only two problems arising in this connection: firstly, the kinetics of appearance of a condensed phase in a flow (when it is not introduced into the flow from outside) and, secondly, the kinetics of the change in the state of the gas during collision of molecules with the surface of these particles (physicochemical kinetics of heterogeneous processes).

In a gasdynamic laser flow the gas temperature drops to such an extent that for certain components the conditions set in in which their condensation occurs. First of all this applies to water vapor, which is the component of a GDL active medium. This process occurs not instantaneously but develops in time. The theory of condensation kinetics is presented in a number of books and papers (for example, in [3.126,127]). Here we shall consider only the basic ideas and give the final results.

The dependence of the saturation pressure p_∞ on temperature below which condensation occurs is given by the Clausius-Clapeyron equation. For a flat phase interface and for a constant latent heat of evaporation L this equation has the form

$$\log p_\infty = -A/T + B \quad , \tag{3.7.1}$$

where A and B are constants given in Table 3.1.

Table 3.1.

Substance	State	A[K]	B[atm]	Temperature range[K]
CO_2	solid	1367	7.025	138-216
N_2	liquid	314.2	4.068	63.2-125
	solid	359.1	4.778	35.0-63.2
H_2O	liquid	2263	6.064	273-395
	solid	2676	7.582	175-273

If the vapor pressure p for a given temperature exceeds p_∞, then we can speak of a supersaturation equal to $s = p/p_\infty$. For the development of condensation there must be the so-called centers of condensation (or, in a more general sense, phase interfaces). If these centers differ in nature from the vapor under consideration (for example, dust particles, solid surfaces of the walls, or in general any condensate of other substance), then we have heterogeneous condensation. With a sufficient amount of foreign admixtures, heterogeneous condensation proceeds very fast even for the values of s slightly exceeding unity.

In homogeneous condensation two stages can be separated. The first stage is determined by the formation of condensation centers, i.e., drops of liquid of such a radius r^* (this radius is called critical), for which further formation of the liquid phase is possible. The growth of these drops constitutes the second stage of condensation. At first the drops may form as a result of density fluctuations in a supersaturated vapor. A drop with dimensions below critical ($r < r^*$) will resorb (evaporate) because the energy of evaporation exceeds the surface energy of the drops being formed. The critical radius r^* is

$$r^* = \frac{2\sigma\mu}{\rho_c RT \ln(p/p_\infty)} \quad , \qquad (3.7.2)$$

where ρ_c is the density of a liquid phase and σ is the surface tension coefficient of a liquid, which is assumed to be independent of r and corresponding to a flat phase interface. The number of molecules g^* in this drop is $g^* = 4\pi(r^*)^3\rho_c/3m$, where m is the mass of a molecule. The basic kinetic parameter is the formation rate J of condensation centers, i.e., the number of drops of the critical size, which are formed per unit volume per unit time. It can be obtained from the balance of the processes of adsorption and evaporation of molecules. According to the Frenkel-Zeldovich theory (see [3.126]),

$$J = \left(\frac{p}{kT}\right)^2 \frac{1}{\rho_c} \left(\frac{2\sigma}{\pi}\right)^{1/2} \exp\left[-\frac{4\pi\sigma(r^*)^2}{3kT}\right] \quad , \qquad (3.7.3)$$

where k is the Boltzmann constant. Then the relative condensation rate is

$$\frac{1}{N}\frac{dN}{dt} = \frac{g^* J}{N} \quad , \qquad (3.7.4)$$

where N is the number of molecules in a gas phase. The value of σ actually depends on the temperature and the radius of the drop [3.127],

$$\sigma_\infty = a - b(T - 273) \quad , \quad (\sigma_r/\sigma_\infty) = (1 + 2\delta/r)^{-1} \quad , \tag{3.7.5}$$

where σ_r, σ_∞ are the values of σ for a certain radius r and for a flat interface ($r \to \infty$), respectively. For water vapors a = 75.6 and b = 0.192, if σ is expressed in erg/cm^2, and $\delta \approx 1$ Å. The account of (3.7.5) can increase the theoretical rate of formation of condensation centers J by several orders [3.128]. One of the principal difficulties of the condensation theory is that it is based on thermodynamic relations for quasi-stationary processes, which is not always correct for kinetic descriptions. Note also that for very high supersaturations s the critical radius r^*, according to (3.7.2), is so small that the drop formed can contain only several molecules. When we use the concept of the surface tension in this case, we also encounter principal difficulties. Despite this, the presented theory proved to be in a good agreement with many experimental results of the study of condensation in supersaturated systems (water vapor, carbon dioxide, etc.).

The second stage of condensation is associated with the growth of condensation centers. The equation which describes the variation of mass of condensing drops can be derived from the balance of the number of condensing and evaporating molecules (see, for example, [3.128]). This stage of condensation proceeds sufficiently fast and results in the appearance of the so-called jumps of condensation in a flow, which are accompanied by a noticeable increase in temperature (as a result of evaporation energy release). In this case the system approaches equilibrium with respect to saturation.

Let us consider the physico-chemical kinetics of heterogeneous processes. Deactivation of molecular vibrations during collisions with a surface is determined by a high probability due to the effect of significant attraction forces, which appear during interaction of arriving molecules with the molecules constituting the condensate surface. The probability of vibrational deactivation P_s during such a collision of a molecule with the surface, according to estimates [3.129] based on the theory [3.130], is from two to three orders higher than during binary collisions of molecules in a gas volume. This probability depends not only on the type of molecules but also on the gas temperature and pressure. At very low gas pressures the solid body surface is free from the adsorbed layer of molecules. As the gas pressure rises, the part of the surface coated with the adsorbed layer increases, and the deactivation probability P_s decreases, approaching the limit P_{sa}, when all the molecules collide only with the adsorbed layer. According to [3.77], for vibrational deactivation of ν_3 of CO_2 on KBr for T = 300 K this occurs at $p > 10$ torr when $P_{sa} = 0.135$, while on a clean surface of the KBr crystal the value of P_s is close to 0.4. It was experimentally established that for T = 300 K the probability of surface deactivation P_s for different molecules is as given in Table 3.2.

Table 3.2.

Molecule, mode of vibration	Surface	P_s	Reference
ν_3, CO_2	teflon, bronze, quartz, molybdenum glass, KBr, pyrex	0.22-0.44	[3.77,131-133]
ν_2, CO_2	NaCl	0.22	[3.134]
ν_3, N_2O	pyrex	0.29	[3.132]
ν_2, N_2O	NaCl	0.37	[3.134]
CO	pyrex	4×10^2	[3.135]
N_2	pyrex, quartz, steel, aluminum, teflon, copper, corundum	$6 \times 10^{-4} - 2.6 \times 10^3$	[3.133,136]

The increase of the gas temperature leads to the decrease of P_s, so, for T = 400 K the deactivation probability on pyrex for ν_3, CO_2 amounts to 0.18; for ν_3, N_2O, $P_s = 0.14$; and for CO, $P_s = 1.86 \times 10^{-2}$ [3.137]. For low temperatures excited molecules are adsorbed on the surface with a sufficiently high probability and can remain in this state for a certain time. As temperature rises, the collisional mechanism of deactivation prevails. This change in the nature of interaction of molecules with a surface makes it possible to explain the observed dependence of $P_s(T)$ [3.77,129,133,138,139]. It should be noted that according to estimates [3.129,140] the surface deactivation probability of CO_2 vibrations on water aerosol particles is close to unity and is lower for nitrogen molecules by three orders of magnitude. These data are required for the analysis of GDL performance, during which water vapor may condense (Sect.5.4).

Similarly, we can consider the effect of collisions with a surface on the development of chemical reactions. When this effect is strong, we can speak about high catalytic activity of the given surface. Of greatest interest in this respect are recombinations of atoms. Experimental data give different catalytic activities. Thus, for T = 350 K recombination of nitrogen atoms on copper proceeds with the probability $P^c = 0.055$, while on tungsten $P^c = 1.4 \times 10^{-3}$; for oxygen atoms on copper $P^c = 0.13$, and on silicon monoxide, below 6×10^{-5}. The probability of surface recombination (especially at low temperatures) depends not only on the temperature but on the pressure as well.

To analyze the effect of heterogeneous deactivation of vibrations and recombination on the development of the processes in the entire mixture volume, we must use kinetic equations, which in the general case are based on the diffusion flux of vibrational energy or of a chemical component to condensed particles. The derivation of diffusion equations from the Boltzmann equation for multicomponent mixtures is given in monographs on the kinetic theory of gases (see, for example, [3.1]). In [3.141] the diffusion equation is simplified. Here we shall give only a simplified treatment of kinetics of heterogeneous relaxation of vibrational energy, applied to the analysis of the processes in gasdynamic lasers [2.50]. Heterogeneous relaxation makes a considerable contribution to the general relaxation of a two-phase flow, when the size of particles is small and the deactivation is determined not by the rate of supply of excited molecules to the particles due to diffusion but by direct collisions with the surface of the particles. Then the heterogeneous relaxation time τ^c does not depend on diffusion properties of a gas and is determined by

$$\tau^c = (\rho_c/\rho)R/3\alpha_c\kappa \tag{3.7.6}$$

for similar particles of radius R, and by

$$\tau^c = 1/\alpha_c\kappa S \tag{3.7.7}$$

for a polydisperse set of particles, where S is the specific surface, that is the surface of particles in a unit volume of gas (the dimension of S is reciprocal length), α_c, ρ_c are the mass fraction and the density of the condensed phase, respectively, and κ is the heterogeneous relaxation coefficient, which is equal to the fraction of molecules in a gas phase that undergoes deactivation as a result of collisions with the surface of particles, i.e., $\kappa = P^c Z^c/N$ or, since $Z^c = N\bar{v}/4$ ($\bar{v}$ is the average thermal velocity of molecules), then

$$\kappa = P^c(kT/2\pi m)^{1/2} \quad , \tag{3.7.8}$$

where m is the mass of a molecule. The quantity τ^c determines the rate of variation of vibrational energy in the relaxation equation of the type (3.4.6)

$$\left(\frac{de_i}{dt}\right)_c = \frac{\bar{e}_i - e_i}{\tau^c} \quad . \tag{3.7.9}$$

In computations this expression must be added to the relaxation equations for vibrational energy in the case of homogeneous relaxation considered in Sect.3.4. In Sect.4.4 we shall give kinetic equations describing heterogeneous processes on the nozzle walls of gasdynamic lasers taking into account diffusion and viscosity of a gas.

4. Relaxation in Nozzle Gas Flow

This chapter begins with the basic equations describing a steady-state inviscid gas flow and considers the features of the mathematical model of a nozzle gas flow. The similarity parameters involved in the description of nozzle flows are developed along with an introduction to the concepts of equilibrium, frozen, and nonequilibrium flows and the principle of instantaneous freezing. In regard to a cooling gas, estimates of the relaxation time and a discussion of recombination processes are presented. The chapter then describes various methods of nozzle contouring and an analysis of a nonequilibrium gas flow in a nozzle. Next, the concept of a boundary layer in a viscous gas flow is introduced along with the results of calculation of the boundary layer width at the nozzle walls in a gasdynamic laser. Viscosity effects on the active medium state are analyzed and operating principles of nozzle arrays in gasdynamic lasers are considered. The chapter closes with a description of the processes in a shock tube with a nozzle, which is also used in the investigation of gasdynamic lasers.

4.1 One-Dimensional Steady-State Inviscid Gas Flow

Up to now we have considered mainly processes in nonmoving media. In gasdynamic lasers, however, the important processes occur in a gas flow. Gas flow in gasdynamic lasers is usually restricted by the walls, that is, it occurs in channels of various shapes[1]. Among these channels of great interest are nozzles, in which gas velocity u increases accompanied by a decrease of pressure p, density ρ, and temperature T. Channels of other shapes and

---

1 Attempts were made to create gasdynamic lasers using jet streams flowing into a free space and some other schemes without channels; some of these schemes are discussed in the next chapter.

functions, such as diffusers, in which the velocity drops and pressure increases, are used in GDLs to increase pressure at the cavity exit, but the analysis of the diffuser flow is beyond the scope of this book (see [1.2, 4.1]).

Equations of gasdynamics describing flowing media follow from the molecular theory of gases[2], [3.1,3]. When solving gasdynamic equations, some simplifications are often introduced. The simplest model of a gasdynamic flow is a one-dimensional steady-state gas flow in the absence of dissipative processes (effect of viscosity, heat conduction, diffusion, etc.). In this case each nozzle cross section flow parameters are assumed to be uniform up to the channel walls (the so-called hydraulic approximation). One-dimensional flow is just an approximation, but it is widely used in the analysis of gasdynamic phenomena (see [3.128]). In this case the following basic equations are used to determine gas parameters.

1) The *continuity equation*, which expresses the law of conservation of mass flux. For a one-dimensional steady-state nozzle flow with the cross-sectional area $S(x)$ this equation is obtained directly from the condition of constant flow rate G, i.e., the mass of the gas flowing through a given channel per unit time,

$$G = \rho u S = \text{const.} \tag{4.1.1}$$

Sometimes the differential form of the conservation law for the mass flux is used,

$$\frac{d\rho}{\rho} + \frac{du}{u} + \frac{dS}{S} = 0 \quad . \tag{4.1.2}$$

2) The *momentum equation* (*Euler equation*), which corresponds to the law of conservation of momentum flux. In essence, this equation is Newton's second law, applied to the gas volume moving under the action of pressure forces. Usually it is written in the differential form

$$\rho u \, du = -dp \quad . \tag{4.1.3}$$

3) The *energy equation*, defines the law of conservation of energy flux. This equation has the simplest form in the absence of energy exchange with the surrounding medium — the channel walls ($dq = 0$) and ignores dissipative

2 For more detail on the questions treated in this chapter we can recommend the textbooks and monographs [3.4,7, 4.2,3].

energy losses. According to the first law of thermodynamics, the quantity of heat dq received by the unit mass of a gas is related to the change of internal energy $d\varepsilon$ of this mass and the work dA done by the gas [for gases $dA = pd(1/\rho)$, where $1/\rho$ is the specific volume]. This relation is expressed by the equation $dq = d\varepsilon + pd(1/\rho)$. From (4.1.3) and $dq = 0$ we obtain $d(\varepsilon + p/\rho) + u\,du = 0$, or

$$dh + u\,du = 0 \quad , \tag{4.1.4}$$

where h is the specific enthalpy. Often the integral form of this equation is used,

$$h + u^2/2 = \text{const.} \quad . \tag{4.1.5}$$

When the active medium moves through a cavity, some part of the gas energy is transformed into radiation (Sect.2.3). Let the unit gas volume lose per unit time the energy W; then in the gas mass occupying the distance dx along the flow the energy $dq = (W/\rho u)\,dx$ is converted into radiation, so that

$$\frac{d}{dx}\left(h + \frac{u^2}{2}\right) = -\frac{W}{\rho u} \quad , \tag{4.1.6}$$

where W is obtained from (2.3.31). Equation (4.1.6) instead of (4.1.4,5) will be used only when describing flow in a cavity.

These basic equations of a one-dimensional steady-state flow should be supplemented with thermal and caloric equations of state (3.1.6) and (3.1.12), and then the specific enthalpy h should be determined. For a perfect gas flow with a constant heat capacity [see (3.1.8)] and for a vibrationally excited gas at equilibrium (3.1.9-11) for $T_i = T$ we obtain simple relations for T, p, and ρ at any point of the flow and in a nonmoving gas at the nozzle entrance (we denote them by T_0, p_0, and ρ_0[3]), depending on the flow velocity u (or the Mach number $\underline{M} = u/a$, where a is the sound velocity); see Appendix J. From these relations it is clear that during acceleration of the nozzle flow the gas is cooled.

To create the active medium in a gasdynamic laser we must provide for fast and extensive cooling. This can be done by accelerating the gas to supersonic speeds exceeding the critical speed ($\underline{M} = 1$). It follows from the relations of the type (J.13) that to obtain a supersonic flow we must have a nozzle which consists of a convergent channel for acceleration to subsonic

---

3 The quantities T_0, p_0, and ρ_0 are usually called *stagnation parameters*.

speeds and a divergent channel for acceleration to supersonic speeds. The nozzle throat cross section S_* for a perfect gas corresponding to $\underline{M} = 1$ is called *critical* for this reason. T_*, p_*, and ρ_*, according to (J.10), are called the *critical* values. When a nonperfect gas flows through the nozzle in which physico-chemical processes or dissipative phenomena occur, then generally the flow velocity can exceed the sound velocity not at the throat but at some other cross section. Usually the difference between these cross sections is small and we can retain the asterisk as the subscript for the *throat* parameters.

For a one-dimensional steady-state flow of an inviscid gas in which chemical reactions occur, all the equations from (4.1.1) to (4.1.5) are valid as well as thermal and caloric equations of state in the form (3.1.6, 12). However, to solve the problem now we must know the law of variation of the gas content. At the limit of infinitely high chemical reaction rates at each point of the flow chemical equilibrium will be maintained. In this case we have an *equilibrium flow*. The calculation procedure for the equilibrium chemical composition of a mixture is given in Appendix G, where to each value of pressure and temperature there corresponds a set of equilibrium values of concentration. Thus we can, according to (3.1.12) and (E.6), determine the dependence of $h = h(p, T)$, $\mu = \mu(p, T)$, which completes the system of gasdynamic equations (with an account of the thermal equations of state). In the general case the problem of an equilibrium nozzle flow of a chemically reacting gas cannot be solved analytically; to do this, the numerical methods are used. The analytical solution can be carried out only in some particular cases, e.g., for singly ionized nozzle flow at equilibrium (see [4.4]).

When describing a *nonequilibrium gas flow*, the basic gasdynamic and thermal equations of state retain their form, the form of the caloric equation (3.1.12) being determined by the nonequilibrium processes occurring in a gas, and, depending on this, some part of the energy subsystems can be considered as in equilibrium (see Appendix D). In the analysis of internal energy relaxation of molecules in a moving medium the most convenient to use are relaxation equations for the specific energy ε_j or the mean energy e_j per molecule of the jth type. Then all the equations given in the previous chapter are valid for the processes in a moving gas as well. To describe the variation of concentration of any component in a flow, we must obtain the continuity equation for this component including kinetics of physico-chemical processes. To do this we must again use the Boltzmann equation and derive the required phenomenological equation (see [3.1,

Chaps.7 and 11], and for a brief discussion [3.141]). The form of this equation for a one-dimensional steady-state flow of inviscid gas on the basis of simple balance considerations is given in Appendix K. Thus for the variation of the mass fraction α_j of the jth-type particle we have, according to (K.3),

$$\frac{d\alpha_j}{dt} = \frac{\mu_j W_j}{\mathcal{N}_A \rho} \quad , \tag{4.1.7}$$

where W_j is the rate of the concentration variation of the jth component as a result of physico-chemical processes, according to (I.7), μ_j is the molecular mass, and $\mathcal{N}_A$ is the Avogadro number. It should be noted that the description of the gas composition variation in a moving gas will be correct only when the variation of the gas density is simultaneously taken into account. It becomes clearer if in (4.1.7) we go over from the mass fraction α_j to the density n_j of the jth-type particles ($\alpha_j = n_j\mu_j/\mathcal{N}_A$). Then for a moving gas we have

$$\frac{d}{dt}\left(\frac{n_j}{\rho}\right) = \frac{W_j}{\rho} \quad . \tag{4.1.8}$$

There is one more circumstance connected with the transition from a nonequilibrium gas flow to an equilibrium one. The flow can be strictly in equilibrium only for infinitely high reaction rates or zero relaxation time. When kinetic equations are used, the equilibrium limit of a nonequilibrium process is the singularity of the type $\infty - \infty$ for equations of the type (3.4.22, 3.6.5,15) involving rate constants, and of the type $0:0$ for the relaxation equations (3.4.8) which include the relaxation time [the numerator on the right-hand side of (3.4.8) is equal to zero, since at equilibrium $\varepsilon = \bar{\varepsilon}$]. Thus, the solution of the problem of the equilibrium gas flow must be carried out with the assumption of the presence of the respective equilibrium[4], i.e., using physico-chemical balance equations.

Let us turn from the form of the equations to their solution. In the analysis of a nozzle flow, three peculiarities appear. The first peculiarity is connected with determining the gas flow rate G: the value of G cannot be determined from the initial state of the gas at the nozzle entrance, since at this region the gas does not move, and the cross-sectional area of the flow $S \to \infty$ (singularity of the type $0 \cdot \infty$). In certain particular cases the

4 The transition from hydrodynamic relaxation equations to equilibrium hydrodynamics is discussed in [Ref.1.8, Chap.6].

gas flow rate G can be found after determining the flow parameters at the nozzle throat as, for example, for a perfect and vibrational equilibrium gas flow (see Appendix J). In the general case, however, these parameters are unknown. Another peculiarity is connected with the transition from the subsonic flow to the supersonic one; at this cross section the solution has a saddle-point singularity. Finally, considerable difficulties can arise in the analysis of quasi-equilibrium flows.

The existence of a saddle point, where the flow velocity is equal to the sound velocity (at the critical "sonic" cross section), can be established in the following way. Let the density of a gas depend on three parameters: pressure p, entropy s, and a certain parameter ξ determining the physico-chemical state of the gas. In particular, to ξ can be ascribed the meaning of concentration (or the set of concentrations) of gas mixture components. Then the density differential can be represented in the form

$$d\rho = \left(\frac{\partial\rho}{\partial p}\right)_{s,\xi} dp + \left(\frac{\partial\rho}{\partial s}\right)_{p,\xi} ds + \left(\frac{\partial\rho}{\partial\xi}\right)_{p,s} d\xi \quad . \tag{4.1.9}$$

By fixing the parameter ξ, we "freeze" the physico-chemical state of the gas (its composition, molecular distribution over vibrational levels, etc.). As was mentioned in Sect.3.1, the sound velocity calculated under this assumption is called frozen. Rewriting (3.1.13), we have

$$a_f = \sqrt{(\partial p/\partial\rho)_{s,\xi}} \quad . \tag{4.1.10}$$

Substituting (4.1.9) into (4.1.2) and taking into account (4.1.3) and (4.1.10), we obtain

$$\frac{du}{u} = \frac{dS/S + [(\partial\rho/\partial s)_{p,\xi}ds + (\partial\rho/\partial\xi)_{p,s}d\xi]}{\underline{M}_f^2 - 1} \quad , \tag{4.1.11}$$

where $\underline{M}_f = u/a_f$. When the flow velocity passes through the value $\underline{M}_f = 1$ (through the value corresponding to the local frozen sound velocity), the numerator on the right-hand side must also be zero, since the flow velocity continuously grows. Thus, for $\underline{M}_f = 1$ the solution for the velocity distribution over the nozzle has a singularity of the type 0 : 0. At this point the same singularity will appear in the equations for other gasdynamic parameters. It is known that for nonequilibrium (and, hence, irreversible) processes the gas entropy s increases ($ds > 0$); if the process is quasi-equilibrium (i.e., is a sequence of equilibrium states), then $ds = 0$. In a particular case of adiabatic variation of a perfect gas state, the gas density ρ, according to (J.1), depends only on pressure p, so that ds = 0 and $d\xi = 0$ (isoentropic process in the absence of physico-chemical transformations), and the expression in the square brackets in (4.1.11) equals zero; hence the transition through the sound velocity occurs at the nozzle throat. If the physico-chemical state of a gas changes ($d\xi \neq 0$), the expression in the square brackets in (4.1.11) differs from zero, and the transition through the frozen sound velocity occurs not at the throat but at some other cross section S_{tr}, where

$$\frac{dS}{S} = -\frac{1}{\rho}\left[\left(\frac{\partial\rho}{\partial s}\right)_{p,\xi} ds + \left(\frac{\partial\rho}{\partial\xi}\right)_{p,s} d\xi\right] \quad . \tag{4.1.12}$$

Carrying out similar transformations of this expression (see [4.2]), we can see that the cross section $S_{tr}(\underline{M}_f = 1)$ is in the supersonic expanding part of the nozzle. The estimates show that usually this shift from the throat is not large and is usually a small fraction of the characteristic linear dimensions of the nozzle throat. A similar shift of the cross section for $\underline{M}_f = 1$ occurs in perfect gas flow when its viscous properties are taken into account, i.e., when ds > 0 for $d\xi = 0$ (see [3.128]). If the gas is in equilibrium (ds = 0), then the transition through the equilibrium sound velocity $a_{eq} = \sqrt{(\partial p/\partial\rho)_{eq}}$ occurs at the throat, where $\underline{M}_{eq} = 1$.

In the general case (including nonequilibrium flow) the flow rate G can be determined in the following way. On the plane $\underline{M}_f$, $\underline{x}$ (Fig.4.1) four regions are separated for integral curves. Two of them correspond to the flow rate $G > G^*$, where G^* is the flow rate for the required solution. These values of G are not realized since they exceed the maximum possible flow rate. For $G < G^*$ two types of solution are possible: a completely subsonic flow in the entire nozzle (below the curve $G = G^*$) and a completely supersonic flow (above the curve $G = G^*$). The curve for $G = G^*$ corresponds to the exact solution with the transition from subsonic to supersonic flow. The value of G^* is usually determined by a *trial-and-error method*. To do this, a zero-order approximation $G^{(0)}$ is first taken (usually from computations for a perfect gas) and then the integration of gasdynamic equations is carried out simultaneously with kinetic equations for a nonequilibrium flow, beginning with a certain cross section where the gas velocity differs from zero. The selection of the G value, approximating G^* with the required accuracy, is performed according to the results of integration near the nozzle throat, i.e., according to the shape of the integral curves in this region. In this case the coordinate x_{tr} that corresponds to the transition through $\underline{M}_f = 1$ is also determined, but this transition also requires an auxiliary method

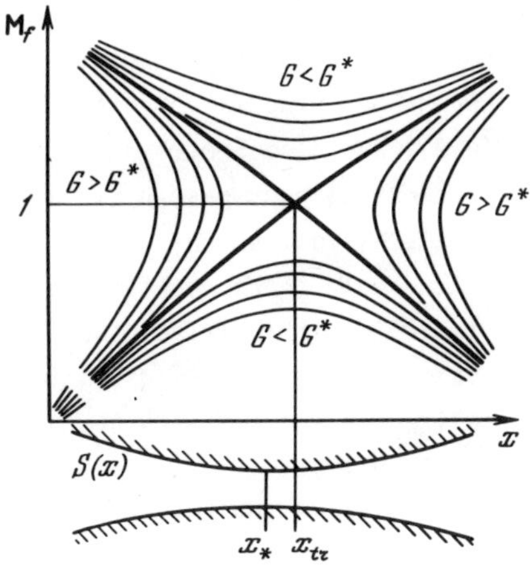

Fig.4.1. Integral curves in the vicinity of a singular point near the nozzle throat

to be used. This can be done in the simplest way by the analytical extension of the results for the nozzle throat $S_*(x_*)$, expanding the parameters into Taylor series near $x = x_*$. Usually it is sufficient to take only linear terms of the expansion, since the singular point is close enough to the throat[5].

There are, however, methods of analysis of a nozzle gas flow which do not involve peculiarities in determining the flow rate for the transition through the sonic section as well. An example is the solution of the so-called *reverse problem* and the *time-dependent method*. In the solution of the reverse problem the nozzle contour $S(x)$ is not given, but is determined from the known distribution of one of the gasdynamic parameters—velocity, pressure, or density [4.6,7]. This distribution is usually found experimentally. The effect of viscosity and other factors in real nozzles in this case are taken into account immediately. At least we can take as an initial distribution the solution of the direct problem for a perfect gas according to (J.4,5,13). In this case it is preferable to take the velocity distribution or density distribution, since in the nozzle flow velocity and density are less dependent on the degree of deviation from equilibrium than are pressure and temperature. Next, the motion equations (4.1.3,5) together with (3.1.6), are solved for the given distribution with respect to the coordinate. The solution of the resulting system of kinetic equations for the flow do not contain singularities. Using (4.1.1), we obtain the nozzle contour (shape) $S'(x)$, which can be compared with the initial contour $S(x)$, and the difference between them can be eliminated using the method of suc-

5 There are also other techniques for calculating G (see [4.5]).

cessive approximations. As a rule, this sequence rapidly converges. Sometimes the inverse method is used only to find the solution for subsonic and transonic regions of the flow (transition through singular points), and then the direct problem is solved (see [4.8]) and also Sect.4.3). Another approach is connected with the use of nonstationary gasdynamic equations to obtain the solution for a steady-state flow as the limiting state ($t \to \infty$) for invariable conditions at the nozzle entrance (time-dependent method). This method also allows us to avoid difficulties connected with the singular points in the steady-state solution. It is given in detail, for example, in [1.2, 4.9] and is realized in the analysis of the nonequilibrium flow in a CO_2 gasdynamic laser [4.10,11].

Let us consider, at last, the difficulties arising in the analysis of *quasi-equilibrium flows*, when kinetic equations are of the form of differential equations with a small parameter at the highest-order derivative. In the chamber in front of the nozzle entrance the gas is at rest in an equilibrium state. In a certain initial region of the nozzle flow deviations from equilibrium are also small, sometimes even in the entire subsonic region and in the part of the supersonic flow. Consequently, in the analysis of the nozzle flow we must inevitably consider quasi-equilibrium states, since deviations from equilibrium may become significant only for a considerable drop of pressure and temperature. In the general form the kinetic equation can be written as

$$\frac{d\xi}{dt} = \varphi(\xi,t) \quad , \tag{4.1.13}$$

where ξ is the generalized parameter of the physico-chemical state of the medium (concentration, energy, etc.). The relaxation time of the process described by (4.1.13) is

$$\tau = (\partial\varphi/\partial\xi)^{-1} \quad . \tag{4.1.14}$$

The integration of (4.1.13) using the widely known Euler or Runge-Kutta method near equilibrium involves considerable difficulties, since even for a smooth variation of ξ we must take a very small integration step $h \leq \tau$, otherwise the solution becomes unstable. We can overcome this difficulty by using the so-called implicit difference schemes of computation, whose algorithm is such that the required value of ξ_{n+1} for the step which follows the known value of ξ_n is implicitly contained in a certain system of algebraic equations [4.12]. The idea of this method is to account for the general nature of the variation of ξ during the transition from point n to n +1 at integration. The essence of this method is given in Appendix L.

In the analysis of relaxing gas nozzle flows other methods are also used, which make it possible to use a sufficiently large stepsite in integrating the system of gasdynamic and kinetic equations (see, for example, [4.13-15]). Thus, representing the required values of ξ_i in the form $\xi_i = \Gamma \exp \alpha_i$ and solving, by the iteration method, the simplified system of difference equations for α_i, we can find the method which works equally well for any extent of the relaxation processes—from the frozen to the equilibrium one [4.14]. An example of this method applied to the analysis of the nozzle flow for a more complicated recombining gas mixture is given in Sect.6.4.

In conclusion we shall pay attention to an important peculiarity of the nozzle flow under consideration for a nonequilibrium excitation of the vibrational degrees of freedom in molecules: the similarity of the flow for the same mixture for equal values of T_0 (the initial temperature at the nozzle entrance) and of the parameter $\lambda = p_0 \ell$ (p_0 is the initial pressure and ℓ is the characteristic nozzle dimension) for the same dependence of the flow area ratio $\bar{S} = S/S_*$ on the coordinate x/ℓ. Indeed, relaxation equations (3.4.19,21) in the general form can be written as

$$u \frac{de_i}{dx} = p \sum \kappa(T) \varphi(e_i, e_j) \quad , \tag{4.1.15}$$

where $\kappa(T)$ is the function of the exchange rate constant or of the relaxation time, which depends only on temperature and the type of molecules, and $\varphi(e_i, e_j)$ is the quantity determined by the mechanism of energy exchange. If in (4.1.15), the conservation equations (4.1.2-4) with (3.1.10), and also in the state equation (3.1.6) we go over to dimensionless variables

$$\bar{x} = \frac{x}{\ell} \ , \quad \bar{S} = \frac{S}{S_*} \ , \quad \bar{p} = \frac{p}{p_0} \ ,$$

$$\bar{\rho} = \frac{\rho}{\rho_0} \ , \quad \bar{T} = \frac{T}{T_0} \ , \quad \bar{u} = \frac{u}{\sqrt{RT_0/\mu}} \ , \tag{4.1.16}$$

then the similarity directly follows from the general form of the resulting equations [4.16]. In the presence of nonequilibrium chemical reactions the number of these parameters increases so greatly that they become useless.

4.2 Deactivation of Molecular Vibrations and Recombination of Atoms in Nozzles

As the flow is accelerated and the gas temperature and pressure drop in the nozzle flow, the rate of physico-chemical processes in a gas, as a rule, considerably decreases. In time a state can set in which the energy exchange between different molecular degrees of freedom (or the chemical reaction) ceases; such a state is usually called *frozen*. Thus in a nozzle the transition from a completely equilibrium flow to a frozen flow takes place. The freezing criterion for the region where the energy exchange or chemical reaction ceases can be obtained in the general form if we introduce the characteristic time t^* during which the value of a certain physico-chemical parameter $\bar{\xi}$ changes considerably as a result of the gas motion and variation of its parameters, t^* is defined by,

$$t^* = \left|\frac{1}{\bar{\xi}} \cdot \frac{d\bar{\xi}}{dt}\right|^{-1} \quad . \tag{4.2.1}$$

If the generalized relaxation time τ of a nonequilibrium process is defined according to (4.1.14), then at first for $\tau << t^*$ we have the equilibrium nozzle flow. Further, as the flow is accelerated and temperature and pressure decrease, τ grows and t^* drops, so that for $\tau \sim t^*$ the relaxation time becomes comparable with the characteristic time of variation of the state of the system. Thus, the noticeable deviation from equilibrium occurs at the nozzle cross section where

$$\left|\frac{1}{\bar{\xi}} \cdot \frac{d\bar{\xi}}{dt}\right| = \frac{1}{\tau} \quad . \tag{4.2.2}$$

If after that the relaxation time becomes much greater than the residence time of a particle in the flow, this deviation leads to freezing of the parameter ξ. If we take for ξ the vibrational energy e, then (4.2.2) can be rewritten in the form

$$\frac{1}{\bar{e}} \cdot \frac{d\bar{e}}{dt} = \frac{1}{\tau_f(p_f, T_f)} \quad , \tag{4.2.3}$$

where the subscript f denotes the region of transition to the frozen state of vibrational energy.

If τ increases very fast, the transition from the equilibrium to the frozen flow occurs practically instantaneously, which enables us to formulate *the principle of instantaneous freezing* [4.17]. Relations (4.2.2,3) illustrate this principle. Its application makes it possible to directly determine the nozzle cross section where the transition from the equilibrium to the frozen flow occurs, which considerably simplifies the analysis of relaxing gas flow in nozzles. In this case, however, it is not always possible to obtain the required accuracy, since the principle is of a criterion nature. It is applicable only for estimates in simple cases, for example, in the analysis of the V-T exchange, whose relaxation time follows the relation $\log \tau \propto T^{-1/3}$. The temperature dependence of the V-V energy exchange probability is less significant, for example, during the exchange between different vibrational modes of the CO_2 molecule. For this reason the principle of instantaneous freezing cannot be recommended as a universal method of analysis of gasdynamic laser flows. This does not mean that the vibrational V-V exchange does not undergo any freezing in the flow. However, the freezing zone in this case can be very extended. It should be stressed that freezing of a certain exchange is caused not only by the temperature drop but also by the decrease of pressure resulting in the reduction of the number of collisions.

In the analysis of gasdynamic laser performance it is very important to know how the rapid decrease of temperature and pressure in a flow affects the distribution of particles over energy levels. If a gas is a mixture of molecules which can be modeled by harmonic oscillators, then, independently of the rate of decrease of temperature and pressure, and absolute values of energy exchange probabilities and mixture composition, the Boltzmann distribution of molecules over the levels will be always retained (if the initial distribution was of the Boltzmann type). The proof of this statement is given in Appendix M. This important conclusion gives the grounds to limit ourselves, when describing the nonequilibrium cooling of the gas composed of harmonic oscillators molecules to consider the relaxation of the average vibrational energy in each mode, i.e., to use the phenomenological equations derived in Sect.3.4, which considerably simplifies the solution of the problem.

Unlike harmonic oscillators, for which the vibrational relaxation time τ is independent of the degree of deviation from equilibrium and direction of the process ($\tau_h^+ = \tau_h^-$ for the same temperature and pressure[6]), in a system of anharmonic oscillators, according to (3.4.13,32), we can expect that

---

6 Further we shall indicate the vibrational relaxation time with a superscript "+" for a gas heating (e.g., in a shock wave) and "-" for cooling (in nozzle flows).

$\tau^+_{an} > \tau^-_{an}$. Indeed, during gas heating and excitation of molecular vibrations the most essential are the transitions between the lowest vibrational levels, while during cooling, on the contrary, the most important are the transitions between the upper levels, for which anharmonicity plays a considerable role (for this reason we can assume that $\tau^+ \approx \tau_h$ and $\tau^- \approx \tau_{an}$); this results in the increase of the deactivation rate. First the possibility of a considerable difference between τ^+ and τ^- was observed in the experiments on measuirng the vibrational deactivation time for diatomic molecules in nozzles and other cooling gas flows; for N_2 the ratio τ^+/τ^- proved to be 15 and higher, and for CO even 10^2 to 10^3 (see [1.10, 4.18-20]).

The analysis of experimental results for CO_2 gasdynamic lasers [4.21] yielded the values of 10 to 10^2 for CO_2-N_2 mixtures. The computation of this ratio using (3.4.13) shows, however, that as a result of anharmonicity the deactivation rate in nozzle flows can be no more than two or three times higher than the rate of excitation. In this connection the question arises, whether or not the assumptions made in the derivation of (3.4.13) are valid for nozzle flows, in particular, the assumption about retaining the Boltzmann distribution of molecules over the levels and the absence of nonresonant V-V transitions. The use of (3.4.32) for τ^- makes it possible to give up these simplifications and to account for the joint effect of the V-T and V-V exchanges (and also deviations from the Boltzmann distribution) on the vibrational deactivation. Then, for a sufficiently high vibrational energy [when the effective "temperature" T_1 of the first level — see (3.4.27) — is comparable with the characteristic vibrational temperature θ] and for low values of the translational temperature T (for $T/T_1 < 0.2$)[7], the deactivation time τ^- can be by several orders of magnitude lower than τ^+. It would seem that we could, taking into account the nonresonant V-V exchange, explain the above experimental results. But experimentally studied freezing of the vibrational energy in a nozzle flow occurs when the deviation from equilibrium is comparatively small, that is, when T/T_1 is still close to unity and the distribution is close to the Boltzmann one, so that in this case (3.4.13) is valid [3.115]. Consequently, τ^+ and τ^- cannot differ by more than a factor of 2 or 3. Accurate experiments [4.20,22] confirmed this for a CO flow; in the experiments with molecular oxygen [4.23a, 4.23b] the ratio $\tau^+/\tau^- \sim 1$ was obtained. The more accurate treatment of the results obtained earlier for N_2 has led to a similar ratio [4.24]. The experiments

---

7 For molecules of the type N_2, CO, O_2 this amounts to less than 300 to 500 K.

on deactivation kinetics for CO_2 [3.68] also confirmed that τ^+ and τ^- are close to each other. It was established in computations [4.25] that the vibrational energy relaxation rate, due to anharmonicity, increases by no more than 30%, this effect decreasing with the parameter $p_0 h_*/2 \times \tan\theta$ (h_* is the throat height and θ is half of the divergence angle of the supersonic part of the nozzle).

The absolute value of the vibrational relaxation time τ affects the flow parameters and the gas state only in the freezing zone, where $\tau \sim t^*$. In the region of equilibrium and completely frozen flow the variations of τ are unimportant. The considerable reduction of the ratio τ^-/τ^+ at the expense of the nonresonant V-V transitions, obtained according to (3.4.32), does not mean at all that this may be a cause of "unfreezing" of the vibrational energy in a cooling gas, since with considerable cooling the absolute value of the V-T exchange probability becomes too low. This "unfreezing" is possible only with an increase of temperature in a channel (due to convergence and shock waves appearing in it) or in channels of very large length.

Thus, inclusion of the effect of anharmonicity on the phenomenological parameters (total vibrational energy, vibrational deactivation time) does not lead to a considerable change in the gas parameters in the nozzle flow. Anharmonicity can affect the molecular distribution over the levels. Usually here we speak about upper levels (for example, in CO molecules—see below). In Sects.2.1 and 3.5 we considered another example of anharmonicity, which was manifested in a peculiar arrangement of the lower levels of twin symmetrical CO_2 vibrational modes due to the Fermi resonance [3.51].

When considering laser transitions between the levels 00^01 and 10^00, 02^00 of the CO_2 molecule, we can, as a rule, limit ourselves to the assumption that all vibrations of this molecule are harmonic. This enables us to consider the molecular distribution over the levels in each mode as the Boltzmann distribution and to put $\tau^- = \tau^+$. This assumption will be adopted when treating processes in a CO_2 gasdynamic laser in Sect.5.1. Similar assumptions will be made in Sect.6.1 when discussing the performance of GDLs on other triatomic molecules.

According to Sect.3.4, during cooling the mixtures in a nozzle after freezing the V-T exchange, in the presence of fast V-V transitions, the quasi-stationary vibrational energy distribution sets in in a gas, for which the values of vibrational temperature T_i, T_j of mixture components (or vibrational modes) i, j are related through (3.4.23,24). It follows from (3.4.23) that in this case we have

$$T_i = \frac{\theta_i}{\theta_j/T_j - (\theta_j - \theta_i)/T} , \qquad (4.2.4)$$

so that with decreasing T the value of T_i can become very large. This "heating" of one of the components was observed in experiments on the cooling of a 5% CO - 95% N_2 mixture in a nozzle, where the vibrational temperature of CO molecules at the nozzle exit at $S/S_* = 30$ greatly exceeded the gas temperature at the nozzle entrance [3.34]. In mixtures of diatomic and triatomic gases, triatomic molecules, which are on combination energy levels (where different vibrational modes are existed simultaneously), can take part in such processes. The quasi-stationary V-V exchange via these combination levels during mixture cooling in a nozzle results in "heating" of certain modes, e.g., in the mixtures of HF-CO_2, DF-CO_2, D_2-CO_2, CO-CS_2 [3.35].

A peculiar nature of quasi-stationary nonequilibrium distribution of molecules over vibrational levels during the nonresonant V-V exchange in a system of anharmonic oscillators (Fig.3.1), mentioned in Sect.3.4, served as the basis for creating gasdynamic lasers on vibrational transitions between highly excited levels in diatomic molecules, in particular, CO molecules. This distribution is realized after a considerable gas cooling in a nozzle, when the probability of direct V-T transitions becomes very low (this kind of freezing has been considered earlier), and there is a possibility of the V-V exchange for which the energy defect is transferred into translational degrees of freedom. In this case anharmonicity plays a great role, and we must find the change in the population of certain molecular levels in a cooling gas. Such an analysis is carried out in considering the CO gasdynamic laser performance (Sect.6.2). In a similar way the population inversion in GDLs operating on twin modes of the CO_2 molecule is created (Sect.5.8).

Let us consider now a nozzle flow in which chemical transformations occur, using as example the simple dissociation and recombination of diatomic homonuclear molecules: $X_2 + M \underset{r}{\overset{d}{\rightleftharpoons}} X + X + M$, where M is any of these particles and letters d and r indicate the direction of the reaction. The state of a gas at the nozzle entrance and the flow in a certain initial region are at equilibrium. Expressing concentrations in terms of mass fractions α_j (Appendix E), for the equilibrium state we obtain, according to the law of mass action (Appendix G),

$$\bar{\alpha} = \sqrt{K_{eq}/(4p + K_{eq})}$$

(the subscript indicating the atomic component is omitted), where K_{eq} is the equilibrium constant (G.5) and p is the gas pressure. During further cooling and drop of pressure the equilibrium in the flow is disturbed; in this case the state of the gas can be described by the chemical kinetics equation (4.1.7). From (E.3, F.1) we obtain

$$\frac{d\alpha}{dx} = \frac{p}{uRT}\left[k_d(1-\alpha)/2 - k_r\,\frac{2p}{RT}\cdot\frac{\alpha^2}{(1+\alpha)}\right] \quad , \tag{4.2.5}$$

where k_d, k_r are the rate constants of dissociation and recombination in molar units—cm^3 mole/s and $cm^6/mole^2$ s, respectively—and u is the flow velocity. Symbolically (4.2.5) can be written in the form

$$\frac{d\alpha}{dx} = R_d - R_r \quad , \tag{4.2.6}$$

where R_d and R_r are the rates of dissociation and recombination. In the conditions of quasi-equilibrium flow these rates are sufficiently high, so that

$$R_d \gg \frac{d\alpha}{dx} \quad , \quad R_r \gg \frac{d\alpha}{dx} \quad . \tag{4.2.7}$$

The drop of temperature and pressure results in a decrease of R_d and R_r. The noticeable deviation from equilibrium begins when R_d, R_r, and $d\alpha/dx$ become commensurable by absolute value. With further decrease of temperature the value of R_d rapidly drops, and the decrease of pressure, according to (4.2.5), results in a considerable reduction of R_r[8]. Thus, in the nozzle flow sooner or later the rates of chemical processes become very low, and the chemical composition of the gas freezes,

$$R_d < R_r \quad , \qquad R_d \;, R_r \to 0 \quad , \quad \frac{d\alpha}{dx} \to 0 \quad . \tag{4.2.8}$$

With a strong effect of the variation of temperature and pressure on the rate of these processes the transition from the equilibrium flow to the frozen flow occurs practically instantaneously [4.17]—when

$$\left(\frac{d\bar{\alpha}}{dx}\right)_f = -R_r \quad . \tag{4.2.9}$$

---

8 In this condition the dissociation rate constant usually depends on temperature [$k_d \propto \exp(-D/kT)$, $D \gg kT$, D is the dissociation energy] to a greater extent than for recombination ($k_r \propto T^{-n}$, $n \approx 1$).

This form of the principle of instantaneous freezing, like (4.2.2,3) makes the analysis of a nonequilibrium nozzle flow of a reacting gas quite universal. Indeed, writing (4.2.9) taking account of (4.2.5),

$$-\frac{u}{2}\cdot\frac{d\bar{\alpha}}{dS}\left(\frac{RT}{p}\right)^2\frac{1+\bar{\alpha}}{\bar{\alpha}^2}=\frac{k_r}{dS/dx}\quad, \tag{4.2.10}$$

we see that the left-hand side of this equation (let us denote it by $\mathscr{F}$) in an equilibrium flow is the function only of the ratio S/S_*, while the left-hand side depends on the rate of change of the nozzle cross section, dS/dx, and on the recombination rate constant at the point of freezing. The value of $\mathscr{F}$ is obtained from the analysis of the equilibrium nozzle flow. By evaluating the right-hand side of (4.2.10), we can immediately find the freezing point and then completely describe the flow in the instantaneous freezing approximation. Examples of the analysis by this method of a gas flow in a recombination laser are given in Sect.6.4.

If several relaxation processes occur in a flow simultaneously (multipass vibrational exchange in polyatomic mixtures, or chemical reactions in a multicomponent gas, or both), the problem becomes so complicated that the principle of instantaneous freezing cannot be applied any longer, since, owing to the difference in the rates of the processes, their freezing occurs at different nozzle cross sections[9]. In this case numerical methods of solution are used. When taking into account the mutual effect of chemical reactions and vibrational relaxation for not too high temperatures (lower than the characteristic vibrational ones), we usually can assume that the characteristic time of a chemical reaction considerably exceeds the vibrational relaxation time. Then during cooling the gas in the nozzle first ceases chemical reactions, freezing the gas composition, and later the vibrational exchange is frozen. In the simplest case of a diatomic gas the boundaries between the completely equilibrized, vibration-equilibrized and completely frozen flow can be determined using the principle of instantaneous freezing. For polyatomic gases and mixtures it is more difficult to distinguish these boundaries in advance, but in the numerical solution it is often possible to simplify the scheme of kinetic processes, assuming that during chemical reactions the gas components are at vibrational equilibrium, and taking into account the possibility of a nonequilibrium vibrational

9 The application of the instantaneous freezing principle for reactions with higher energy content makes it possible to approximately solve certain problems about multicomponent gas flows in nozzles [4.26,27].

exchange only after freezing the chemical composition of the mixture. In the next approximation we must take into account the absence of chemical and vibrational equilibrium and use (3.6.10) and (3.6.11) with the rate constant according to (3.6.8). For the flow of the CO_2-N_2-H_2O mixtures such calculations are made in [4.28].

4.3 Nozzle Contouring. Non-One-Dimensional Gas Flow

To develop GDL nozzles we must solve two problems: 1) rapidly cool a gas to create an active medium, and 2) provide the required oscillation parameters in a cavity. This requires that the medium must be accelerated to a high velocity and be homogeneous. In a nozzle gas flow the most intensive cooling occurs after passing through the so-called *accelerating region* of the supersonic part of the nozzle; here higher cooling rates correspond to greater divergence angles (until shock waves appear and the flow is separated from the nozzle walls). To obtain a homogeneous active medium and supply it to the cavity, the nozzle should have a *uniformation region*, at the end of which the transverse velocity component must be close to zero and the flow uniform. The creation of a nozzle which consists of the accelerating and the uniformation regions and provides at the exit a sufficiently uniform flow is a typical problem in the development of wind tunnel nozzles, intended for modelling of supersonic flight. For this purpose the nozzle contour is given a special shape. Evidently, this problem cannot be solved with one-dimensional approximations. For a gasdynamic laser performance, as a rule, flows with a plane symmetry are required, therefore we shall further consider plane two-dimensional flows and flows in two-dimensional nozzles[10].

The analysis of a nozzle flow can be carried out, as was noted in Sect. 4.1, according to two different approaches — the direct and the reverse. In the *direct problem* the nozzle contour is given, and we determine the distribution of all gasdynamic parameters in the flow. In the *reverse problem*, on the contrary, we have the distribution of one of the parameters (usually at the nozzle axis of symmetry) and find the shape of the nozzle contour and all the other parameters. Hence, nozzle contouring is the reverse problem.

We shall begin the discussion of this problem with the most simple case of a two-dimensional steady-state vortex-free isoentropic flow of an inviscid

10 In some GDL designs axiosymmetrical nozzles are also utilized (Sects. 4.4 and 6.2).

and perfect gas. This model to a certain approximation can be used to describe a GDL nozzle flow (for example, by introducing the effective value of the ratio of specific heats γ). In this case for the flow velocity w we can write the following motion equations [4.3]

$$(u^2 - a^2)\frac{\partial u}{\partial x} + uv\left(\frac{\partial u}{\partial y} + \frac{\partial v}{\partial x}\right) + (v^2 - a^2)\frac{\partial v}{\partial y} = 0 \quad , \tag{4.3.1}$$

$$\frac{\partial u}{\partial y} - \frac{\partial v}{\partial x} = 0 \quad , \tag{4.3.2}$$

where u, v are x and y components of the velocity w (the x axis is directed along the nozzle axis, and the y axis, perpendicular to it) and a is the sound velocity. The main difficulty in the solution of this problem is the nonlinear nature of (4.3.1). Gasdynamic equations can be solved by the approximate linearization of (4.3.1) or with the substitution of variables that transforms (4.3.1) into a linear equation. In recent years numerical methods for the solution of nonlinear equations have been successfully developed. According to (4.3.1), the steady-state gas flow in the subsonic part of the nozzle is described by equations of elliptical type, and in the supersonic part, of hyperbolic type. The change of the type of equations complicates the analysis of the flow in the transient transonic region of the nozzle. Only in the particular case of a specially arranged flow will the transient surface be plane and perpendicular to the direction of the flow, i.e., when at the center of the nozzle flow (at the point on the axis of symmetry where the flow velocity is equal to the sound velocity) the flow velocity gradient is zero. In this case the solution of the problem for the gas flow in subsonic and supersonic parts of the nozzle can be found separately. If the flow velocity gradient at the nozzle center differs from zero, the transient surface is curved (Fig.4.2). Then the analysis of the subsonic and the supersonic parts of the nozzle flow must be carried out simultaneously, which is difficult because the motion of a gas in the transonic region is described by equations of a complex elliptic-hyperbolic type. For this reason, in contouring the nozzle for simplicity the transient surface is assumed to be plane, and the two parts of the nozzle are considered separately. However, it is very difficult to realize the plane transient surface because of nozzle manufacturing inaccuracies and the effect of viscosity.

High-velocity gradients are required to achieve rapid cooling of a gas in gasdynamic lasers, so that the problem of designing nozzles with curved

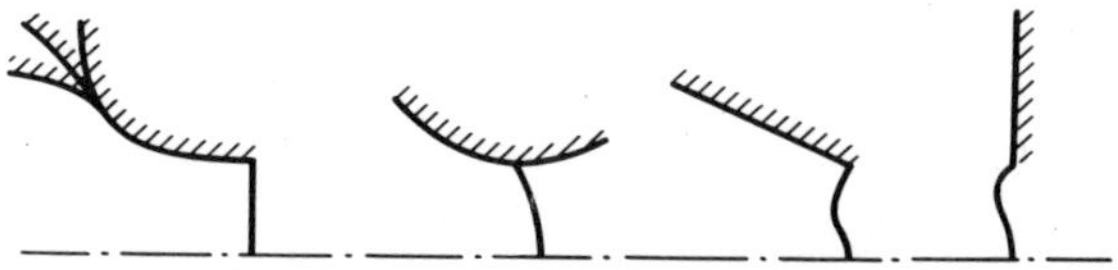

Fig.4.2. Sonic transient lines for different shapes of the subsonic part of a nozzle

transient surfaces is of current interest. To solve this problem the velocity potential $\varphi(u = \partial\varphi/\partial x,\ v = \partial\varphi/\partial y)$ is sometimes expanded in a series in powers of x and y near the nozzle center x = y = 0 (the Meyer method) and the distribution of u and v in the transonic region is found. The expansion coefficients are determined by the nozzle shape near the throat and by the gas properties. The disadvantage of this method is that the radius of convergence of this series with respect to the y axis is unknown beforehand. In 1904 Chaplygin put forward a suggestion which was of fundamental importance for the development of gas dynamics. He transformed (4.3.1) to the linear form by substituting for the variables x, y the variables φ (velocity potential) and ψ (function of flow defined so that $\partial\psi/\partial x = -\rho v/\rho_0$, $\partial\psi/\partial y = \rho u/\rho_0$, ρ_0 is the density of a nonmoving gas), and the transition from the physical plane of the flow to the plane of the velocity hodograph. The variable ψ numbers the stream lines, the tangent to which indicates the direction of the velocity vector at every point. In a steady-state flow these lines coincide with the trajectories of particles in the flow. Thus, the family of lines $\psi(x, y)$ = const. gives a graphic picture of a two-dimensional nozzle flow. The solution of the Chaplygin equation obtained as a result of this transformation makes it possible to determine the shape of the transient lines of a two-dimensional nozzle for various types of a subsonic part of the nozzle (Fig.4.2). Using these methods, we can determine the contour of a subsonic part of the nozzle. In wind tunnels the subsonic contour of a nozzle is usually found using the Vetoshinsky empirical equation

$$y = y_*\left[1 - \left(1 - \frac{y^2}{y_{en}^2}\right)\frac{(1 - x^2/\ell^2)^2}{(1 + x^2/3\ell^2)^3}\right]^{-1/2}, \tag{4.3.3}$$

where y_{en} and y_* are the ordinates of the contour at the nozzle entrance (x = 0) and at its throat ($x = \ell$); $\ell = (3 \text{ to } 4)y_*$. Sometimes the subsonic contour is given by the shape of an arc of a certain radius R; for $R \geqq 4h_*$ (h_* is the throat height, $h_* = 2y_*$) the transient surface is close to a plane. The gas flow rate G through the nozzle is equal to the flow rate G_0 for a one-dimensional flow (4.1.1) only for a plane transient surface. For a curved surface in a two-dimensional nozzle the flow rate is

$$G = G_0\left[1 - \frac{\gamma + 1}{360}\left(\frac{h_*}{R_m}\right)^2\right] , \tag{4.3.4}$$

where R_m is the radius of curvature at the nozzle throat.

The shape of the transient surface (or its projection—the sonic line) is determined by the nature of the gasdynamic parameter variation in the transonic region of the nozzle. With a plane transient surface harmful shock waves in this region are absent. For the flows with finite acceleration at the nozzle center (i.e., with a curved sonic line) the continuous flow without shock waves in the transonic region is possible only under certain conditions for the velocity distribution near the nozzle center. For example, for a linear velocity variation, when the sonic line has the shape of a parabola concave downstream, the flow is continuous if the ratio of velocity gradients in subsonic and supersonic flows lies between 0.25 and 1. In this case, with a sufficiently steep contour of the subsonic part of the nozzle, the entrance of the supersonic part should not be too flat. For this reason the nozzle contour should have an inflection point at the throat. A sharp increase in the slope of the curved transient surface, however, may result in local zones of stagnation in the flow and corresponding increase of pressure, and also in shock waves appearing already because of the flow overexpansion.

The experimental study of the flow field in the transonic part of a gas-dynamic laser nozzle [4.29] showed that despite the great variety of subsonic nozzle contours (the variation of the angle between the subsonic contour and the flow axis from 20° to 90°), the flow pattern at the beginning of the supersonic region changes little. In all cases, however, in the peripheral zone of the flow weak shock waves are observed due to the flow overexpansion (similar shock waves were observed in [4.30]).

The analysis of the flow in the supersonic part of the nozzle can be conveniently carried out using the *method of characteristics*, which consists in the determining of the curves (characteristics) at every point of which certain relations are fulfilled, which follow from the motion equations for a gas (4.3.1,2). Introducing the Mach angle α ($\alpha = \arcsin 1/\underline{M}$) and angle β formed by the flow velocity vector with the x axis ($u = w\cos\beta$, $v = 2\sin\beta$),[11] we can easily obtain for (4.3.1,2) the equation for the

11 Equation $dy/dx = \tan\beta$ describes stream lines.

characteristics (see, for example [4.3,31a]),

$$\frac{dy}{dx} = \tan(\beta \pm \alpha) \quad , \tag{4.3.5}$$

and the relation on the characteristics,

$$\nu(\underline{M}) \mp \beta = \text{const.} \quad , \tag{4.3.6}$$

where $\nu(\underline{M})$ is the Prandtl-Meyer function

$$\nu(\underline{M}) = \sqrt{(\gamma + 1)/(\gamma - 1)}$$
$$\times \arctan\sqrt{(\underline{M}^2 - 1)(\gamma + 1)/(\gamma - 1)} - \arctan(\sqrt{\underline{M}^2 - 1}) \quad . \tag{4.3.7}$$

The upper sign in (4.3.5,6) corresponds to the characteristics of the first family, and the lower sign, to the second one. The physical meaning of the characteristics is that they are the lines along which weak perturbations propagate in a supersonic flow (the Mach lines). To apply the method of characteristics we must in (4.3.5) pass over from the derivatives to the ratio of finite differences and define the distribution of the $\underline{M}$ values ($\underline{M} > 1$) over a certain transonic line, and step by step carry out the analysis of the velocity field in the entire flow using (4.3.5-7). The selection of the initial (transonic) distribution of $\underline{M}$ is determined by the solution of the problem for the sub- and transonic regions of the flow. Often it is assumed that $\underline{M}$ is a constant a little greater than unity (say, $\underline{M} = 1.001$) along the line across the flow near the nozzle throat. Evidently, this assumption is close to reality only for the flows with plane transient surfaces.

In the nozzle with an inflection point in the accelerating region the flow makes a turn around this point.[12] In this case the characteristics of the second family emerge from point A and go to the nozzle axis, while the characteristics of the first family, on the contrary, go from the axis to the nozzle walls (Fig.4.3a). The analysis of the flow turn using the method of characteristics is continued until the number $\underline{M}$ at the nozzle axis reaches the value $\underline{M}_{ex}$, which is required at the nozzle exit. The analysis of the uniformation region—from the limiting characteristic AB of the accelerating region to the last characteristic of the uniformation region (all

12 A supersonic semilimited flow round the exterior convex angle is called the Prandtl-Meyer flow, for which a number of simple relations was obtained (see [4.1,31b]).

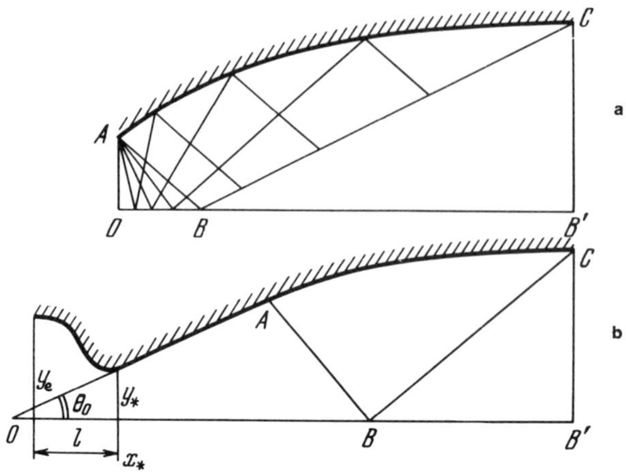

Fig.4.3a,b. Nozzle contouring

values of the parameters on this characteristic are constant) — consists basically in determining the supersonic nozzle contour. For a two-dimensional nozzle with a uniform flow at the exit all the characteristics of the first family in the uniformation region (to the right of AB) are straight lines. Hence, to determine the nozzle contour coordinates we can use the condition of a constant flow rate through the characteristics of the first and the second families (in the region from the nozzle contour to their intersection). The uniformation region should not be too short, since the rapid contraction of the supersonic flow results in the convergence of the characteristics and their intersection, pointing to the ambiguity of the solution at this point, i.e., the appearance of a shock wave. The rate of variation of gasdynamic parameters in such a nozzle can be estimated using (J.4,5) and the Mach number distribution. The example of this distribution over the nozzle axis on the accelerating region OB is given in Fig.4.4a for different values of γ [4.32]; on the uniformation region BB' the value of $\underline{M}$ along the axis is constant. From (4.3.4) it follows that the distance BB' = $(H/2)\sqrt{\underline{M}^2 - 1}$, where H is the height of the nozzle exit. The total length $\bar{L}$ of this nozzle related to H is shown in Fig.4.4b, and the tables of contour coordinates are given in [4.32]. The supersonic nozzle with an inflection point has minimal length and provides for the highest cooling rate; for this reason such nozzles are widely employed in gasdynamic lasers [1.22, 4.33-36].

The problem of nozzle contouring is considerably simplified if in the accelerating region the flow is radial, when all the variables, as in a one-dimensional flow, depend only on one coordinate (here the r coordinate —

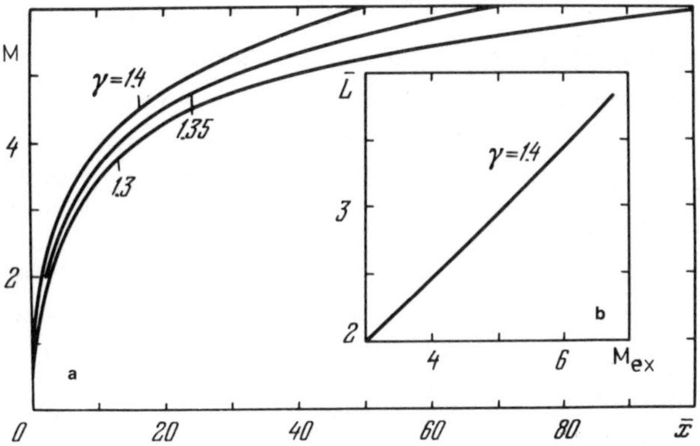

Fig.4.4. Mach number distribution on the axis of the accelerating part of a nozzle with the inflection point (a) and reduced length $\bar{L}$ of this nozzle (b) (the curves for γ = 1.3, 1.35, and 1.4 practically coincide). Reduced coordinate $\bar{x}$ is normalized with respect to $h_*/2$, nozzle length $\bar{L}$, to the height H of the nozzle exit

the radius of the cylinder), see Fig.4.3b. In this case there is a simple relation between the Mach number and the radius r [4.31a],

$$\frac{r}{r_*} = \frac{1}{\underline{M}}\left(\frac{\gamma - 1}{\gamma + 1}\underline{M}^2 + \frac{2}{\gamma + 1}\right)^{(\gamma+1)/2(\gamma-1)} , \tag{4.3.8}$$

where r_* is the radius of the cylindrical surface of transition through the sound velocity value. Here the supersonic accelerating region has the shape of a wedge inner surface (wedge nozzle), and the parameters at every point of the limiting characteristic AB depend only on the distance r from the conditional center O of the source. The contour of the uniformation region of the nozzle, as before, can be easily obtained from the equality of the gas flow rates across the lines AB and CB'. Let θ_0 be the angle between the plane of the accelerating part and the nozzle axis (< AOB); then for a certain value of the angle of inclination θ of the uniformation part contour ($0 \leqq \theta \leqq \theta_0$) the contour coordinates are

$$x = \frac{H}{2}\,\frac{\cos\theta}{\theta_0}\,\frac{r}{r_B}\,\{1 + (\theta_0 - \theta)[\sqrt{M^2 - 1} - \tan\theta]\} ,$$

$$y = \frac{H}{2}\,\frac{\cos\theta}{\theta_0}\,\frac{r}{r_B}\,\{\tan\theta + (\theta_0 + \theta)[1 + \tan\theta\sqrt{M^2 - 1}]\} , \tag{4.3.9}$$

where $\theta = \nu(\underline{M}_{ex}) - \nu(\underline{M})$, $\underline{M}$ is the Mach number at the wall, $\nu(\underline{M})$ is determined according to (4.3.7), and r/r_B is the ratio of the current value of the radius of the cylindrical source according to (4.3.8) and the value r_B at point B. The measurement of pressure distribution in such nozzles showed [4.37] that for $\underline{M}_{ex}$ = 2 and 3 the flow in the accelerating region is close to radial. Behind the uniformation region in sufficiently long nozzles (the total length is more than several times as large as H/2) for $\underline{M}_{ex}$ = 1.25 to 4 the nonuniformity across the flow does not exceed $\Delta\underline{M} = \pm 0.01$.

The nozzle contour for isoentropic flows and for nonequilibrium flows can be determined from the solution of the reverse problem for the entire nozzle in the x, ψ coordinates with a given distribution of velocity (or pressure) over the nozzle axis ($\psi = 0$) [4.38]. The initial distribution should provide for the continuity of the flow without shocks and the uniformity of a gas-dynamic field at the nozzle exit with a minimal nozzle length. This distribution can be obtained semi-empirically—either from the analysis of the experimental results or by the selection of certain analytical expressions which fulfill these conditions. The system of motion equations for a two-dimensional gas flow in these coordinates is

$$\frac{\partial y}{\partial \psi} = \frac{1}{\rho u} \quad , \quad \frac{\partial (py)}{\partial \psi} = \frac{p}{\rho u} - y\frac{\partial v}{\partial x} \quad , \quad \frac{1}{2}\frac{\partial w^2}{\partial x} + \frac{1}{\rho}\frac{\partial p}{\partial x} = 0 \quad , \tag{4.3.10}$$

where w is the overall flow velocity. These equations are supplemented by the equation for the stream lines,

$$v = w\frac{\partial y}{\partial x} \Big/ \sqrt{1 + \left(\frac{\partial y}{\partial x}\right)^2} \quad , \quad u = \sqrt{w^2 - v^2} \quad . \tag{4.3.11}$$

For a given stream line the energy conservation equation (4.1.5) and also the state and kinetic equations can be written. Further, the computational algorithm for all the quantities along the stream line with a given velocity (or pressure) distribution can be compiled. This algorithm completely coincides with the methods of the solution of the problem for a one-dimensional flow of a nonequilibrium gas. For the analysis of a two-dimensional flow we must know the geometry of the stream lines (to find y) and the velocity (or pressure) distribution on them. To do this the entire field x, ψ under consideration is divided by straight lines (in the general case nonequidistant), which form a rectangular network. Proceeding from the known values of parameters on the layer ψ_n (usually beginning from the axis, $\psi = 0$), the values of y, p, u, v, and ρ on the layer ψ_{n+1} can be found by the iterative method. Then other required values can be found by integrating kinetic equations over each stream line as described in Sect.4.1. The boundary conditions are usually found from equilibrium conditions at the nozzle entrance.

The great advantage of the solution of the reverse problem is that the analysis of sub-, tran-, and supersonic flow regions can be carried out using a unique scheme, which is very convenient for operating on a computer. Thus the shape of the nozzle contour with very small radius of curvature at the throat (practically with a point of inflection) is determined.

The shape of the subsonic and transonic part of the nozzle contour practically does not depend on the physico-chemical state of the gas. In the supersonic part of the flow the difference between the stream lines for a frozen and an equilibrium flow becomes noticeable. For a nonequilibrium flow, unlike the perfect and equilibrium gas flow, the shape of the nozzle contour depends on p_0 and h_*, and it cannot be represented in the relative coordinates. Hence, strict speaking, each nonequilibrium gas flow regime has its own nozzle contour. In practice, however, the problem of nozzle contouring is usually solved for a perfect gas flow and then a correction, which takes into account the effect of viscosity but not of the deviation from equilibrium, is introduced (see Sect.4.4). Note that in the solution of the reverse problem we obtain, in addition to the nozzle contour, the field of all the characteristics of a nonequilibrium flow.

The direct analysis of a non-one-dimensional nonequilibrium flow can be successfully carried out using different techniques. Here we shall give only some aspects of the solution of the direct problem applied to nonequilibrium expansion flows. These are formed, for example, during supersonic flow round the acute angle (the Prandtl-Meyer flow). The elements of this flow are realized in the turn of the flow around the nozzle point of inflection. The Prandtl-Meyer flow itself is important as a way of rapid gas cooling. In the flow, around the vicinity of the inflection point, three regions can be conveniently separated (Fig.4.5): the first region is near the wall, where as a result of a very fast cooling the state of a gas has no time to change during the transition through the expansion wave (here the characteristic length of relaxation $\ell_r = a_0\tau \gg r$, r is the distance from the inflection point and a_0 is the sound velocity in the oncoming flow); the second region is at such a distance r for which $r \sim \ell_r$; and the third region is far from the wall, where $r \gg \ell_r$ and where during the transition the state of the gas remains at equilibrium. The straight line OA is the first line of the perturbation propagating from the inflection point O with a velocity equal to the frozen sound velocity a_f (4.1.10); the straight line OA' corresponds to the propagation of first perturbations with the equilibrium sound velocity a_e. The limiting characteristic of the first family, OB, is now curved, since the gas properties vary along this charac-

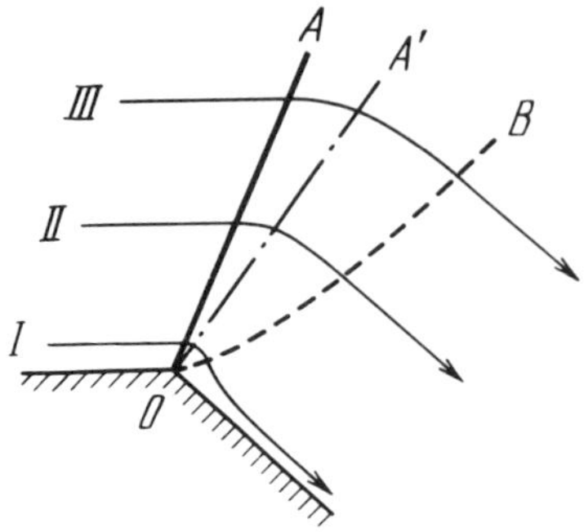

Fig.4.5.
Prandtl-Meyer flow

teristic.[13] An important feature of a two-dimensional nonequilibrium flow is its vorticity, determined by the nonequilibrium nature of physico-chemical processes in it [4.2]. Analytically this means that the right-hand side of (4.3.2) differs from zero. Vorticity of a nonequilibrium flow may result in certain anomalies in the gasdynamic parameter distribution. Thus, behind the inflection point in a nonequilibrium flow there appears a positive gradient of pressure due to the reflection of expansion waves from the vortex zone. The presence of the positive pressure gradients near the walls facilitates the undesirable separation of the flow from the nozzle walls (Sect.4.4).

In the first analyses of a two-dimensional nonequilibrium flow in nozzles with inflection points the flow field in axiosymmetrical nozzles was determined by the method of characteristics, beginning from the straight transonic line with constant parameters and the velocity vector parallel to the nozzle axis. The solution was carried out by numerical methods and took much time. Considerable saving of time was achieved by using the above-mentioned method of networks in the x, ψ plane both in the direct and the reverse problem on the nozzle flow. The difficulties associated with the transition through the sound velocity in the direct problem can be avoided by going over to equations of nonstationary gas dynamics and by using the time-dependent method. Then the complicated system of elliptic-hyperbolic equations in two variables x, y transforms into a hyperbolic system in three variables x, y, t, for which a number of different schemes for finding a numerical solution have been developed [4.12]. The application of this method for the analysis of nozzle flows, however, is expedient only for sub- and transonic

13 In a nonequilibrium flow the characteristic directions are determined by the frozen sound velocity [4.2].

parts of the flow, since for a supersonic flow the equations are of the hyperbolic type already without this transformation, and the introduction of a new variable complicates the scheme and increases the computation time. In the solution of the direct problem for a given nozzle contour shock waves can appear in a supersonic flow. A sufficiently effective method of solution must provide for the analysis of the entire flow taking account of shock waves without their separation. Such is the *shock-capturing method*, in which the equations are written in the form of integral relations for certain cells into which the supersonic part of the flow is divided with a certain period [4.39,40]. Methods of analysis of not only two-dimensional, but also three-dimensional nozzle flows are given in detail in [4.12]. In conclusion it should be noted that account of the nonequilibrium nature of the nozzle flow can be taken in such a way that first the solution for the equilibrium flow or for a perfect gas flow is obtained and then, by determining the arrangement of stream lines, kinetic equations are solved along these lines. Of course, this approach is valid only when the effect of physico-chemical processes on gasdynamic parameter distributions is small.

4.4 Effect of Viscosity. Gas Flow in Nozzle Array Wakes

The basic notion which enables us to successfully solve the majority of problems of viscous gas dynamics is that the effect of viscosity is strong only in the vicinity of the walls (in the boundary layer), while the parameters in the remaining part of the flow (the core) experience only indirect influence of viscosity. The shear stress τ_{sh} (the force of internal friction in a gas per unit shear area) is proportional to the flow velocity gradient

$$\tau_{sh} = \mu \frac{du}{dy} , \qquad (4.4.1)$$

where μ is the temperature-dependent dynamic viscosity coefficient of the gas. Let us assess the boundary layer thickness δ in which the variation of velocity across the flow mainly takes place. The inertia force of a moving gas acting on a gas layer of height δ is approximately equal to $\rho\delta(du/dt) = \rho\delta u(du/dx) \approx \rho\delta u^2/\ell$, where ℓ is the characteristic length. By equating this force to the shear stress ($\tau_{sh} \approx \mu u/\delta$), we obtain

$$\rho \frac{\delta u^2}{\ell} \approx \mu \frac{u}{\delta} \quad , \qquad \text{from which} \quad \frac{\delta}{\ell} \approx \frac{1}{\sqrt{\underline{\underline{Re}}}} \quad , \tag{4.4.2}$$

where Re is the Reynolds number[14], which characterizes the ratio of inertia forces in a flow to friction forces: $Re = ul/\nu$, ν is the kinematic viscosity ($\nu = \mu/\rho$). According to (4.4.2), for sufficiently high values of Re the boundary layer thickness is considerably smaller than the characteristic nozzle dimensions. The equations describing the flow (Prandtl equations) are obtained from the general equation of viscous gas dynamics [4.3,41], where with the account of $Re \gg 1$ the terms containing $1/Re$ to higher powers are ignored. For a two-dimensional flow in a boundary layer the continuity, momentum, and energy equations are, respectively, of the form

$$\frac{\partial \rho u}{\partial x} + \frac{\partial \rho v}{\partial y} = 0 \quad ,$$

$$\rho\left(u \frac{\partial u}{\partial x} + v \frac{\partial u}{\partial y}\right) = - \frac{\partial p}{\partial x} + \frac{\partial}{\partial y}\left(\mu \frac{\partial u}{\partial y}\right) \quad , \quad \frac{\partial p}{\partial y} = 0 \quad , \tag{4.4.3}$$

$$\rho\left(u \frac{\partial h}{\partial x} + v \frac{\partial h}{\partial y}\right) - u \frac{\partial p}{\partial x} = - \frac{\partial Q}{\partial y} + \mu\left(\frac{\partial u}{\partial y}\right)^2 \quad ,$$

where h is the enthalpy and Q is the heat flux associated with the energy transfer due to thermal conductivity and diffusion of particles. It is essential that the gas pressure does not change across the boundary layer. Thus, to fulfill (4.4.3), the boundary layer should be relatively thin ($\delta \ll 1$); but on the other hand, the boundary layer thickness must be greater than the mean free path of molecules.

Unlike rocket engines, in whose nozzles the boundary layers are typically turbulent, gasdynamic laser nozzle flows usually exhibit laminar boundary layers [4.36]. As Re increases, a transition from a laminar layer to a turbulent one occurs on a plate—usually when $Re \approx 10^5$ to 10^6 [4.1,41]. Pulsations in the flow in front of the nozzle entrance cause earlier appearance of a turbulent layer; a considerable acceleration of the nozzle flow, however, may result in a reverse transition to the laminar flow [4.42].

---

14 In addition to the Reynolds number, of great importance in the dynamics of viscous and heat-conducting gas is the Prandtl number, $Pr = c_p\mu/\lambda$ (λ is the thermal conductance). It characterizes the degree of similarity of velocity and temperature fields in a gas. For perfect monoatomic gases $Pr = 0.67$, for diatomic gases $Pr = 0.72$, for triatomic gases $Pr = 0.8$. The account of physico-chemical processes results in a certain change of Pr.

The flow in a laminar layer is described by (4.4.3); for a turbulent boundary layer a similar system of equations includes certain coefficients (turbulent "viscosity" μ_t, diffusion D_t, etc.), which require additional relations for their determination [4.43,44].

In the analysis of GDL performance two problems arise when considering gas viscosity: nozzle contouring for a viscous gas flow and determining the state of the active medium in a boundary layer. For the first problem we shall mention two types of influence of the viscous boundary layer on the flow core. First, the boundary layer distorts the outer inviscid flow as a result of its displacement and reduction of the core cross section and, second, perturbations propagate through the region of low flow velocities near the wall upstream, which is never observed in inviscid supersonic flows. The most important result of the second type of interaction of the boundary layer with the outer flow is separation of the flow from the nozzle walls.

In GDL nozzle contouring, of great importance is the final result of the effect of viscosity on the flow parameters; for this reason it is expedient to use integral methods accounting for friction. One of these methods consists in the replacement of the actual flow velocity distribution in a boundary layer by a certain set of approximate curves which depend on one or more parameters. These parameters are selected so that certain integral conditions are satisfied. The most widely spread are the so-called Karman integral conditions, which follow from the balance of momentum as applied to the boundary layer flow [4.3,41]. For this purpose such boundary layer parameters as the thickness of displacement δ^* and the thickness of the momentum loss δ^{**} are introduced. The value of δ^* is determined by equating the amount of gas flowing through the layer of height δ^* (with a velocity and density of the outer flow) to the decrease of the gas flow rate through the boundary layer (the subscript e indicates the flow core),

$$\rho_e u_e \delta^* = \int_0^\infty (\rho_e u_e - \rho u) dy \quad , \quad \text{or} \quad \delta^* = \int_0^\infty (1 - \rho u/\rho_e u_e) dy \quad . \tag{4.4.4}$$

The value of δ^{**} is obtained by equating the momentum of the layer of thickness δ^{**} to the momentum lost in the boundary layer due to friction for the actual flow rate,

$$\rho_e u_e^2 \delta^{**} = \int_0^\infty \rho u (u_e - u) dy \quad ,$$

or

$$\delta^{**} = \int_0^\infty (\rho u/\rho_e u_e)(1 - u/u_e) dy \quad . \tag{4.4.5}$$

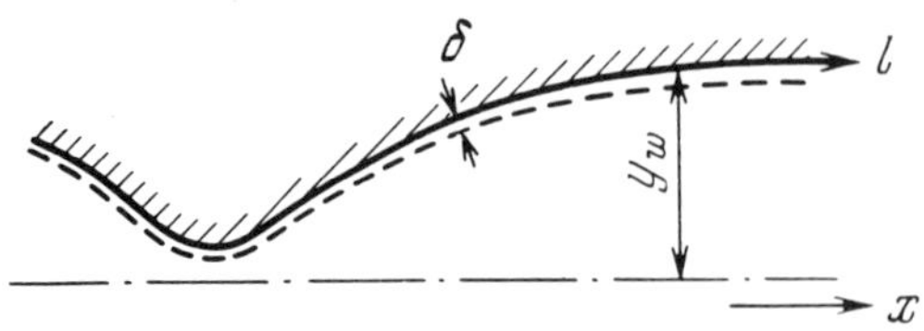

Fig.4.6.
Boundary layer at nozzle walls

Let us introduce the quantities $H^* = \delta^*/\delta^{**}$, the form factor of the boundary layer, and $\zeta = \tau_w/\rho_e u_e^2$, the local friction coefficient at the wall [τ_w is the shear stress according to (4.4.1); the subscript w refers to the conditions at the wall]. Then, assuming the gas to be perfect and integrating the momentum equation for a viscous gas with respect to y, we arrive at the equation which describes the variation of δ^{**} in a two-dimensional nozzle,

$$\frac{d\delta^{**}}{d\ell} + \frac{2 + H^* - \underline{M}^2}{1 + (\gamma - 1)\underline{M}^2/2} \cdot \frac{\delta^{**}}{\underline{M}} \cdot \frac{d\underline{M}}{d\ell} + \frac{\delta^{**}}{y_w} \cdot \frac{dy_w}{d\ell} = \zeta \quad , \tag{4.4.6}$$

where ℓ is the distance along the wall, y_w is the coordinate of the nozzle wall (Fig.4.6), and $\underline{M}$ is the Mach number for the flow core, whose distribution over the nozzle is assumed to be known. Thus, to solve the problem we must know the values of H^* and ζ. These functions can be approximately represented as functions of $\underline{M}$, the wall temperature T_w, and the Reynolds number $\underline{\underline{Re}}$, obtained for a laminar or a turbulent flow near the plane plate (see [4.45]). After integrating (4.4.6), we obtain the distribution of δ^{**} and $\delta^* = H^*\delta^{**}$ in the entire nozzle. Moreover, it is found that the layer thickness is minimal in the transonic region and that is grows considerably with the distance from the throat. For this reason in the analysis of the boundary layer in the supersonic part of the nozzle it can be assumed to a good approximation that at the throat $\delta^* = \delta^{**} = 0$.

The results of computations of the laminar layer parameters according to (4.4.6) for a two-dimensional wedge or axiosymmetrical nozzle flow with the temperature at the entrance $T_0 = 1800$ K and the wall temperature $T_w = 300$ K are given in Fig.4.7 depending on the nozzle area ratio S/S_* [4.36]. Here the upper curve shows the product of the current values of δ^{**}/H and $\sqrt{p_0 h_* \sin\theta}$, where H is the nozzle height, h_* is the throat height, and θ is the angle between the contour and the nozzle axis (for axiosymmetrical nozzles H, h_* should be replaced by the diameter). This product is almost independent of the flow symmetry and γ, and for $S/S_* > 10$, also of the area ratio. The second curve makes it possible to determine δ^*, and, finally, on the third curve δ^{***} denotes the integral thickness of the energy loss,

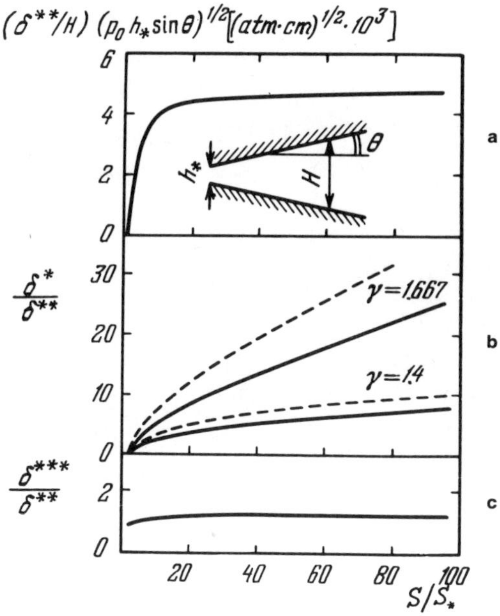

Fig.4.7a-c. Laminar boundary layer for two-dimensional wedge nozzle walls (solid lines) and an axiosymmetrical cone nozzle (dashed lines). In (a) and (c) these curves practically coincide

$$\delta^{***} = \int_0^\infty (\rho u/\rho_e u_e)(1 - h/h_e)dy \quad , \tag{4.4.7}$$

where h is the enthalpy of the gas along the stream line according to (4.1.5). This quantity can be computed using the equation similar to (4.4.6) [4.41]. The computations were carried out for nitrogen (γ = 1.4) and helium (γ = 1.667) on the assumption that $\underline{Pr}$ = 1 and that there is a linear dependence of viscosity on temperature. For the recalculation of these results for other gases, we must multiply the ordinates of the first curve by $\sqrt{\nu_0/a_0}$ (ν_0 and a_0 are kinematic viscosity and the sound velocity at the nozzle entrance) and divide by the same value for N_2 (or He). The Prandtl number and the temperature dependence of viscosity weakly affect these results, and the variation of T_0 results in the almost proportional variation of δ^{**}.

The thickness of displacement δ^* for a laminar boundary layer in a two-dimensional nozzle with the inflection point (see Fig.4.4) can be assessed using the relation $\delta^* \approx 5x\ \underline{Re}_0$, where $\underline{Re}_0 = h_* u_m/2\nu_0$, ν_0 is the kinematic viscosity for the conditions at the nozzle entrance, u_m is the maximum flow velocity $u_m = \sqrt{2h_0}$, and h_0 is the stagnation enthalpy [4.32]. Here x is the distance from the nozzle throat. This relation was obtained for $\underline{Pr}$ = 0.72 and for a heat-insulated wall. When nozzle walls are cooled, the thickness of the boundary layer noticeably decreases. In experiments with conical nozzles it was found that $\delta^* = 0.49x\,\underline{Re}_r^{-0.3}$ [4.46], where $\underline{Re}_r = ux/\nu_r$, and the kinematic viscosity ν_r is taken for the temperature $T_r = T + 0.5(T_w - T) + 0.22(T_a - T)$; here T_a is the so-called equilibrium wall temperature

$$T_a = T_0 \frac{1 + (\gamma - 1)r\underline{M}^2/2}{1 + (\gamma - 1)\underline{M}^2/2} \quad ; \qquad (4.4.8)$$

r is the recovery coefficient close to 0.89 [4.41].

For CO_2 gasdynamic lasers (Chap.5) the relative area of the nozzle occupied by the boundary layer for $\underline{M}_{ex} = 4$ is about 15% for $T_w/T = 1$ and 5% for $T_w/T = 0.2$. For the parameters averaged over the flow cross section for $T_w/T_0 = 1$ this results in an increase of gas temperature approximately by 7% and of pressure, approximately by 25%, while the flow velocity decreases by 5%. To compensate the effect of the boundary layer, the nozzle walls at the exit sometimes are made diverging [4.37].

The boundary layer is formed not only on the nozzle walls, but also on its plane side surface and on the walls of a GDL cavity. The calculations for the layer thickness at these planes both for an accelerating flow in the nozzle region and for a flow with constant parameters in the cavity behind the nozzle can be carried out using (4.4.6). The most simple is the dependence of δ^* on the distance x for the cavity flow with a laminar layer ($\delta^* \propto \sqrt{x}$). Thus, for a CO_2 gasdynamic laser the displacement thickness δ^* at the side wall of the cavity varies, according to [4.47], is

$$\begin{aligned} \delta^*/h_* &= 1.13\sqrt{5.25 \times 10^{-3}(x/h_* - 13) + 0.024} \\ &\quad \text{for} \quad x \geqq 13h_* \quad \text{at} \quad \underline{M}_{ex} = 4.7 \; , \\ \delta^*/h_* &= 3.5\sqrt{1.6 \times 10^{-3}(x/h_* - 30) + 0.0154} \\ &\quad \text{for} \quad x \geqq 30h_* \quad \text{at} \quad \underline{M}_{ex} = 5.8 \; , \end{aligned} \qquad (4.4.9)$$

where h_* is the nozzle throat height and x is counted from the throat ($x = 13h_*$ and $x = 30h_*$ correspond to the nozzle exit for the indicated values of $\underline{M}_{ex}$). The increase of the boundary layer thickness at the plane wall surface of a laser results in the reduction of the flow cross section and creates compression and shock waves, which disturbs the uniformity of the cavity flow. In the linear approximation the relative distortion of the gas density is related to the slope θ_w of the outer boundary of the layer in the following way [4.47]

$$\Delta\rho/\rho \approx M_{ex}\theta_w \quad . \tag{4.4.10}$$

The value of θ_w at the side wall increases, reaching 20 mrad (1.15°) at the nozzle exit, and then decreases. Hence, according to (4.4.10), we can expect that in the cavity density perturbations with $\Delta\rho/\rho \approx 0.05$-$0.10$ will appear. These perturbations can be decreased by making the side walls concave [4.47], or simply arranging them at an angle θ to the flow axis, beginning from the nozzle throat. Experiments [4.34] showed that for $\theta = 0.75°$ the wedge divergence of the side walls makes it possible to reduce density perturbations at the nozzle exit on the average from 5% to 1.8%.

To achieve high negative temperature gradients for fast freezing of physico-chemical processes in GDLs, sometimes nozzles with a very small throat height h_* are used. If in this case h_* becomes comparable with δ^*, the separation of the boundary layer loses its sense, and the problem of the viscous gas flow should be solved for the flow as a whole. For the initial pressure p_0 of about several atmospheres and a temperature T_0 of $\sim$ 1000 K the flow conditions set in for h_* of about 0.1 mm and below. Gas-dynamic equations for the flow in such slender channels for relatively small $\underline{Re}$ numbers can be written, assuming that the order of magnitude of the transverse and longitudinal nozzle dimension ratio and that of the ratio of the transverse velocity components and gradients of variables to their longitudinal components are the same [4.48]. Then these slender channel equations are solved together with physico-chemical kinetics equations, in particular, for the vibrationally nonequilibrized nitrogen in [4.49]. The contouring of the supersonic nozzle, in which the boundary layer thickness is comparable with the transverse flow dimensions, is carried out, for example, in [4.50].

For a high nozzle area ratio the gas pressure drops so that the flow may become overexpanded; in this case the pressure p_{ex} at the nozzle exit becomes less than the pressure in the cavity. This results in the appearance in this region of a system of oblique and straight shock waves, which distort the flow field. The perturbations of pressure appearing in the boundary layer propagate upstream, which may result in the separation of the flow from the nozzle walls far from the exit. At the separation point the shear stress τ_w is vanishing, and behind this point a return flow appears. During flow separation the pressure increases jumpwise—$\Delta p_s = p_s - p_e$ (p_e is the pressure in the flow before the separation point)—and then the pressure changes little up to the nozzle exit ($p_s \approx p_{ex}$). The ratio $\Delta p_s/p_e$ depending on $\underline{M}$ is determined experimentally. It was found that in rocket engine nozzles $p_s/p_e \approx 2$ to 5. In experiments studying the flow separation in small two-dimensional nozzles with a laminar layer and not very high Reynold numbers (which is typical of gasdynamic lasers) it was shown [4.51] that flow separation occurs at a considerably lower ratio p_s/p_e equal to 1.2 to 1.7. The flow separation from the nozzle walls results not only in considerable inhomogeneities appearing in the cavity, but it disturbs the design regime of the flow and makes it impossible to create an active medium in gasdynamic lasers.

Let us go over to the analysis of the effect of viscosity on the active medium state in GDLs. For this we must supplement the Prandtl equations (4.4.3) with the continuity equations for individual components of a non-equilibrium gas mixture in a flow. These equations are the generalization of (4.1.9) for the case of a two-dimensional flow in a boundary layer and include the diffusion transport of the jth component across the layer [3.141],

$$\rho\left(u\,\frac{\partial\alpha_j}{\partial x} + v\,\frac{\partial\alpha_j}{\partial y}\right) = -\frac{\partial}{\partial y}\,(\rho\alpha_j V_j) + m_j W_j \quad , \tag{4.4.11}$$

where α_j is the mass fraction, m_j is the mass of a molecule, W_j is the rate of formation (I.7), and V_j is the diffusion rate of the jth component. Here the diffusion flow is mainly determined by the concentration gradients, and for this reason barodiffusion and thermal diffusion can be ignored. In this case the diffusion rate V_j is written in the most simple form for a binary mixture [3.1,141] (Fick's law),

$$\alpha_j V_j = -D_j(\partial\alpha_j/\partial y) \quad , \tag{4.4.12}$$

where D_j is the diffusion coefficient, which is close to μ/ρ and μ is the viscosity coefficient (in calculations the experimentally determined values of D_j are often used). If a mixture contains many components ($j > 2$), then for the sake of simplicity we can combine them into two groups with respect to their diffusion properties and then use (4.4.12) or more complex expressions for V_j [3.1,3]. Vibrational relaxation processes in a boundary layer with a constant chemical composition are described by (4.4.11) and (4.4.12), where instead of concentration the average number of vibrational quanta e_i of the ith type appear (see [4.52]),

$$\rho\left(u\,\frac{\partial e_i}{\partial x} + v\,\frac{\partial e_i}{\partial y}\right) = \frac{\partial}{\partial y}\left(\rho D_i\,\frac{\partial e_i}{\partial y}\right) + \rho\left(\frac{de_i}{dt}\right)_r \quad , \tag{4.4.13}$$

where D_i is the diffusion coefficient and $(de_i/dt)_r$ is the relaxation rate. For vibrationally excited molecules D_i is somewhat lower than for unexcited ones due to the V-V exchange. For example, this difference for CO_2 molecules does not exceed 10% [3.77].

The boundary conditions for these equations are determined by the state of an inviscid flow at the flow core and by the values of parameters at the nozzle walls. The conditions at the wall in the presence of heterogeneous processes were considered in Sect.3.7. Sometimes it is assumed for estimates that near the wall there is an equilibrium state at the wall temperature. For the solution of the boundary layer equations in partial derivatives it is expedient to carry out a transformation to new coordinates, which results in ordinary differential equations. Often the Lees-Dorodnitsin transform is used, where for a two-dimensional flow the following coordinates are introduced:

$$\xi(x) = \int_0^x \rho_e u_e \mu_e dx \quad , \qquad \eta(x,y) = Y(y)/s(x) \quad , \tag{4.4.14}$$

where

$$Y(y) = \int_0^y (\rho/\rho_e)dy \quad , \qquad s(x) = \sqrt{2\xi(x)}/\rho_e u_e \quad .$$

These transformations are given in most monographs and textbooks on viscous gas dynamics (for example, [3.141, 4.3]). The system of equations thus obtained for a boundary layer in the general case is solved numerically.

For simple estimates of the gas state in a boundary layer we can introduce a parameter—the Damkeler number $\underline{\underline{Da}}$—which is equal to the ratio of two characteristic times: the lifetime τ_d for a particle in a boundary layer, determined by diffusion, and the reaction time τ_r (the relaxation time). Approximately $\tau_d \approx \delta/V_j$, and the diffusion rate, according to (4.4.12), can be estimated by the value of $V_j \approx D/\delta \approx \mu/\rho\delta$. Thus, $\tau_d \approx \rho\delta^2/\mu$. Using (4.4.12), for constant values of the parameters along the flow we obtain

$$\underline{\underline{Da}} = \ell/u\tau_r \quad . \tag{4.4.15}$$

When $\underline{\underline{Da}} >> 1$, the flow in the boundary layer is close to equilibrium, and when $\underline{\underline{Da}} >> 1$, to the frozen flow. For $\underline{\underline{Da}} \approx 1$ we must solve the problem for a nonequilibrium flow.

GDL nozzles have small dimensions: the throat height usually amounts to 0.2-2 mm, and at the exit, to several centimeters. For this reason to obtain high output power it is expedient to use in lasers several nozzles, which form a multiple nozzle array (Fig.4.8). This increases the gas flow rate

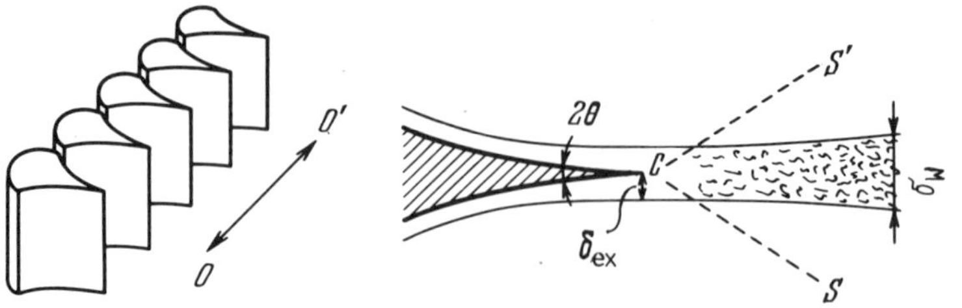

Fig.4.8. Nozzle array and flow behind the nozzle edge. OO'-optical cavity axis; CS, CS'-position of a shock wave

through the cavity in comparison with a one-nozzle design. The difference in the flows with a single nozzle and with a nozzle array is observed only at the array exit, where behind each nozzle edge shock waves appear and a wake is formed as a result of the leakage of boundary layers from the nozzle walls. In this connection two problems arise when the nozzle arrays are used. First, the shock waves and wakes distort the optical homogeneity of the medium in the cavity as a result of the gas density variation (Sect. 2.5). Second, in the shock waves and wakes the change in the laser level population is observed, and thus the state of the active medium changes.

Shock waves behind the array edge appear because the angle between two supersonic flows from two neighboring nozzles differs from zero, since nozzle edges have a finite angle $2\theta_{ex}$. Usually the nozzle edge angle is small, so that the shock waves are rather weak and form with the flow an angle close to $\alpha = \arcsin(1/\underline{M})$. Further they intersect, are reflected from the cavity walls, and form a rhomboid system of flow perturbations in the cavity. Away from the nozzle array the density perturbation decreases, so that [4.53]

$$\frac{\Delta\rho}{\rho} = \frac{\theta_{ex}\underline{M}^2}{(\underline{M}^2-1)^{1/2} + \theta_{ex}x\underline{M}^4/2H(\underline{M}^2-1)} \quad , \tag{4.4.16}$$

where x is the distance from the nozzle array edges and H is the exit height of a nozzle. This density perturbation, as was pointed out in Sect. 2.5, results in a change of the refractive index. Besides, the shock waves affect the state of the active medium because of increase of the gas temperature and pressure.

In a viscous wake behind the nozzle array edges the flow soon becomes turbulent. In the transient region at first the wake as a whole begins to oscillate at a frequency close to $f = 0.05\,u_e/\delta^{**}$, where u is the flow velocity and δ^{**} is the momentum loss thickness in a laminar boundary layer at the nozzle edges [4.54]; then other pulsation frequencies appear, and the flow in the wake loses its stability. This transition to a turbulent regime occurs when $\underline{Re}_*$ (δ^{**}) is about 100 (the characteristic dimension here is equal to δ^{**}). It should be noted that general regularities for this wake are almost the same as for an incompressible gas flow at subsonic velocities [4.55]. Further we shall consider the flow parameters averaged with respect to time for all micropulsations.

The basic laws of variation of the wake parameters are determined by the conservation of the momentum loss thickness δ^{**}(4.4.5) and the energy loss thickness δ^{***} (4.4.7) in the wake; in this case the lower limit of the respective integrals (4.4.5) and (4.4.7) must be taken at $-\infty$, or these values must be doubled. The values of $2\delta^{**}$ and $2\delta^{***}$ will be exactly constant if the flow is not bounded, the upsteam transfer of perturbations is absent, and there is no loss in the shock waves appearing at the nozzle edge. These effects somewhat change the values of $2\delta^{**}$ and $2\delta^{***}$ near the edge; in the region of the turbulent wake these values are practically constant [4.55].

Let us determine the wake thickness δ_w as the distance between the points in the flow for which the difference $u_e - u^0$, where u^0 is the flow velocity at the wake center line, is equal to 1% of $u_e - u_0$. Let $(\delta^{**})_{ex}$ be the momentum loss thickness at the nozzle exit. Then just behind the edge the wake thickness δ_w is close to $2(\delta^{**})_{ex}$, and in the initial region, whose length equals several $(\delta^{**})_{ex}$, the value of δ_w slightly decreases, forming a wake throat, after which the wake expands. After the transition to the turbulent regime in the far wake [at a distance larger than $50\text{-}100\delta^{**}{}_{ex}$] $\delta_w = \sqrt{\kappa\delta^{**}{}_{ex}x}$, where κ is the empirical factor close to $\sim$2.5-3 [4.55]. The reading of the x coordinate is taken from the point of transition to the turbulent flow regime; in estimates this point can be approximately identified with the coordinate of the nozzle exit. The characteristic scale of medium inhomogeneity in the cavity flow is obviously commensurable with the wake thickness. In a turbulent wake the transverse integral scale of turbulent pulsations Λ amounts to $0.1\text{-}0.2\delta^{*}{}_{ex}$ [4.36].

The active medium state in a wake can be determined if we know the gas temperature and flow velocity distribution across and along the wake. Often it is assumed that the gas pressure in a cavity flow in the absence of strong shock waves is practically constant everywhere. The degree of deviation from statistical equilibrium in a gas does not strongly affect the wake structure and the distribution of temperature and pressure in the wake. To a good approximation we can assume that in a far wake across the flow these quantities obey the Gauss distribution [4.41]

$$\Delta u' = a_u \exp[-(2Y/b_u)^2] \quad , \quad \Delta T' = a_T \exp[-(2Y/b_T)^2] \quad , \tag{4.4.17}$$

where $\Delta u' = (u_e - u)/u_e$, $\Delta T' = (T - T_e)/T_e$, a_u, a_T are the values of $\Delta u'$ and $\Delta T'$ at the wake center line, and b_u, b_T are the quantities related to the wake thickness. The Y coordinate is related to y through the Lees-Dorodnitsin transform (4.4.14), which in a particular case of a two-dimensional turbulent wake retains its meaning, if instead of molecular viscosity of a gas, μ, we shall take the turbulent viscosity μ_t. In the near wake at the nozzle array edge the distribution (4.4.17) is less precise, thought it can also be used in the analysis, if we introduce the respective corrections to a_u, b_u, a_T, and b_T. The dependence of these quantities on the x coordinate is determined from the integral relations resulting from the conservation of δ^{**} and δ^{***} in the wake (see [4.52]). The value of b_u is practically equal to a half of the total wake thickness δ_w, and $b_T = b_u(\underline{Pr}_t)^{-0.5}$, where $\underline{Pr}_t$ is the turbulent Prandtl number usually equal to 0.7. Thus, the thermal wake is only slightly thicker than the dynamic one. According to [4.52],

$$a_u = \frac{s_{ex}}{s_{ex} + (4\delta^{**}x/\underline{Re}_t)^{1/2}} \quad , \tag{4.4.18}$$

where s_{ex} is the value of s(x) according to (4.4.14) at the nozzle exit and $\underline{Re}_t$ is the turbulent Reynold number equal to 13 according to experiments [4.55]. In the far wake

$$a_T = \frac{\sqrt{\underline{Pr}_t}}{1 - \sigma_e}\left(2\sigma_e - \frac{\delta^{**}}{\delta^{***}}\right)a_u \quad , \tag{4.4.19}$$

where σ_e for a perfect gas is equal to $1 - T_e/T_0$. Using these relations for determining the gas temperature and flow velocity in the physico-chemical kinetics equations and proceeding from the known solutions for an inviscid flow, we can determine the active medium state in a turbulent wake behind the nozzle array. An example of this solution is given in Sect.5.6. The variation of gas density in a wake, which affects the optical homogeneity of the active medium, is obtained from the temperature distribution using the state equation for averaged quantities in the form (3.1.6), if the gas pressure in the wake is assumed to be approximately constant.

In the turbulent wake theory many hypotheses and empirical parameters are inevitably used to determine μ_t, $\underline{Re}_t$, $\underline{Pr}_t$, etc. Therefore of great importance is the experimental study of the flow field behind the nozzle array. These investigations were carried out, for example, in [4.34] on an array

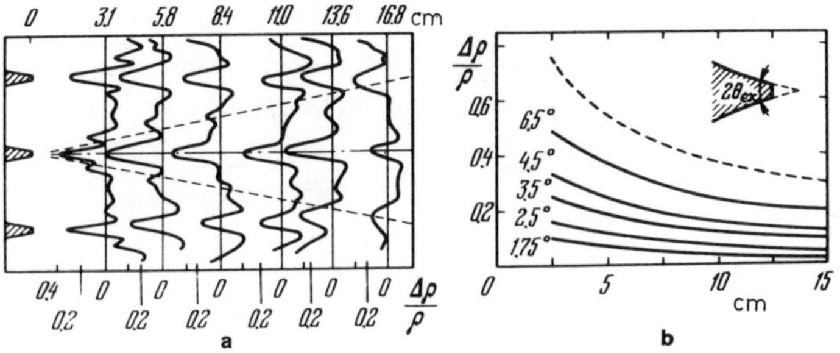

Fig.4.9a,b. Gas density distribution in a flow behind the nozzle array. (a) Typical density distribution; (b) density perturbations $\Delta\rho/\rho$ as a function of distance x along the flow for different angles $2\theta_{ex}$ formed by the nozzle edges (solid lines correspond to the effect of shock waves, dashed lines, to a turbulent wake)

of contoured nozzles, each of which had the throat height $h_* = 1.35$ mm and the area ratio $S/S_* = 24.5$. The variation of the gas density in a wake is maximal at the nozzle exit (here it can reach 60 to 80%) and then it decreases as the gas moves away from the nozzle edges and the wake expands, according to the theory, in inverse proportion to $\sqrt{x}$ (Fig.4.9). This perturbation exceeds the change in density due to shock waves when the edge angle is below 6.5°.

The detailed study of the flow-field properties in gasdynamic laser nozzle wakes is carried out in [4.56], where a possibility of the increase of pressure at the wake center line (up to 40% in comparison with the pressure at the nozzle array edge) was found. For distances $x/H > 1.2$, where the wake can be considered turbulent, there is good agreement between experimental data and the wake parameters calculated using the linearized turbulent wake theory according to [4.57]. This agreement, however, is not observed for the transition region of the wake. These facts point to a complex nature of the processes in a GDL cavity flow. As a matter of fact, there is no completely reliable theoretical description of this flow, and for this reason every newly created GDL of this type must be experimentally investigated to obtain a sufficiently homogeneous flow field.

To decrease the effect of inhomogeneities introduced into the flow by two-dimensional nozzle arrays, in [4.58] a screen nozzle, which consists of a large number of very short axiosymmetrical nozzles with small (fractions of a millimeter) throat diameters d_* and divergence angles of 20 to 40°, was suggested. The analysis (with the account of boundary layers and mixing of flows from each nozzle) and experiments showed that at a distance of about 100 diameters of the nozzle exit the flow field is practically homogeneous (the density pulsations are below 1%). This nozzle screen is made in the form of a plate of a thickness about 0.9 cm with closely arranged conical orifices.

4.5 Shock Tube with a Nozzle

Laboratory studies of the processes in GDLs are conveniently carried out using shock tubes, in which large enough volumes of a heated and compressed gas can be obtained when a shock wave propagates through a given mixture, which is then cooled in the nozzle. The operation principle of a shock tube is very simple: the shock wave is formed as a result of the fast action on the gas under investigation of another (driving) gas after the rupture of a diaphragm, which separates the gas under study in the low-pressure chamber from the driving gas in the high-pressure chamber. Expansion waves propagate simultaneously in the opposite direction. A detailed description of shock tube performance under different operating conditions can be found in [1.8, 4.59,60]. Wave processes in a shock tube are described by the x-t diagram shown in Fig.4.10, where OA is the shock wave path and OB', ..., OB" are expansion waves. The shock wave is followed by the contact surface OC, which separates the driving gas from the gas under investigation. When the shock wave reflects from the end plate of the low-pressure chamber, the reflected shock wave AC is formed; thus the gas under investigation is compressed and shock-heated twice: first by the incident shock wave (the transition from region 1 to region 2 in Fig.4.10) and then by the reflected wave (transition $2 \rightarrow 5$). The shock-heated gas flows through the nozzle mounted at the end of the tube.

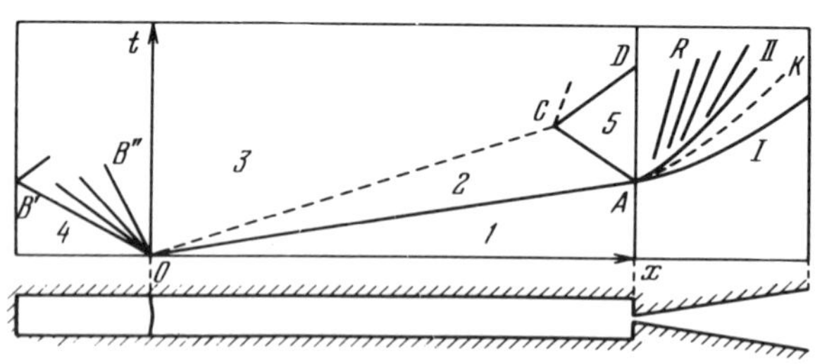

Fig.4.10. Development of processes in a shock tube with a nozzle

The temperature and pressure (and, hence, the composition) in region 5 at the nozzle entrance are usually calculated from initial pressure p_1, temperature T_1 of the gas with a given composition in the low-pressure chamber, and the velocity V of the incident shock wave. In this analysis use is made of the conservation laws for the mass, momentum, and energy fluxes during the transition through the front, first, of the incident, then of the reflected shock waves, and also the equations of state and the law of mass action at equilibrium. The procedure of this analysis is given, for example, in [4.59-61]. The high temperatures and pressures facilitate the rapid attainment of statistical equilibrium in the gas behind the front of the reflected shock wave. Behind the front of the incident wave, however, equilibrium in some cases has no time to set in because of the low rate of relaxation processes.

To assess the possibility of utilizing a shock tube for modelling steady-state processes in gasdynamic lasers, we must know the time during which the gas in a heated state is in front of the nozzle entrance (the working time τ' of a shock tube). In the absence of viscosity this time is determined by the interval between the moment of the wave reflection (t_A in Fig.4.10) and the arrival to the end plate of perturbation CD—a shock wave or an expansion wave after the interaction of the reflected wave with a contact surface. The nature of perturbation CD is determined by the relation [4.62]

$$\frac{(\gamma_3+1)/(\gamma_3-1)+p_2/p_3}{(\gamma_2+1)/(\gamma_2-1)+p_2/p_3} \lessgtr \frac{\mu_2(\gamma_2-1)T_3}{\mu_3(\gamma_3-1)T_2} , \qquad (4.5.1)$$

where γ_i, μ_i are the ratios of specific heats (the gas is assumed to be perfect) and the molecular mass of the gas in the ith region. If the left-hand side of (4.5.1) is greater than the right-hand side, then an expansion wave propagates to the direction of the end plate after interaction with the contact surface; in the opposite case—a shock wave. If there is equality in (4.5.1), we may speak about "joining" the conditions at the contact surface: in this case CD is just a weak sonic perturbation, which does not change the gas state in region 5 (for $\gamma_2 = \gamma_3$ and $p_2 = p_3$ this corresponds to the coincidence of the values of sound velocities $a_2 = a_3$). For $T_1 = T_4$ and driving the mixture of 10% CO_2-40% N_2-50% He by helium the joining conditions are fulfilled for the Mach number of the incident wave ($\underline{M} = V/a_1$) $\underline{M}_a = 2.2$; when nitrogen is driven by helium, $\underline{M}_a = 3.4$, and by hydrogen, $\underline{M}_a = 6$. In these conditions the working time τ' is sufficiently large—it is limited only by the arrival of the reflected expansion wave OB' to the end of the shock tube. In practice, however, the joining conditions cannot be realized often (one of the reasons is that Mach numbers different from $\underline{M}_a$ are used), and the working time τ' is limited by the time t_D ($\tau' = t_D - t_A$).

The account of viscosity considerably reduces τ' for two reasons. First, the real motion of the contact surface, taking account of the boundary layer at the shock tube walls, noticeably differs from the path OC for an ideal shock tube. Because of the drift of a part of the gas into the boundary layer the contact surface is accelerated, and the time required for it to reach the point x_C decreases almost by half [4.59]. Then, assuming that $t_C - t_A \approx t_D - t_C$ and $\gamma_5 = \gamma_2 = \gamma$, for the working time of a shock wave we obtain

$$\tau' = (L/V)(\gamma - 1)/2\gamma \quad , \tag{4.5.2}$$

where L is the length of the low-pressure chamber (more precise estimates of τ' are given in [4.63]). The other cause of the reduction of τ' is the interaction of the reflected wave with the boundary layer at the tube walls, which appears during the propagation of the incident wave. This interaction results in the phenomenon similar to the flow separation from the wall. As a result the gas accumulates in the near-wall region, and the reflected shock front splits into two parts (*bifurcation* takes place—see [4.60,61]). This effect, as the front moves away from the end plate, is first manifested in stagnation, and then, after a noticeable contraction of the unperturbed flow core, in the acceleration of the flat part of this wave. For this reason temperature T_5 and pressure p_5 coincide with the rated values obtained in the assumptions of a one-dimensional inviscid flow only immediately after reflection (usually during about 200 μs [4.64,65]). Then, due to the reflected wave, the stagnation pressure p_5 drops (sometimes to 30% and below) [4.66]. Further, as the wave is accelerated, these values may again approximate the rated values (sometimes the pressure may become higher than the rated one [4.60,61]). If the equality in (4.5.1) is not fulfilled and perturbation CD is a shock wave (for $\underline{M} > \underline{M}_a$), then as a result of multiple reflection of the shock waves in this region the pressure and temperature of the gas at the end plate increase somewhat [4.67]. Thus, because of the complicated nature of these described phenomena, it is expedient to compute the pressure and temperature from the velocity of the reflected wave front, or to directly measure temperature and pressure of the gas in front of the nozzle entrance.

The bifurcation of the reflected wave front results in a burst of the cold driving gas to the end plate, which limits the working time τ' of a shock tube. This time is large (more than 2 to 3 ms) only for low values of the incident shock wave velocity V ($\underline{M}$ < 2 to 2.3), and it rapidly decreases with an increase of $\underline{M}$, amounting to about 1 ms for $\underline{M}$ = 2.8, 400 μs for $\underline{M}$ = 4, and about 200 μs for $\underline{M}$ = 6 [4.67,68]. As a result of mixing the cold driving gas with the gas under investigation the temperature drops and the composition of the gas, which then flows into the nozzle, changes.[15] Disregard of this phenomenon may sometimes lead to a wrong interpretation of experimental results for gasdynamic lasers. The revision [4.70] of the results of studies [4.71,72] of a CO_2 gasdynamic laser performance with a shock tube may serve as an example. It is essential that the gas pressure in this case usually changes little, so that pressure control of the shock tube working time can be ineffective.

During operation with a nozzle part of the working time is spent to start the nozzle, i.e., for its transition to a steady-state regime, which corresponds to the temperature and pressure of the gas in front of the nozzle entrance. Without a diaphragm in front of the nozzle, during the start-up, first the primary shock wave I propagates through the gas in the nozzle (Fig.4.11), then the contact surface K, behind which the gas entering the nozzle moves, and further, secondary shock wave II propagates upstream (and, as a rule, is removed by the flow). This wave is formed as a result of a pressure drop in an expanding flow in comparison with the pressure of the gas behind the primary shock wave. Finally, the start-up is completed by a nonstationary expansion wave R, in which the steady-state distribution of pressure and other parameters at the nozzle flow set in. If the nozzle operates in underexpansion conditions, when pressure at the nozzle exit is low (this can be provided by a diaphragm in front of the nozzle and by evacuating this volume separately), both shock waves are removed by the flow from the nozzle considerably earlier than the nonstationary expansion wave. In this case the start-up of the nozzle proceeds most quickly (an "ideal start-up", to which a dot-and-dash line in Fig.4.11 corresponds). With an increase of the counterpressure at the nozzle exit (overexpansion regime) the intensity of a secondary shock wave (line II in Fig.4.11) considerably increases, which leads to its slowing down in

15 Often this burst and mixing is of a pulsation nature, which leads to the respective pulsations in amplification and GDL oscillations at the shock tube nozzle exit [4.69].

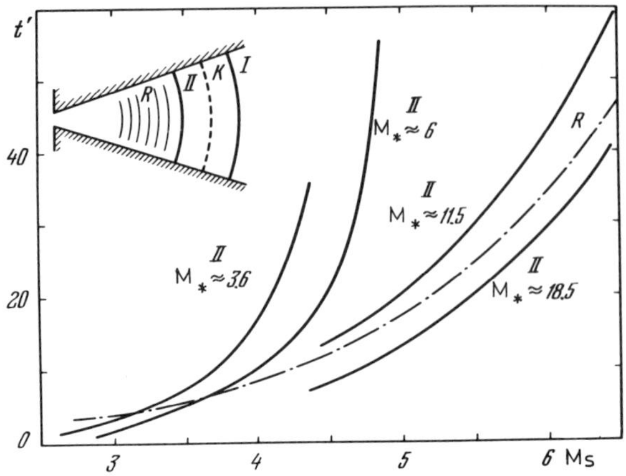

Fig.4.11. Reduced time $t' = ta_5/\ell$ of perturbations propagating during starting a nozzle at a shock tube end face [ℓ-characteristic nozzle dimension for a wedge nozzle equal to $h/(2\tan\theta)$, a_5-sound velocity of a gas at the nozzle entrance]

the nozzle and increases the start-up time, or even makes it impossible to obtain the rated steady-state conditions of the flow. Figure 4.11 provides some results for nozzle start-up time. Values of the reduced time t' corresponding to the propagation of shock waves I, II and expansion wave R in the nozzle for a perfect gas ($\gamma = 1.4$) are given as a function of $\underline{M}_S$ for a steady-state flow in different nozzle cross sections for different intensities of the primary shock wave (the Mach number $\underline{M}_*$) at the nozzle throat [4.73,74]. For problems considered in this book in the majority of cases $\underline{M}_*$ is close to 4 to 6. For more precise calculations of $\underline{M}_*$ we can use the relations resulting from the theory of shock wave propagation in channels with varying cross section (see, for example, [4.60,75]). The gas flow rate through a nozzle during its performance leads to a decrease of pressure p_5 and temperature T_5 in front of the nozzle entrance, this decrease, however, is very small.

In spite of the difficulties mentioned here, shock tubes are widely and successfully used in the development of gasdynamic lasers. Respective reviews can be found in [4.76-78].

5. Infrared CO_2 Gasdynamic Laser

This chapter considers the processes in the CO_2 gasdynamic lasers and gives the relaxation equations for the active medium of this type of laser in various approximations. The mathematical expression for the gain coefficient is compared to experimental results. Also presented are the results of experimental studies both of the output power measurements and the optimization problem for small-signal gain and output power taking into account the previous results for a two-dimensional active medium flow in a nozzle and for a flow in an unstable resonator. Next, various kinds of fuel are analyzed for their effectiveness on CO_2 gasdynamic laser performance. The chapter then estimates the effect of impurities on laser parameters and considers supersonic mixing gasdynamic lasers. Estimates of the laser efficiency and a discussion of various ways of increasing gasdynamic laser efficiency are presented. The chapter concludes with the results of the development of a twin-mode laser and suggestions concerning the creation of an active medium in expansion waves and shock waves.

5.1 Kinetics of Processes in an Active Medium. Gain Coefficient

Great progress in the development of gasdynamic lasers (GDLs) has been achieved in devices producing infrared radiation at a wavelength of about 10 μm, using transitions between vibrational levels of CO_2 molecules in mixtures with nitrogen or other gases, e.g., helium, or water vapor. The output power for these lasers has reached tens and hundreds of kilowatts using different methods of mixture heating (by shock waves, combustion, etc.) in different laser constructions with supersonic nozzles and nozzle arrays.

A mathematical model of the processes in a CO_2 gasdynamic laser can be created on the basis of physico-chemical kinetics, relaxation gas dynamics, and quantum electronics, considered in the previous chapters. The most detailed account of transitions between all vibrational levels of molecules during vibrational energy exchange in CO_2 and its mixtures with other components was given in Sect.3.5. For mixtures with a CO_2 content higher than 3% to 5% (precisely these mixtures are utilized in gasdynamic lasers) this model gives results similar to those obtained using simpler kinetic schemes based on microkinetic relaxation equations (Sect.3.4). For this reason the required accuracy for describing the processes in CO_2 gasdynamic lasers can be obtained with the following assumptions [3.19,82, 4.69]. Let all the rotational degrees of freedom of a molecule be at equilibrium with the translational ones. All modes of molecular vibrations are approximated by a harmonic oscillator model; then, according to Sect.4.2, among the molecules belonging to the same i^{th} mode, regardless of the nature of the temperature variation, the Boltzmann distribution with a vibrational temperature T_i sets in. Let us ascribe, as before, the subscripts i = 1, 2, 3 to the three modes of the CO_2 vibrations (symmetrical valence, symmetrical deformation, and antisymmetrical valence vibrations) and i = 4 to vibrations of the N_2 molecule. The Fermi resonance in the CO_2 molecule results in the equality $T_1 = T_2$ (Sect.3.5). It is expedient to assume the transition (001) $\rightleftharpoons$ (030) to be the main exchange channels between the third and the second modes. Suppose that in the mixture, in addition to CO_2 and N_2, there can be He atoms (i = 5) or H_2O molecules (i = 6). The time of the V-T exchange for H_2O molecules is very small, so we can assume that vibrations of these molecules are at equilibrium with the translational motion. Thus, the kinetic scheme adopted here includes the following processes:

$$\text{V-T}: CO_2(\nu_2) + M \rightleftharpoons CO_2 + M \quad , \qquad N_2(v) + M \rightleftharpoons N_2 + M \quad ;$$

$$\text{V-V}: CO_2(\nu_3) + M \rightleftharpoons CO_2(3\nu_2) + M \quad ,$$

$$CO_2(\nu_3) + N_2 \rightleftharpoons CO_2 + N_2(v) \quad , \tag{5.1.1}$$

where M = N_2, CO_2, He, or H_2O; the mode of excited vibrations and the number of quanta ν_2 participating in the V-V exchange transition are indicated in the brackets. Proceeding from Sect.3.4, we can write for (5.1.1) the relaxation equations for the vibrational energy expressed in terms of the average number of quanta e_i in the given mode according to (3.1.9). Assuming $\theta = 2\theta_2$ (θ_i is the characteristic vibrational temperature), for $T_1 = T_2$ we obtain $e_1 = e_2^2/(2e_2 + 1)$. After summing up the equations for the vibrational energy of the first and the second modes taking account of the degeneracy of the latter ($\theta_1 e_1 + 2\theta_2 e_2$), we obtain

$$\frac{de_2}{dt} = \frac{(2e_2+1)^2}{6e_2^2+6e_2+1}\left[\frac{p}{\tau_2}(\bar{e}_2 - e_2) + \frac{3}{2}\varphi_{32}\right] ,$$

$$\frac{de_3}{dt} = -\varphi_{32} - \xi_{N_2}\varphi_{34} , \quad \frac{de_4}{dt} = \xi_{CO_2}\varphi_{34} + \frac{p}{\tau_4}(\bar{e}_4 - e_4) , \tag{5.1.2}$$

where

$$\varphi_{32} = pk_{32}\{e_3(e_2+1)^3 - e_2^3(e_3+1)\exp[-(\theta_3 - 3\theta_2)/T]\} ,$$

$$\varphi_{34} = pk_{34}\{e_3(e_4+1) - e_4(e_3+1)\exp[(\theta_3 - \theta_4)/T]\} ;$$

and τ_2 and τ_4 are the values of the time of the V-T exchange for CO_2 (ν_2) and N_2, respectively: $\tau_2 = [\sum \xi_j/\tau_2(j)]^{-1}$, k_{32} and k_{34} being the rate constants for the V-V and V-V' exchange, $k_{32} = \sum \xi_j k_{32}(j)$, and ξ_j is the molar fraction of the jth component.

The up-to-date values of τ and k for the processes involving CO_2 were given in Sect.3.5. The relaxation times τ_4 for collisions of the N_2 molecule with different partners are given in [3.25,26, 5.1-3]. It should be noted that the V-T deactivation of N_2 during collisions with N_2 and CO_2 for not very high temperatures has very low probability and in all the cases is masked by the deactivation during collisions $N_2 + H_2O$ or $N_2 + He$. The coefficients in the expression $\log\tau_i(j) = a_i(j)T^{-1/3} + b_i(j)$ and $\log k_{32}(j) = c(j)\,T^{-2/3} + d(j)T^{-1/3} + e(j)$ recommended for computations are given in Table 5.1. The values of τ and k obtained using these coefficients have dimensions s atm and $s^{-1}atm^{-1}$, respectively, i.e., are related to the atmospheric pressure of a gas partner in collision.

Table 5.1.

j	CO_2	N_2	He	H_2O
a_2	16.26	19.49	12.32	- 21.0
b_2	- 7.66	- 7.84	- 8.29	- 5.46
a_4	104	95.5	56.2	27.65
b_4	-11.2	-10.76	- 9.85	- 9.242
c	178.3	378.3	184.7	352.3
d	-58.98	-111.0	-59.47	-87.26
e	10.19	13.0	9.50	12.05

The recommended value of k_{34} is given by the formula $\log k_{34} = 0.732 \times 10^{-6}\, T^2 - 2.094 \times 10^{-3}\, T + 7.662$, where k_{34} is given in $s^{-1}atm^{-1}$. Let us recall that the quantity k_{32} appearing here is related to k_{32}^0 for a linear equation of the V-V exchange (3.5.8) by (3.5.9).

The simplest model of a gasdynamic laser flow is a one-dimensional steady-state flow of an inviscid gas in a single supersonic nozzle. A system of equations describing this flow of the mixture under consideration consists of continuity equation (4.1.1), momentum equation (4.1.3), and energy equation (4.1.5), and also thermal and caloric equations of the gas state (3.1.6, 10). The last of these equations includes the quantities e_i determined from relaxation equation (5.1.2). The solution of this system of equations was discussed in the previous chapter. As a result of this solution we obtain the distribution of all gasdynamic parameters and the values of the vibrational temperature T_i along the nozzle axis. The set of these quantities allows us to find the population of any vibrational level of molecules in any cross section of the flow.

Let us go over to the computation of a small-signal gain. The equation for the small-signal gain for an isolated line in terms of the population N_m of the upper and N_n of the lower level was obtained in Sect.2.2. In this case the upper laser levels are the rotational levels of the first vibrational state of the antisymmetrical valence mode of vibrations (00^01) of the CO_2 molecule. A CO_2 laser oscillation is observed for the transitions $00^01 \rightarrow 10^00$ and $00^01 \rightarrow 02^00$, so that the lower laser levels are the rotational levels of the corresponding vibrational states (see Fig.2.3). The population of the upper laser level N_m, corresponding to the rotational quantum number J_m, according to (D.4) and (D.9) (taking into account that the symmetry factor σ for the CO_2 molecules is 2), is given by

$$N_m = \frac{n_{CO_2}}{Q} \cdot \frac{2\theta_{001}}{T}\, g_m \exp\left[-\frac{\theta_3}{T_3} - \frac{\theta_{001}}{T} J_m(J_m + 1)\right] \quad , \qquad (5.1.3)$$

where n_{CO_2} is the number of CO_2 molecules per unit volume and Q is the partition function with respect to vibrations in the harmonic oscillator approximation (D.11),

$$Q = [1 - \exp(-\theta_1/T_1)]^{-1} \times [1 - \exp(-\theta_2/T_2)]^{-2} [1 - \exp(-\theta_3/T_3)]^{-1} \quad , \qquad (5.1.4)$$

θ_{001} is the characteristic rotational temperature for the vibrational state 00^01, $\theta_{001} = hcB_{001}/k$ (B_{001} is the rotational constant, h and k are the Planck and the Boltzmann constants, c is the light velocity), and g_m is the statistical weight of the rotational level ($g_m = 2J_m + 1$). A similar expression can be written for the population of the lower laser level N_n, say, in the state 10^00,

$$N_n = \frac{n_{CO_2}}{Q} \cdot \frac{2\theta_{100}}{T} g_n \exp\left[-\frac{\theta 1}{T_1} - \frac{\theta_{100}}{T} J_n(J_n + 1)\right] \quad , \tag{5.1.5}$$

where all the quantities have the same meaning as previously (in particular, $g_n = 2J_n + 1$). The constants appearing here are $\theta_1 = 1997$ K, $\theta_2 = 960$ K, $B_{001} = 0.3866\ \mathrm{cm}^{-1}$, and $B_{100} = 0.3897\ \mathrm{cm}^{-1}$. In the CO_2 spectrum the most intensive is the transition $J_{m_0} = 19 \to J_{n_0} = 20$ in the P branch, which will be further denoted as the P20 ($00^01 \to 10^00$) transition.

The optical small-signal gain k_0 can be obtained from (2.2.10) in the form

$$k_0 = \frac{\lambda^2 A_{mn}}{8\pi^{3/2}c}\left(N_m - \frac{g_m}{g_n} N_n\right) \frac{a}{\Delta_c} H(a,\omega) \quad , \tag{5.1.6}$$

where the same notations as in Sect.2.2 are adopted, except for Δ_c, for which the prime is omitted (from here on Δ_c is the collisional half-width of the line in cm^{-1}).

In addition to the populations of the levels N_m and N_n, the expression for k_0 includes the spontaneous emission probability A_{mn} and the collisional half-width Δ_c, which determines the ratio of the collisional half-width of the line to the Doppler half-width ($a = \Delta_c \sqrt{\ln 2}/\Delta_D$) and thus the value of the Voigt function $H(a,\omega)$ for a certain distance ω from the center of the line (Sect.2.2). The values of A_{mn} and Δ_c are measured in optical experiments involving the investigation of the light absorption for different gas pressures. For low pressures, where collisional broadening is insignificant, we can determine A_{mn_0} by measuring the absorption coefficient. For the transition P20 ($00^01 \to 10^00$) in early studies [5.4] the value of $A_{mn} = 0.213$ ($\pm$ 11%)s^{-1} was obtained. In subsequent studies this value was determined more precisely (see [5.5]); the mean value according to the most reliable results [5.6-8] is $A_{mn} = 0.187\ \mathrm{s}^{-1}$. With the change of the number J this value changes insignificantly [5.5,7]. Similar measurements for higher pressures, when collisional broadening becomes important, allowed the determination of the value of Δ_c. The analysis of these results for the transition P20 ($00^01 \to 10^00$) for T = 300 K (see [5.5]) leads to the following values of

$\Delta_c^{\,1}$:

Collision	$CO_2 + CO_2$	$CO_2 + N_2$	$CO_2 + He$	$CO_2 + H_2O$
Δ_c^0 [$cm^{-1}atm^{-1}$]	0.096	0.07	0.061	0.042 .

With the increase of the rotational quantum number Δ_c decreases (within several tens of percentage points).

The lines in the bands $00^01 \rightarrow 10^00$ and $00^01 \rightarrow 02^00$ for this temperature do not overlap even on the optical path of about 50 cm atm; the influence of other vibrational transitions in the CO_2 molecule on the contour of these bands was observed, however, which is especially noticeable when the gas temperature increases [5.10,11]. In Sect.2.1 it was shown that to a first approximation $\Delta_c \propto 1/\sqrt{T}$; in [5.10,12,13] it was established, however, that for the line P20 ($00^01 \rightarrow 10^00$) the experimental results in the temperature range up to 1600 K are better explained not only by the temperature variation of the optical collision diameter σ, but also by the increase of the interfering effects from other bands. Thus, for example, the line R23 ($01^11 \rightarrow 11^10$) is separated from the line P20 ($00^01 \rightarrow 10^00$) only by 0.01 cm^{-1}, and for this reason the increase in the population of the levels 01^11, 11^10 with the temperature results in the noticeable distortion of the line contour of P20 ($00^01 \rightarrow 10^00$) [5.14]. For other lines this dependence may be different [5.10]. Thus, in a gas mixture the collisional half-width Δ_c for two different assumptions on its temperature dependence is either

$$\Delta_c = p\sqrt{(300/T)}\sum(\Delta_c^0)_j\xi_j \quad , \qquad \text{or} \qquad \Delta_c = p(300/T)\sum(\Delta_c^0)_j\xi_j \quad , \qquad (5.1.7)$$

where p is the gas pressure in atmospheres, ξ_j is the molar fraction of the jth component, and $(\Delta_c^0)_j$ are the values of Δ_c for T = 300 K, indicated above. If the gas temperature at the GDL nozzle exit is close to 300 K, the results of computations are independent of the choice of the dependence (5.1.7).

The equation for the small-signal gain (5.1.6) can be simplified. Thus, we can consider that the rotational state constants of the upper and lower laser levels are the same ($B_{001} \approx B_{100}$, $J_m \approx J_n$); for not very high temperatures [such that $\exp(-\theta_2/T_2) \ll 1$] in (5.1.3,4) $Q \approx 1$. The most important simplification is associated with the disregard of the Doppler broadening

---

1 For collisions $CO_2 + H_2O$ the estimate is given with respect to the broadening efficiency for $CO_2 + H_2$ [5.9], but taking account of the mass of the H_2O molecule.

[5.15]. For typical conditions at the GDL nozzle exit ($T = 300$ K, $p = 0.01$ atm, the mixture of 10% CO_2 - 40% N_2 - 50% He) this gives the value of the small-signal gain exaggerated by more than 60%; with an increase of the gas pressure this difference is reduced (to 30% for $p = 0.015$ atm and to 10% for $p = 0.035$ atm). Making these assumptions and proceeding from (5.1.6), we obtain for the small-signal gain at the center of the line P20 ($00^01 \to 10^00$) (for $A_{mn} = 0.187\ s^{-1}$ and $\Delta_c \propto 1/T$)

$$k_0 = 97.5 \frac{\xi_{CO_2} \exp(-223/T)}{T \sum (\Delta_c^0)_j \xi_j} \left[\exp\left(-\frac{\theta_3}{T_3}\right) - \exp\left(-\frac{\theta_1}{T_1}\right)\right] cm^{-1} \quad . \qquad (5.1.8)$$

Thus, if the line contour is close to the dispersion one, the small-signal gain is independent of pressure. In the other limiting case, i.e., for low pressures and high temperatures, when the role of the Doppler broadening is great, the small-signal gain is proportional to pressure. The temperature dependence of the small-signal gain has a peak [for (5.1.8), at $T = 223$ K], which is due to the peak population of the rotational levels under consideration, depending on the temperature.

Of greatest importance is the dependence of the small-signal gain on the vibrational temperatures T_3 and T_1 ($T_1 = T_2$), which determine the population inversion $\Delta N = [N_m - (g_m/g_n)N_n]$. For $\Delta N = 0$, $k_0 = 0$; this corresponds to the ratio T_3/T_1, which is practically equal to $\theta_3/\theta_1 = 1.69$. For $T_3/T_1 > 1.69$ the gain coefficient is positive. Evidently, the value of k_0 is maximum for the highest possible T_3 and the lowest possible T_1. An upper limit to T_3 is provided by the vibrational energy before entering the nozzle, while the lower limit to T_1 is given by the translational gas temperature in a flow. Figure 5.1 gives the idea of the dependence of k_0 on the vibrational temperature T_3 and T_1, where k_0 is given for the center of the line P20 ($00^01 \to 10^00$), which has a dispersion contour, depending on T_3 and T_1 for 10% CO_2 content in the mixture in the assumption that $T_1 = T_2 = T$. The dashed line in this figure corresponds to the equilibrium state ($T_3 = T_1$).

The small-signal gain k_0 and the absolute value of the population inversion ΔN determine the state of an active medium entering the cavity. For this reason, before discussing the GDL performance as a whole, we shall consider first the results of the solution of the problem about creation of the GDL active medium. We shall then compare these results with experiment, which will enable us to estimate the accuracy of the described model of the processes in the GDL. The distribution curves for the gas pressure and temperature over the nozzle; the vibrational temperature T_2, T_3, and T_4;

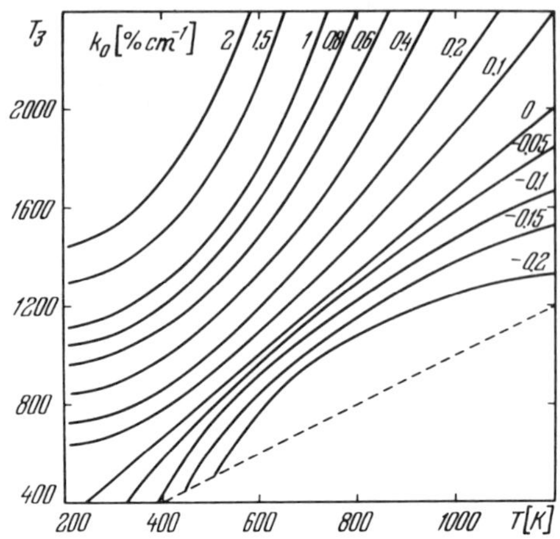

Fig.5.1. Small-signal gain k_0 for collisional broadening of the line P20 ($00^01 \rightarrow 10^00$) with 10% CO_2 content in the mixture. The half-width Δ_c is taken for 10% CO_2 - 40% N_2 - 50% He mixture, $\Delta_c \propto T$

and also the population inversion ΔN and the small-signal gain k_0, obtained as a result of the joint solution of the system of equations of gas dynamics and kinetics and the state equations [(4.1.1,3,5, 3.1.6,10, 5.1.2)] are given in Fig.5.2 for the transition P20 ($00^01 \rightarrow 10^00$). In this example we considered a flow in a two-dimensional nozzle, whose subsonic contour had the shape of a circle with 4 mm radius and 2 mm throat height, and its supersonic contour had the shape of a diverging wedge with a half-angle equal to 15°. The computations have been carried out for the mixture of 10% CO_2 - 40% N_2 - 50% He and 15% CO_2 - 83% N_2 - 2% H_2O for different values of pressure p_0 and temperature T_0 at the nozzle entrance.[2]

The subsonic flow differs little from the equilibrium flow if the initial pressure is higher than 5 atm. The fast change of gas temperature and pressure in the critical region of the flow first of all freezes the V-T exchange with nitrogen molecules, since the vibrational relaxation time τ_4 is large. Then the V-T exchange for CO_2 molecules slows down. When carbon dioxide is cooled in a nozzle in the absence of nitrogen and other admixtures, the values of T_2 and T_3 are frozen at approximately the same level and the inversion is absent, which is due to similar characteristic times of the V-T and V-V exchange in CO_2 (Sect.3.5). The presence of nitrogen in the laser mixture changes the nature of freezing of antisymmetrical vib-

---

2 Some of the results discussed here had been obtained for somewhat different values of kinetic and optical parameters than those indicated above. The difference in the results, as a rule, is not large.
Here and further, concentrations of gas mixture components are expressed in molar fractions.

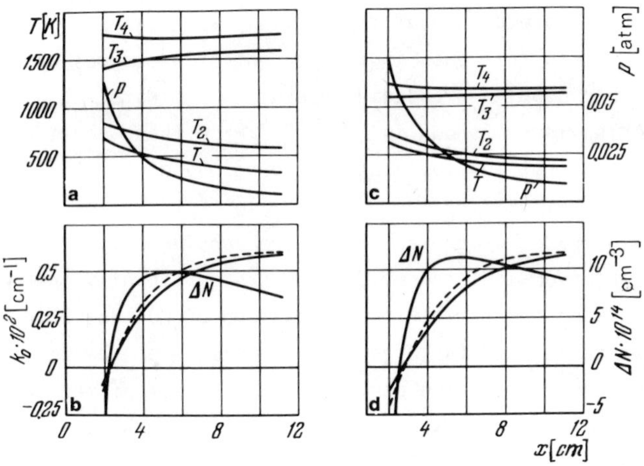

Fig.5.2. Distribution of temperature T, T_i and pressure (a, c), absolute inversion ΔN and small-signal gain k_0 (b, d) in a two-dimensional wedge nozzle for 10% CO_2 - 40% N_2 - 50% He mixture, T_0 = 2540 K, p_0 = 5.5 atm (a, b); 15% CO_2 - 83% N_2 - 2% H_2O, T_0 = 1700 K, p_0 = 5.15 atm (c, d). Dashed lines correspond to $\Delta_c \propto 1/T$, solid lines, to $\Delta_c \propto 1/\sqrt{T}$

rations of CO_2 molecules. The intensive and continuous exchange between the levels ν_3 of CO_2 and vibrationally excited nitrogen provides the pumping of vibrational energy from nitrogen to the third mode of CO_2 vibrations; for this reason after the drop T_3 passes through a minimum and then becomes close to T_4. The value of this minimum depends on the conditions in the gas and the rate of cooling the mixture. In particular, for sufficiently fast exchange CO_2 $(\nu_3) \rightleftharpoons N_2$ the minimum is not observed. It should be noted, however, that the probability of the CO_2 $(\nu_3) \rightleftharpoons N_2$ exchange is not high enough to provide the simultaneous freezing of the CO_2 (ν_3) vibrations and the equality $T_3 = T_4$: the pumping of vibrational energy from N_2 to CO_2 takes a certain time, which is slightly smaller than the characteristic time of the nozzle gas flow. Complete freezing of the symmetrical CO_2 vibrations in the GDL nozzle does not occur, and T_2 is only slightly higher than the translational gas temperature.

After the difference between T_3 and T_2 has increased and T has dropped, conditions set in in the flow, for which the absolute value of the population inversion ΔN of the states 10^00 and 00^01 and the small-signal gain coefficient k_0 become positive. As the gas expands in the nozzle, the inversion ΔN increases and passes through a maximum: the decrease of ΔN is connected with the pressure drop in the flow. The gain coefficient k_0 in the examples under consideration increases along the nozzle until the gas pressure can still provide for the prevalence of the collisional broadening of lines (as was pointed out, k_0 in this case is independent of pressure) and then somewhat decreases, when Doppler broadening begins to dominate (this part of the dependence for k_0 is not shown in Fig.5.2). From these considerations it is clear that the use of any temperature dependence of Δ_c (5.1.7) changes the result insignificantly.

The presence of helium or water vapor in the mixture considerably affects the value of the population inversion: these admixtures significantly increase the deactivation rate of the symmetrical vibrations of CO_2 and bring T_2 closer to T. Helium, which is used in large amounts because of the small value of enthalpy of a monatomic gas, also promotes the increase of the cooling rate in the mixture. Water vapor is so effective that it is sufficient to add only several percent H_2O to a CO_2-N_2 mixture.

The theoretical model considered here required experimental verification. To do this, we must compare the theoretical values of the small-signal gain with the measured values for the same initial conditions, gas composition, and nozzle contour, which can be realized most simply in experiments with shock tubes (Sect.4.5). To measure the small-signal gain k_0, use is made of the test radiation from the low-power CO_2 electric-discharge laser, directed across the flow. Such measurements could result in a systematic error in the interpretation of k_0, which is usually related to the center of the line. This error depends both on the spectral composition of the laser test radiation and on the contour of the lines in the laser under study [5.16]. For a single-mode test laser oscillation (which is close to reality for low-power CO_2 lasers) the error due to the test frequency shift from the center of the line by ω, expressed in terms of the Doppler width $\Delta\nu_D$ (2.1.21), is

$$\delta = 1 - H(a,\omega)/H(a,0) \quad , \tag{5.1.9}$$

where $H(a,\omega)$ is the Voigt function (2.1.20), and a is the ratio of the collisional and the Doppler widths of the line (2.1.21). If the pressure in the GDL under study is high enough and the line contour is close to the dispersion one, then from (5.1.9) and (B.4) it follows that

$$\delta = \omega^2/(a^2 + \omega^2) \quad . \tag{5.1.10}$$

The value of ω in low-power electric-discharge CO_2 lasers with the Doppler contour usually is not higher than unity, and in gasdynamic lasers, as a rule, $a = 50$-$100\,p$, where p is expressed in atmospheres, so that for $\omega \leq 1$ and $p \geq 0.02$ atm $\delta \leq 20$ to 50% and it rapidly decreases with an increase of pressure. These quantities characterize the precision which can be expected when comparing experimental results with the theoretical solution for such an experiment. The experimentally obtained small-signal gain k_0 also includes other random and systematic errors; on the other hand, unavoidable errors in kinetic and optical parameters used in computations make the theoretical value of k_0 insufficiently accurate as well (see below).

The comparison of the theoretical small-signal gain with the experimental values for the 10% CO_2 - 40% N_2 - 50% He and 15% CO_2 - 83% N_2 - 2% H_2O mixtures obtained in a shock tube of a large diameter [4.69] with p_0 ranging from 2.5 to 7 atm, T_0 = 1600 to 2100 K and with a confidence level of 0.95 leads to the conclusion that (for $A_{mn} = 0.21\ s^{-1}$) the difference between the theoretical and the experimental values does not exceed approximately 25%. A similar result was obtained in [5.17,18] for the same mixtures in a wider temperature range (T_0 = 1200 to 2500 K) and for higher pressures (up to 23 atm); the kinetic scheme used in these investigations is rather complicated due to the account of the exchange between ν_1 and ν_2 modes of the CO_2 vibrations and the attempt to separate the energy exchange channels (001 → 030) and (001 → 110) [3.79], which, however, does not affect much the theoretical results for k_0. Good agreement of the theoretical value of k_0 with the experimental results was obtained in experiments involving the inversion formation in a gas mixture flow behind the slit at the end of a shock tube [3.80]; in this analysis relaxation equations similar to (5.1.2) were used.

The model of the processes in a CO_2 gasdynamic laser considered here is not very complicated. However, it can still be simplified. The first step in this line is the *linearization* of the system of kinetic equations (5.1.2) without changing the number of relaxing degrees of freedom of molecules in this gas mixture. This is achieved by assuming the vibrational temperature T_2 of the symmetrical modes is close to the translational gas temperature (this is evident from the results of the analysis) and that $\theta_3 - \theta_4 \ll T$. We expand the expression for φ_{32} in (5.1.2) into a power series of $\bar{e}_3 - e_3$, where $\bar{e}_3$ is the equilibrium value of e_3, and limit ourselves to the first term of the expansion. We also assume that in the first equation of (5.1.2) the fraction in front of the brackets equals unity and $\theta_3 = \theta_4$. Thus we obtain the system of linearized kinetic equations

$$\frac{de_2}{dt} = \frac{p}{\tau_2}(\bar{e}_2 - e_2) - \frac{3}{2}pk_{32}^0(\bar{e}_3 - e_3) \quad ,$$

$$\frac{de_3}{dt} = pk_{32}^0(\bar{e}_3 - e_3) - \xi_{N_2}pk_{34}(e_3 - e_4) \quad ,$$

$$\frac{de_4}{dt} = \xi_{CO_2}pk_{34}(e_3 - e_4) + \frac{p}{\tau_4}(\bar{e}_4 - e_4) \quad . \tag{5.1.11}$$

The rate constant of the exchange, k_{32}^0, appearing here, is related to k_{32} in (5.1.2) through (3.5.9). This system was used in the analyses [2.20, 4.10, 5.19]. In [2.20, 4.10] it was assumed that $\bar{e}_3 \approx e_3\,(T_2)$, since $T_2 \approx T$. Equations (5.1.11) can be simplified further if we take into account that the rate of V-T exchange for N_2 is low in comparison with the rates of other processes; hence the last term in the third equation is sometimes ignored [5.20,21].

The next step in simplifying the kinetic scheme of exchange is to take account of only two relaxing "blocks" of vibrational states—$CO_2(\nu_1) + CO_2(\nu_2)$ and $CO_2(\nu_3) + N_2$—where in addition to $T_1 = T_2$ it is assumed that the N_2 molecules are at vibrational equilibrium with the third mode of the CO_2 vibrations ($T_3 = T_4$). Then, assuming that $\theta_4 = \theta_3$, we always have $e_4 = e_3$, and the system (5.1.11) is reduced to

$$\frac{de_2}{dt} = \frac{p}{\tau_2}(\bar{e}_2 - e_2) - \frac{3}{2}pk_{32}^0(\bar{e}_3 - e_3) \quad , \quad \frac{de_3}{dt} = pk_{32}^0(\bar{e}_3 - e_3) \quad . \tag{5.1.12}$$

The analysis shows that the effect of the V-V exchange (the second term on the right-hand side of the first equation) on the vibrational state of the symmetrical modes is small. Then, ignoring the effect of the V-V exchange on e_2 (despite the violation of the balance in the vibrational exchange scheme among CO_2 molecules, appearing in this case), we can propose a simpler model (the *Anderson model* [1.2, 4.11]). Introducing the vibrational energy of each block (per gram of the mixture) in the form $\varepsilon_I = \xi_{CO_2}R(\theta_1 e_1 + 2\theta_2 e_2)/\mu$ and $\varepsilon_{II} = R(\xi_{CO_2}\theta_3 e_3 + \xi_{N_2}\theta_4 e_4)/\mu$ (μ is the molecular mass of the mixture and R is the gas constant), we arrive at two simple kinetic equations (assuming that $\theta_1 = 2\theta_2$ and taking into account the V-T exchange for N_2),

$$\frac{d\varepsilon_I}{dt} = \frac{p}{\tau_I}(\bar{\varepsilon}_I - \varepsilon_I) \quad , \quad \frac{d\varepsilon_{II}}{dt} = \frac{p}{\tau_{II}}(\bar{\varepsilon}_{II} - \varepsilon_{II}) \quad , \tag{5.1.13}$$

where

$$\tau_1 = \tau_2 \quad , \quad \tau_{II} = (\xi_{CO_2} + \xi_{N_2})/(\xi_{CO_2}k_{32}^0 + \xi_{N_2}/\tau_4) \quad , \text{ and}$$

$$\tau_i = [\sum \xi_j/\tau_i(j)]^{-1}(i = 2,4) \quad , \quad k_{32}^0 = \sum \xi_j k_{32}^0(j) \quad .$$

The difference between T_2 and T is small, so in some cases we can ignore the first equation assuming that $T_1 = T_2 = T$ everywhere and put forward the scheme with *one relaxing block* of the vibrational states of CO_2 (ν_3) + N_2. Then to describe vibrational exchange kinetics we need only one equation

$$\frac{d\varepsilon_{II}}{dt} = \frac{p}{\tau_{II}} (\bar{\varepsilon}_{II} - \varepsilon_{II}) \quad . \tag{5.1.14}$$

This kinetics model is valid only for a high He content of more than 30%, water vapor content of more than 1%, a pressure p_0 above 5 atm, and for a temperature T_0 below 2000 K.

Simple schemes of relaxation processes in gasdynamic lasers are obtained using the instantaneous freezing principle (Sect.4.2) in the form (4.2.3), applied to the vibrational energy of the second block of vibrational states of CO_2+N_2. The obtained values of $(T_3)_f$ at the freezing point f can be used to determine the relation between T_2 and T in the further flow of the mixture [5.22a]. The characteristic relaxation time τ_m of the upper laser level in a cooling mixture, according to (4.1.4), for linearized equations of the type (5.1.13) is $\tau_m = \tau_{II}/p$; if we assume that $\xi_{N_2} >> \xi_{CO_2}$ and ignore the V-T exchange for N_2, then $\tau_m = \xi_{N_2}/\xi_{CO_2}pk^0_{32}$ [5.19]. Proceeding from (4.2.3), for the freezing point we have

$$\left| \frac{u\tau_m}{\bar{\varepsilon}_{II}} \left(\frac{p_0}{p}\right) \frac{d\bar{\varepsilon}_{II}}{dt} \frac{dT}{d\bar{S}} \right| = \lambda \quad , \tag{5.1.15}$$

where $\bar{S} = S/S_*$, λ is the similarity parameter (Sect.4.1), which for a wedge nozzle is $\lambda = p_0h_*/(2 \tan \theta)$, h_* is the throat height, and θ is half of divergence angle of the supersonic part of the nozzle. According to [5.22a], (5.1.15) is more accurate if the right-hand side is multiplied by the factor 0.8. The results of computation for $(T_3)_f$ according to (5.1.15) for the CO_2-N_2-He mixture are given in [5.22a].

The simplest assumption is that the vibrational temperature T_3 freezes at the nozzle throat and T_2 always remains at equilibrium with the gas temperature T [5.22b]. For pressure $p_0 > 5$ to 10 atm this gives an upper estimate of the population inversion and the small-signal gain k_0 for the theory of the maximum gain [5.22b]. For the most typical values of the quantities in (5.1.15) we obtain the estimate

$$p_0\tau_m \approx 10^{-5}h_* \quad , \tag{5.1.16}$$

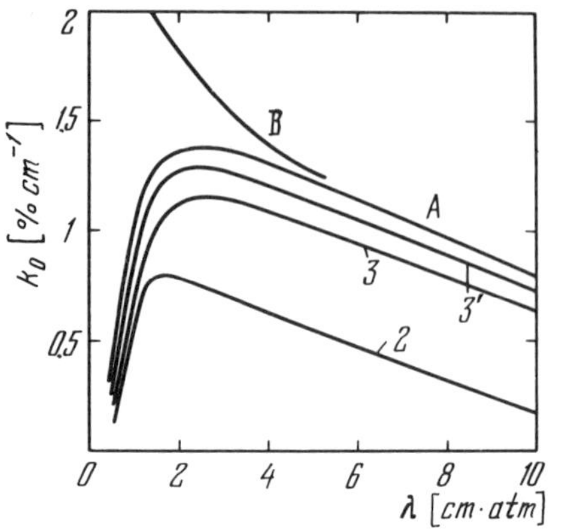

Fig.5.3. Small-signal gain k_0 as a function of $\lambda = p_0h_*/(2\tan\theta)$ for $S/S_* = 30$ according to analysis for different schemes of vibrational exchange kinetics. 3 —— (5.1.2); 3' —— (5.1.11); A —— (5.1.13); B —— (5.1.14); 2 —— the same as 3, but taking account of the V-V exchange through the second level ν_2 of CO_2

where h_* is expressed in centimeters, p_0 in atmospheres, and τ_m in seconds, so that (on the basis of the data in Fig.3.5 for k_{32}^0) we should have $h_* \lesssim 1$ mm, which determines the characteristic dimensions of the CO_2 gasdynamic laser nozzle.

The selection of the scheme is dictated by the problem to be solved and the accuracy required. Note, however, that for almost all schemes mentioned here the authors found adequate agreement with experimental results for the gain coefficient k_0. Systematic errors, misalignment of the compared conditions, and difference in the initial data make it impossible to effectively use experimental results to assess relative accuracy of these models. This can be done by comparing the results of the theoretical solution of the same problem with similar initial conditions, gas composition, and nozzle contour in different models. An example of such a solution is given in Fig.5.3 for the flow of the 10% CO_2 - 40% N_2 - 50% He mixture in a two-dimensional wedge nozzle with T_0 = 1500 K. The subsonic flow was assumed to be at equilibrium; the gain coefficient was computed by (5.1.8) including only the collisional broadening. For comparison, the results of the analysis for the V-V exchange ($\nu_3 \rightleftharpoons \nu_1$, ν_2) via the second level of CO_2 (ν_2) are given in the same figure. The relative difference in the results with application of nonlinear system (5.1.2), linearized equations (5.1.11), Anderson's model (5.1.13), and the model of a single relaxing "block" (5.1.14) is small only for sufficiently high values of the similarity parameter λ ($\lambda \geq 3$ atm cm). It becomes significant with decreasing λ (see also [5.23]). The difference in the channels of the V-V exchange in CO_2, on the contrary, noticeably affect the value of k_0 for large λ. The linearization conditions for the CO_2-N_2-H_2O mixtures are fulfilled better, since T_2 and T are close in value. This results in the smaller difference between the results for different models. Thus, the analysis of these mixtures in the wide range of initial parameters showed that the maximum difference between the results obtained according to (5.1.2) and (5.1.13) in the population inversion of the levels under consideration amounted to 15%, and in temperature T_3, 4% for T_0 = 2500 K and λ = 10 atm cm. This difference decreases with an increase of λ and a decrease of T_0. The comparison of different kinetic models according to the results of computation of the gain coefficient is carried out also in [5.23,24].

To assess the possibility of simplifying the relaxation mechanism, it is useful to know the contribution of each process of (5.1.1) to the results of analysis. This contribution is determined by the "sensitivity" of the small-signal gain k_0 to the variation of the rate constants of the V-V exchange and the time of the V-T relaxation. Let us illustrate the contribution of a certain channel to k_0 using as examples of two variants of the gas-dynamic laser flow analysis according to Anderson's kinetic model (5.1.13) (the first variant: 7% CO_2 - 92% N_2 - 1% H_2O mixture, T_0 = 1200 K, p_0 = 20 atm, the nozzle area ratio S/S_* = 20; the second variant: 7% CO_2 - 89.5% N_2 - 3.5% H_2O mixture, T_0 = 1800 K, p_0 = 35 atm, the nozzle area ratio S/S_* = 50) [1.2, 5.25a]. The results are shown in Table 5.2.

Table 5.2.

Process	Partner in collision	Variant I		Variant II	
		$\beta \times 10^4$	$\Delta k_0/k_0$ [%]	$\beta \times 10^4$	$\Delta k_0/k_0$ [%]
V-T (CO_2)	CO_2	3.6×10^{-1}	0.2	1.9×10^{-1}	0.1
-"-	N_2	2.9×10^{-1}	1.2	1.7×10^{-1}	0.7
-"-	H_2O	4.5	14.2	5.1	15.3
V-T (N_2)	N_2	4.3×10^{-3}	0.1	1.8×10^{-2}	0.1
-"-	H_2O	1.8	1.8	9.0	23.5
$\nu_3 \rightleftharpoons \nu_1, \nu_2$	CO_2	1.5	0.4	1.0	0.3
-"-	N_2	4.4	22.0	3.0	15.0
-"-	H_2O	5.0	25.0	6.2	31.0

In the Table β is a measure of the relative change of k_0 for a 1% variation of the rate of the respective process. It is clear from Table 5.2 that the small-signal gain k_0 has low sensitivity to the V-T deactivation by collisions N_2+N_2, CO_2+N_2, CO_2+CO_2. Despite the low water vapor content, the role of collisions involving H_2O molecules is significant. As was noted in Sect.3.5, the respective exchange rate constants, as a rule, are known so far with a low accuracy. On the basis of these values of β, we can assess the accuracy of calculation of the small-signal gain k_0 if we specify the accuracy in the exchange rate with respect to each channel. Table 5.2 gives the relative error $\Delta k_0/k_0$ (in percent) thus obtained in [5.25a]. Note that this error is close to the difference between the theoretical and experimental results, which was obtained in [4.69].

In conclusion it should be noted that in [5.26] a closed-form correlation is given for the peak small-signal gain obtained as a result of analysis and statistical treatment of many results of theoretical and experimental investigations of the GDL parameters for CO_2-N_2-H_2O mixtures. This correlation is used for approximate engineering estimates of k_0 at the nozzle exit, if the initial gas parameters, nozzle area ratio, and the throat height are known.

5.2 Small-Signal Gain Optimization

To choose optimal operating conditions for a gasdynamic laser we must know how different factors—mixture composition, initial conditions (pressure p_0 and temperature T_0 at the nozzle entrance), nozzle contour, etc.—affect its characteristics. Let us determine the conditions for obtaining the highest small-signal gain. This problem had been successfully solved already in the first studies of CO_2 gasdynamic lasers, in which mixture composition, nozzle contours, and operating conditions providing sufficiently high values of k_0 had been established. Further the number of parameters and the range of their variation were considerably increased. At first the workers tried to find optimal conditions by varying in a theory or experiment a certain parameter separately, other parameters remaining constant; later the problem of simultaneous optimization of many parameters was solved. We shall consider here the results of variation of separate factors determining k_0, main attention being paid to experimental investigations.

In the analysis of the effect of a gas mixture composition we shall first limit ourselves to mixtures containing only nitrogen, helium, and water vapor; other mixtures will be treated in Sect.5.4. The measurements of the small-signal gain k_0 in a mixture emanating from a slit nozzle at a shock

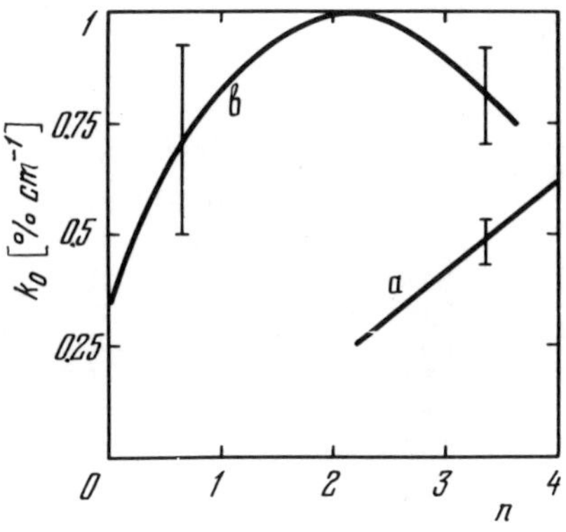

Fig.5.4. Effect of the relative helium-nitrogen content on small-signal gain k_0 in the mixtures of CO_2:He = 1:n for T_0 = 2550 K, $(p_0)_{CO_2}$ = 7 atm (a) and CO_2:N_2:He = 1:n:2.9 for T_0 = 2350 K, $(p_0)_{CO_2}$ = 4.3 atm (b). Vertical lines show the spread of experimental results (as in other figures). Here and further, concentrations are given in mole fractions

tube [5.27] have shown that the addition of helium to CO_2 results in positive values of k_0 (Fig.5.4). Thus, it is possible in principle to create a CO_2-He gasdynamic laser without nitrogen, since helium sufficiently deactivates the lower laser CO_2 level and to a lesser extent its upper level (for the ratio $\xi_{He}/\xi_{CO_2} > 10$ the deactivation of the upper level becomes noticeable). The addition of nitrogen to this mixture, however, leads to a considerable increase of the small-signal gain (Fig.5.4), which is connected with the transition of the vibrational energy from nitrogen to the upper laser level of CO_2. Too large an amount of nitrogen decreases the cooling rate of the mixture and thus depopulates the upper laser level.

Already in the first works on GDLs [1.15] the possibility of obtaining inversion in a binary mixture containing only CO_2 and N_2 had been envisaged. The small-signal gain in such mixtures is determined by the V-T deactivation rate of the ν_2 mode of CO_2 during collisions with N_2 molecules. The small-signal gain is positive only when $\tau_2(N_2) < \tau_2(CO_2)$, and nitrogen deactivates the lower laser level in the CO_2 molecule. This inequality was observed in [3.26]; thus the analysis of the flow of the binary CO_2-N_2 mixture based on $\tau_2(N_2)$ according to [3.26] indicated the possibility of obtaining inversion (for example, [3.81]). According to more precise data [3.57], however, $\tau_2(N_2) > \tau_2(CO_2)$, which excludes the possibility of obtaining inversion [5.28]. Indeed, in the overwhelming majority of experiments the gain (or oscillations) in these mixtures was not observed. The facts indicating the inversion in the binary N_2-CO_2 mixture, which were presented in [4.71, 5.17,29], are apparently erroneous due to the presence of uncontrolled impurities.

When measuring a small-signal gain k_0 at the transition P16 (001 → 100) in the CO_2-N_2-He mixture in a two-dimensional nozzle flow for fixed values

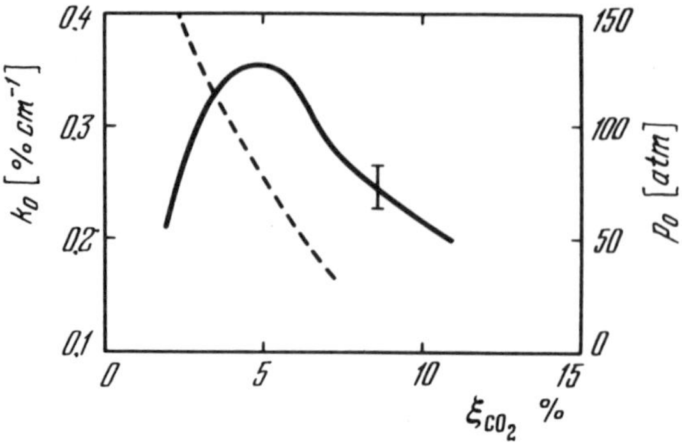

Fig.5.5. Effect of CO_2 concentration on small-signal gain k_0 with 40% He in the mixture (T_0 = 1500 K, p_0 = 82 atm). The dashed line shows the variation of optimum values of ξ_{CO_2} with initial pressure p_0

of T_0, ξ_{He}, and the flow area ratio, the peak ξ_{CO_2} was observed, providing the maximum small-signal gain k_0, and it was established that the optimal values of ξ_{CO_2} decrease with increasing pressure (Fig.5.5) [5.30].

Already the first studies of CO_2-N_2-H_2O mixtures [1.24] revealed that for T_0 = 1000 K and p_0 = 5 atm the optimal H_2O content is close to 1 to 2%, and that with an increase of the water vapor concentration up to 10%, k_0 considerably decreases, which is connected with the growing deactivating effect of H_2O molecules. Raising the initial temperature and the cooling rate (in nozzles with small throat heights and with inflection points, and also in jets) resulted in increased k_0 in these mixtures and made the drop of k_0 for large values of ξ_{H_2O} not so extreme.[3] Figures 5.6 and 5.7 give the idea of this dependence. Shown in Fig.5.6 are the results of measurement of k_0 in a two-dimensional wedge [5.32] and profiled [5.31] nozzles, and in Fig.5.7, in a jet from a slit [5.33]. In the analysis of mixtures containing water vapors in a wide range of initial temperatures (1500 K $\leq T_0 \leq$ 3200 K) it was found that even with 50% H_2O the small-signal gain k_0 in the jet remains positive and for T_0 = 2200 K it amounts to 3×10^{-4} cm^{-1} (in the mixture of CO_2:N_2:H_2O = 1:4:5) [5.33].

The dependence of the small-signal gain on the *initial temperature* for fixed values of other parameters has a maximum. As the temperature grows, the level population and the absolute value of inversion also increases. The deactivation rate of the upper laser level and the role of the Doppler broadening of the line, however, also increase, which leads to a decrease of k_0. For very high temperatures (above 2500 to 3000 K), the CO_2 dissociation begins [5.38,39]. The position and the magnitude of this peak are determined

3 See, for example, [5.31-37].

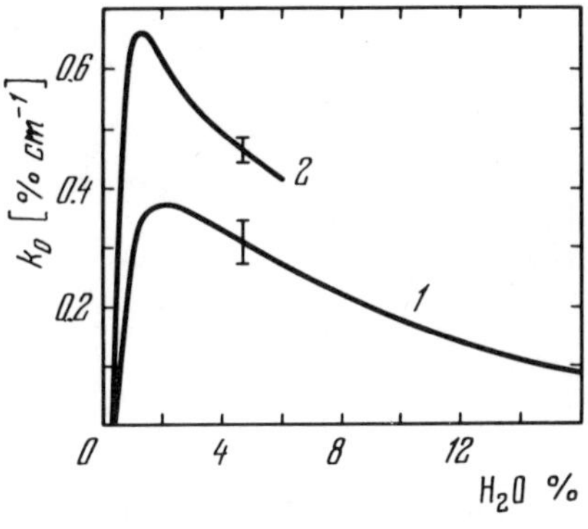

Fig.5.6. Variation of small-signal gain k_0 with concentration of water vapor added to the mixture of $CO_2:N_2$ = 1:5.7 (1-- wedge nozzle, T_0 = 1800 K, p_0 = 5.3 atm) and $CO_2:N_2$ = 1:6 (2-- contoured nozzle, T_0 = 1550 K, p_0 = 10 atm)

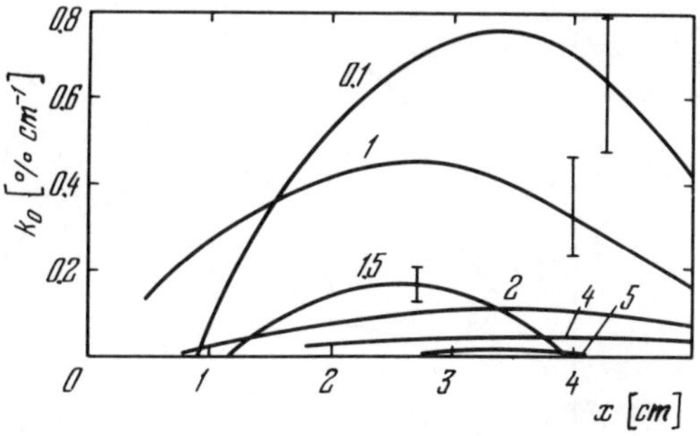

Fig.5.7. Small-signal gain distribution in $CO_2:N_2$ = 1:4 mixture when water vapor is added (numbers at the curves show the $\xi_{H_2O}:\xi_{CO_2}$ ratio); h_* = 0.8 mm, T_0 = 2250 K. x — the distance from the slit

by the specific conditions of the experiment; the general idea of the dependence k_0 (T_0) is given in Fig.5.8 for different mixture compositions and pressure p_0, different nozzle contours, and different lines of the CO_2 spectrum.

Going over to the analysis of the dependence of k_0 on the *initial pressure* and *nozzle geometry*, we shall recall that in a one-dimensional flow of the same relaxing mixture for the same value of T_0 there is the similarity for constant values of parameters $\lambda = p_0\ell$ and $\bar{S}$, where ℓ is the characteristic nozzle dimension and $\bar{S} = S/S_*$ is the area ratio of the flow (Sect.4.1). For a wedge nozzle $\ell = h_*/(2 \tan \theta)$. In some other cases it is convenient to take the length of the supersonic part of the nozzle L as its characteristic dimension. According to (5.1.6), the small-signal gain k_0 in the general case depends on p, increasing with pressure. For sufficiently high p, however, and in the absence of the effect of the Doppler broadening, k_0 does not explicitly depend on pressure [see (5.1.8)]. For these conditions the results of investigation of k_0 for the same values of ξ_j, T_0, and $\bar{S}$ characterize all GDL nozzles with respect to the parameter $\lambda = p_0\ell$.

In all experiments it was found that k_0 first rapidly increases with p_0 (this is connected with an increase in the absolute number of CO_2 molecules in a mixture and then gradually decreases as a result of the effective collisional deactivation of the upper laser level and a decrease of T_3. This

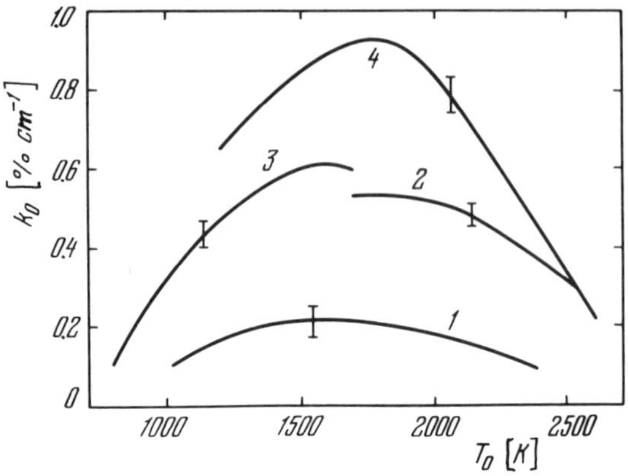

Fig.5.8. Variation of small-signal gain k_0 with the initial temperature. 1 -- [5.17]; 2 -- [5.40]; 3 -- [5.22b]; 4 -- [5.38,39]

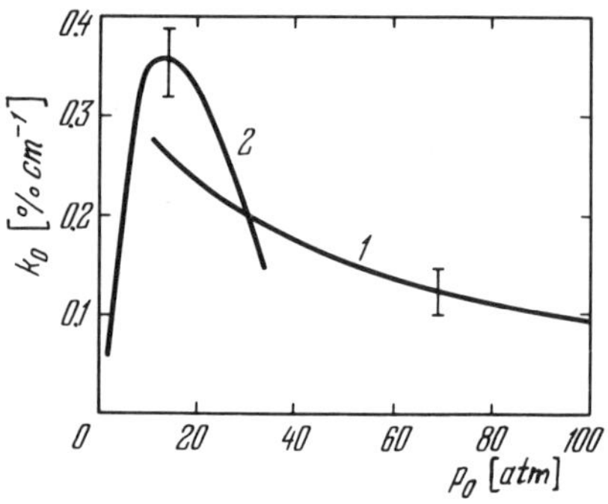

Fig.5.9. Variation of small-signal gain with initial pressure for helium-containing mixtures (1) (2.6% CO_2 - 54.3% N_2 -43.1% He) and a mixture with water vapor (2) (12.6% CO_2 - 81% N_2 - 6.4% H_2O)

dependence (Fig.5.9) was obtained for mixtures containing helium [4.69, 5.21] and water vapor [4.10, 5.41] for a wedge nozzle with $\theta = 15^0$ ($h_* = 2$ mm for the mixture designated by 1 and $h_* = 0.5$ mm for mixture 2). Taking into account the similarity with respect to the parameter $p_0h_*/(2 \tan \theta)$, we can conclude that for a fixed pressure p_0, k_0 has a peak also in the dependence on $h_*/(2 \tan \theta)$, which determines the cooling rate of a gas in the nozzle. If the nozzle divergence angle is small and the throat height h_* is large, then the cooling rate of the gas is low and the small-signal gain can even be negative (small deviations from equilibrium). With an increase of the divergence angle and a decrease of h_* the population inversion appears and the small-signal gain k_0 grows. With very narrow slits and too large divergence angles, however, all the vibrational degrees of freedom of the CO_2 molecule, including symmetrical vibrations, may be frozen, and the population inversion may be not so significant. If after such a rapid expansion of the flow (in the transonic region) the nozzle contour is made gently sloping (contoured nozzle with an inflection point, Sect.4.3), then the deactivation of the lower laser level can be achieved. Such nozzles provide the highest values of the small-signal gain and the output power.

The experimental data showing the effect of different factors on a small-signal gain k_0, presented here, were subject to theoretical analysis using different kinetics models (Sect.5.1). This made it possible to carry out the numerical solution of the *variational problem of optimization* with respect to a certain set of parameters. Let us consider the results of the solution of this problem in [3.81, 5.42,43], obtained by the configuration method [5.44], which unlike other optimization methods, e.g., the gradient method, effectively overcomes difficulties connected with the presence of "gullies" in a multidimensional space of the parameters being optimized. These parameters are molar fractions ξ_j of the components, initial temperature T_0, initial pressure p_0, the product $\lambda = p_0\ell$, and also the parameters determining the form of the function $\bar{S}$ depending on the reduced distance x/ℓ along the nozzle center line. The pressure p_0 is a separate optimization parameter due to the explicit dependence of k_0 on pressure when the Doppler broadening of the lines is taken into account. For low temperature and high pressure this dependence becomes weak. In this case the effect of pressure can be taken into account only through the parameter λ; this approach, however, restricts the generality of the solution. For a graphical representation of the nature of the found optimum and for a quantitative estimate of non-optimal conditions in the process of solution it is expedient to have the topographical arrangement of the small-signal gain levels. This is important for determining experimental conditions and "steepness" of the variation of parameters in the vicinity of the optimum. These levels should be plotted for a pair of parameters being optimized, other parameters having the optimal values.

First we consider the solution of the optimization problem for two-dimensional wedge nozzles. In this case it is expedient to take for the initial parameters ξ_j, T_0, p_0, L (L is the length of the supersonic nozzle) for a certain value of the characteristic length $\ell = h_*/(2 \tan \theta)$. The results for the same values of ξ_j, T_0, and p_0 will also be valid for other nozzles having the same reduced length $X = L/\ell$ and for sufficiently high p_0, when collisional broadening dominates for all the wedge nozzles with the same parameter $\lambda = p_0\ell$. Thus, in triple mixtures five parameters are optimized simultaneously (with $\sum \xi_j = 1$). First we fix $\ell = 0.373$ cm, which determines the nozzle with $h_* = 2$ mm, $\theta = 15^0$. This nozzle, in particular,

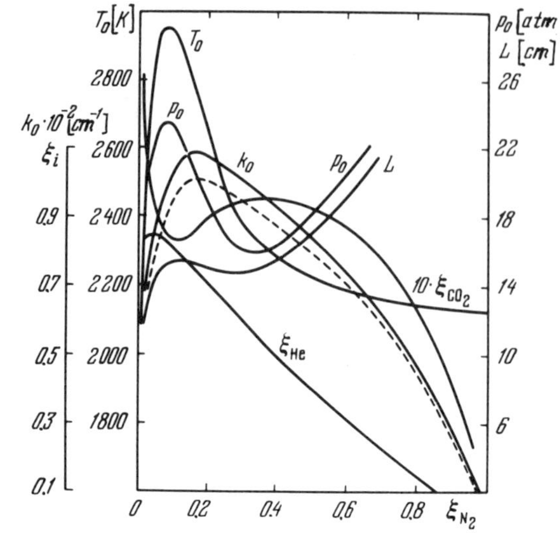

Fig.5.10. Optimum parameters for a flow of a CO_2-N_2-He mixture in a two-dimensional wedge nozzle; $\ell = h_*/(2 \tan\theta) = 0.373$ cm

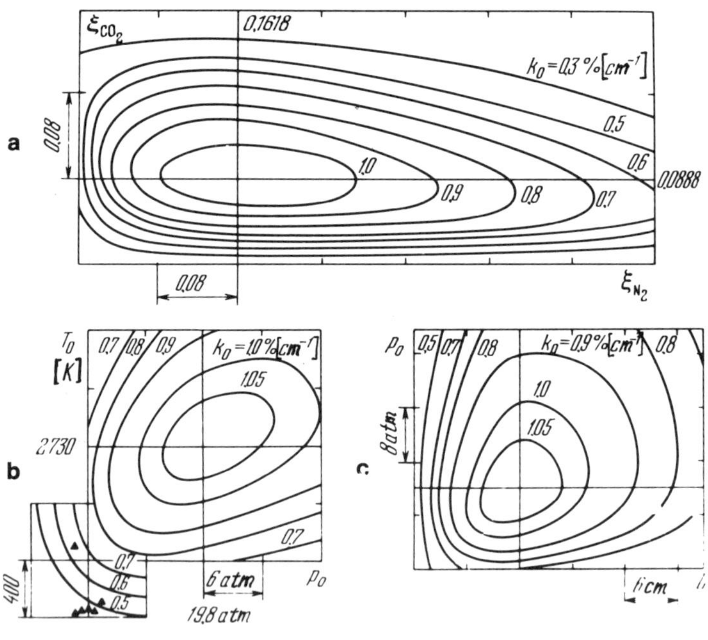

Fig.5.11. Small-signal gain levels near the optimum $k_0 = 1.08\%\ cm^{-1}$ in a nozzle with $\ell = 0.373$ cm for a CO_2-N_2-He mixture

was utilized in experiments [4.69, 5.21].[4] Figure 5.10 shows the results of the analysis for a CO_2-N_2-He mixture depending on the nitrogen concentration. Each point on this plot (and on similar plots below) is the optimum for a fixed value of the abscissa. The dash line, corresponding to the optimum k_0 for $\Delta_c \propto 1/T$, is given here for comparison. Figure 5.11 shows the small-signal gain levels for a combination of parameters $\xi_{CO_2} - \xi_{N_2}$, $T_0 - p_0$, $p_0 - L$. Black triangles indicate the values of the parameters in the experiment [4.69]; it is evident that they are far from the optimum. In the vicinity of the optimum k_0 is more sensitive to the variation of ξ_{CO_2} than to the variation of ξ_{N_2}. It is important that the deviation of pressure p_0 from the optimum by 6 to 8 atm or of temperature by 400 K decreases k_0 only by approximately 10%. By decreasing the characteristic nozzle dimension $\ell = h_*/(2 \tan \theta)$, we can achieve a considerable increase of the small-signal gain k_0. The results of optimization for the mixtures CO_2-N_2-He and CO_2-N_2-H_2O for different ℓ's (up to ℓ = 0.01 to 0.02 cm) in Figs.5.12 and 5.13 are given in the form of the dependence on the reciprocal quantity $a = 1/\ell$ (cm^{-1}). As ℓ decreases, the optimum pressure p_0 grows and the nozzle length L decreases; the optimum gas composition ξ_j and temperature T_0 change.

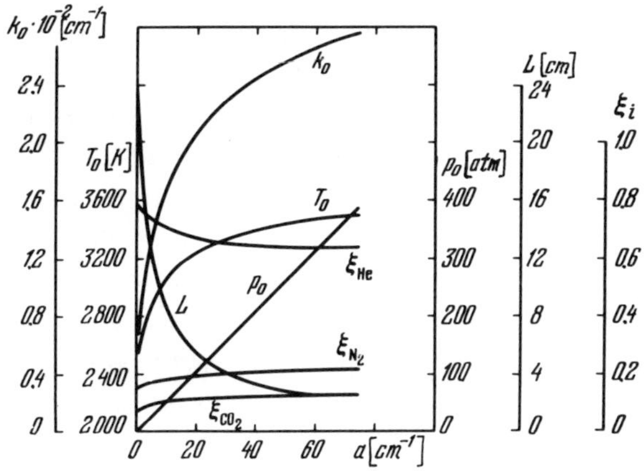

Fig.5.12. Variation of optimum parameters with a decrease of characteristic size of a two-dimensional wedge nozzle ℓ ($a = 1/\ell$) for CO_2-N_2-He mixtures

4 In these and subsequent calculations it was assumed that $A_{mn} = 0.21\ s^{-1}$, $\Delta_c \propto 1/\sqrt{T}$; other values are given in [3.81, 5.42,45]. The values of τ_2 (N_2) are taken from the review [3.26]; according to the remarks made at the beginning of this section, this leads to certain inaccuracies for CO_2-N_2 mixtures with a low helium content and for pure binary mixtures CO_2-N_2.

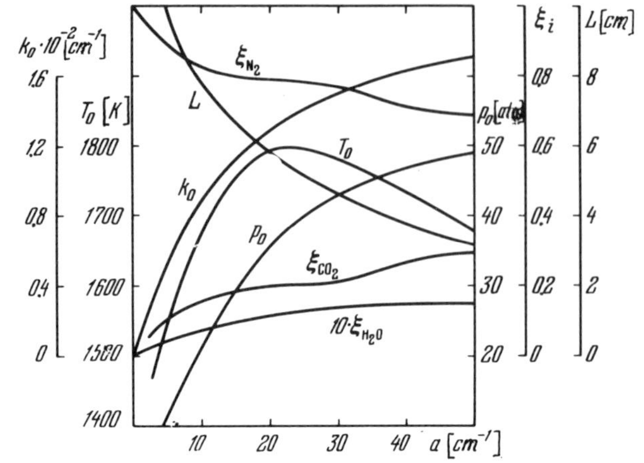

Fig.5.13. The same as Fig.5.12 for CO_2-N_2-H_2O mixtures

Let us now consider the results of optimization for an arbitrary nozzle contour, when in the solution we must find the optimal law of variation of the nozzle cross section. In this case again we shall limit ourselves to the one-dimensional flow approximation, the two-dimensional flow in a gasdynamic laser being treated later. We shall optimize only the contour of the supersonic part of the nozzle, assuming that in the subsonic flow the deviations from equilibrium are small. For a characteristic dimension we shall take the required optimal length of the supersonic part of the nozzle. The dependence of the cross-sectional area of the nozzle, S, on the distance x from the throat is given in the dimensionless form (S/S_*), so that (passing over to a plane nozzle with cross-sectional area S to height h) $h/h_* = 1 + \omega(x/L)$, where ω is the required function, for which $\omega(0) = 0$. The form of this function can be determined by a certain (not large) set of parameters. One of the possibilities of defining these parameters is to choose the derivatives α_i of this function at certain i^{th} grid points along the supersonic nozzle length L. Between these points a certain, e.g., parabolic, approximation can be found, whose coefficients are determined by the values of function ω_i and its derivatives α_i at the grid points. Let us divide the supersonic part of the nozzle along its length in the ratio 1:3:5 and let i = 0 for x = 0, $\alpha_i \geq 0$ (the flow is not contracted), and $\alpha_3 = 0$, so that at the nozzle exit the contour is parallel to the x axis, which is typical for gasdynamic lasers. Then the required function $\omega(x/\ell)$ is determined by three parameters (α_0, α_1, α_2), which are related to the slope of the nozzle contour at the grid points by

$$\tan\theta_i = \alpha_i h_*/2L \quad (i = 0, 1, 2) \quad , \tag{5.2.1}$$

and $\theta_3 = 0$. Instead of these parameters we can take some others, for instance, we can introduce the quantity $\bar{S}_{ex} = \omega(1)$, which determines the degree of the flow expansion at the nozzle exit. The quantity L, which also determines the nozzle shape, enters into the similarity parameter $\lambda = p_0 L$.

The results of optimization with respect to seven parameters for a CO_2-N_2-He mixture are shown in Fig.5.14 by solid lines. For comparison, the dashed line here shows the results of optimization for the wedge nozzle flow, the respective quantities being denoted by primes. The peak small-signal gain increases with pressure. For the same pressure p_0, k_0 noticeably exceeds k_0', and on the contrary, T_0 is less than T_0'. This is important for setting up the experiment, since high temperatures are difficult to obtain. The content of CO_2 and N_2 in the optimal mixture is practically the same ($\xi_{CO_2} \approx \xi_{N_2} \approx 20\%$). The optimal divergence angle, which is characterized by α_0, for a nozzle of arbitrary shape is several times as large as that for a wedge nozzle (this provides higher values of k_0); further the cross-sectional area of such a nozzle varies insignificantly (α_1 and α_2 are almost by two orders of magnitude lower than α_0). The optimal area ratio of this nozzle flow $\bar{S}_{ex}$ increases with pressure p_0.

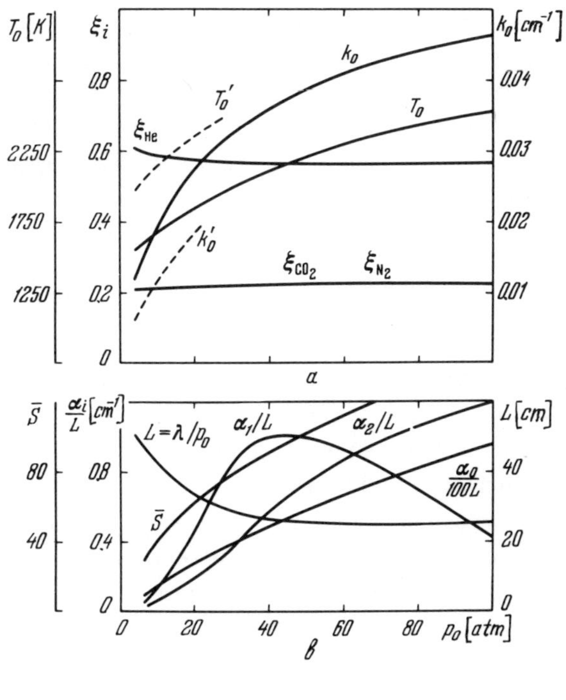

Fig.5.14. Optimum parameters for a CO_2-N_2-He mixture flow in a nozzle of arbitrary shape

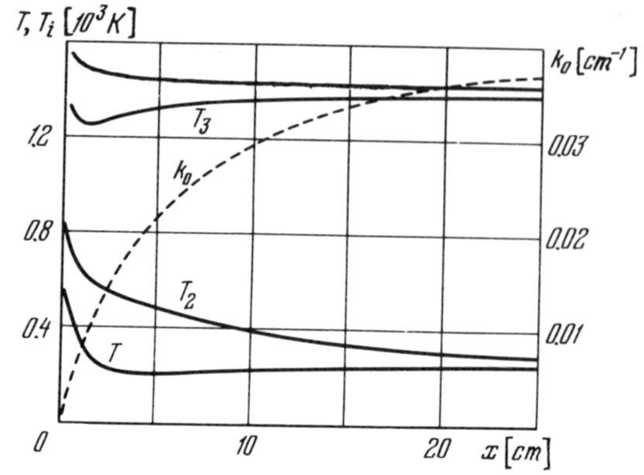

Fig.5.15. Distribution of vibrational temperature T_i and translational temperature T and small-signal gain k_0 along the optimum nozzle axis for p_0 = 50 atm

The mechanism of formation of a considerable population inversion and high small-signal gain in a nozzle for the optimal variation of the cross-sectional area is shown in Fig.5.15. During the rapid gas expansion immediately behind the nozzle throat not only the values of T_4 and T_3 are frozen, but to a certain extent T_2 as well. Further, in a channel with an almost uniform cross-sectional area and with the length that corresponds to the deactivation time of the energy of symmetrical modes of CO_2 vibrations, T_2 drops and becomes close to T. The value of T_3 here increases somewhat due to the energy exchange $CO_2(\nu_3) \rightleftharpoons N_2$ and does not decrease, since the deactivation time of the upper laser level of CO_2 at these conditions is sufficiently large. The length of this channel can be reduced by slightly heating the flow, e.g., using shock waves, and increasing the V-T exchange rate without noticeable change of the V-V transition rate. Note that the solution of the optimization problem for a nozzle in which the contraction of the flow is possible in the supersonic part ($\alpha_i < 0$) indeed results in the optimum contour either with $\alpha_1 < 0$ or $\alpha_2 < 0$. The increase of k_0 after narrowing is, however, insignificant [5.46]. On the other hand, the narrowing of the supersonic flow leads to the formation of shock waves, in which the gain coefficient may not increase but drop [5.47,48].

The slope of the optimum contour at the beginning of the supersonic part of the nozzle is rather large, which can result in generation of shock waves, separation of the flow, and other phenomena that limit attainment of peak values of the small-signal gain. For the optimum nozzles described here the effect of shock waves may turn out to be significant when the half divergence angle θ_0 of the supersonic part of the nozzle is greater than 20^0. Putting $\theta_0 \leq 20^0$, we must also limit the throat height h_* ($h_* \lesssim 0.75$ mm for

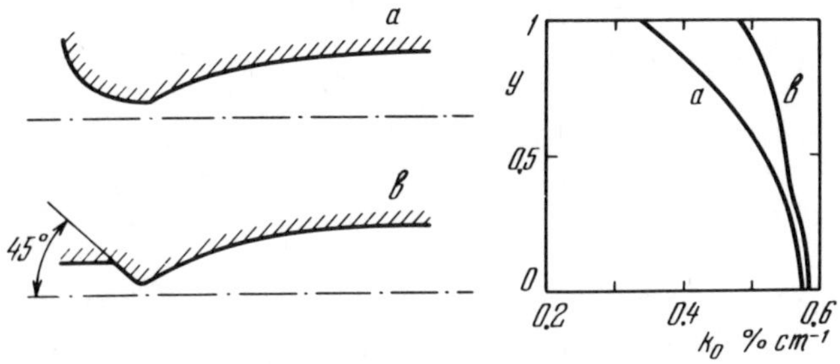

Fig.5.16. Small-signal gain distribution across the flow for nozzle contour a and b in 12.7% CO_2 - 79.4% N_2 - 7.9% H_2O mixture, T_0 = 1700 K, p_0 = 10 atm

p_0 = 10 atm, $h_* \lesssim 0.15$ mm for p_0 = 50 atm, etc. [5.42]) to provide the optimal conditions. In the general case we must carry out the analysis of a non-one-dimensional GDL nozzle flow. The methods of this analysis were considered in Sect.4.3. Here we shall only mention the results [5.49a,49b] and continue the treatment of this problem in the next section. To determine the field of a parameter in a two-dimensional nozzle [3.135, 5.49] the time-dependent method was used for the sub- and transonic regions of the flow and the net-characteristic method, for the supersonic region taking account of the vibrational exchange kinetics according to (5.1.2). Figure 5.16 shows the distribution of the small-signal gain across the flow at the exit of two plane nozzles ($\underset{\sim}{M}_{ex}$ = 4 for γ = 1.29) with different contours of subsonic parts [5.49a]. The value of k_0 varies considerably over the flow cross section, but upon averaging with respect to the coordinate across the flow it is only 10 to 20% lower than k_0 for a one-dimensional flow. Other examples of such an analysis will be given in Sect.5.3 (see also [5.50,51]).

The solution of the optimization problem gives the values of the initial temperature T_0, for which chemical reactions cannot be ignored. The account of chemical reaction kinetics in the nozzle flow of a CO_2 gasdynamic laser has shown that, as a rule, in the subsonic region the chemical composition is close to equilibrium. Freezing of the chemical composition occurs shortly behind the throat so that, further downstream, the concentrations of the components differ little from the values of ξ_j^* at the throat (more precise data were obtained in [4.28]). In this approximation a simple correction can be introduced into the optimum values of the initial concentration obtained earlier by equating them to ξ_j^* and then determining ξ_j', which provides for the equilibrium formation of ξ_j^* during the transition from T_0 to T_*. In this case we can assume that the effectiveness of the CO and O_2 molecules formed in small amounts as a result of the CO_2 dissociation differs little from the effectiveness of nitrogen (Sect.5.4).

5.3 Output Power. Power Optimization

Let us go over to the analysis of the basic characteristic of a gasdynamic laser—the output power. First we shall get acquainted with the results of experimental studies of the dependence of the output power P on certain parameters, and then discuss the methods of computing P and the results of optimization. In addition to the mixture composition, initial conditions (T_0 and p_0), and the nozzle geometry in this case we must take into account the properties of a cavity mounted at the nozzle exit, since these properties can in the long run determine the effectiveness of conversion of energy into radiation. The absolute value of the power registered in an experiment is determined, apart from other factors, by the mass flow rate of the mixture, and to a lesser exlent by the relative form of the output power dependencies.

Limiting ourselves to mixtures containing carbon dioxide, helium, and water vapor, we shall consider the results of power measurement [1.23, 2,19, 5.22b,52-55]. In these experiments it was found that the dependence of radiation power on the CO_2 content passes through a peak, which depends on the initial temperature T_0 and other factors. As temperature increases, the peak with respect to ξ_{CO_2} shifts from 16 to 20% for $T_0 \approx 1100$ K to 8 to 10% for $T_0 = 1500$ K. For $T_0 \approx 1500$ K it is preferable to deal with mixtures which contain more helium than nitrogen [5.54]. The variation of radiation power P depending on the water vapor content is similar to the results of the study of the dependence k_0 (ξ_{H_2O}) [1.24]; it is confirmed by the results of [5.55a]. For a fixed $\xi_{H_2O} = 0.5\%$, P decreases with an increase of ξ_{CO_2} for $\xi_{CO_2} \geqq 10\%$ [2.19, 5.52].

The typical dependence of the output power on temperature T_0 has a peak (see, for example, [1.23, 4.30,35, 5.54]); its position, however, depends on other factors, in particular, on the initial pressure p_0. In the 10% CO_2 - 40% N_2 - 50% He mixture for $p_0 = 115$ atm this peak is observed for $T_0 \approx 3000$ K (Fig.5.17). In experiments [5.22b] with mixtures of similar composition for $p_0 = 8.5$ atm the peak power P was not attained up to $T_0 = 2200$ K. On the other hand, in [5.55b] a power of 5 kW was reported on CO_2-N_2-H_2O mixtures for lower temperatures ($T_0 = 650$ K) and the radiation threshold was established for 450 K $< T_0 <$ 750 K, depending on the losses in the cavity

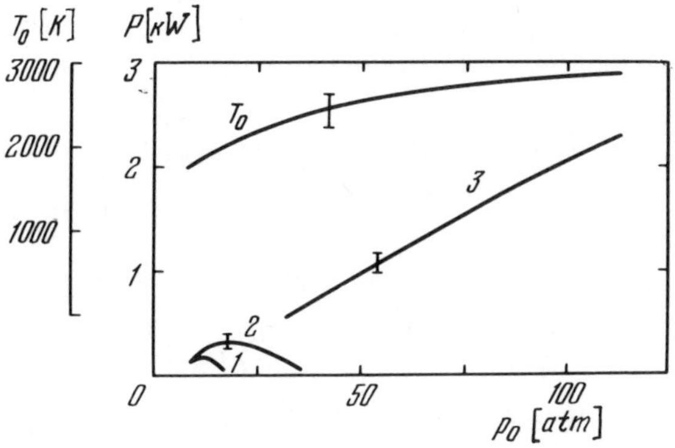

Fig.5.17. Effect of initial pressure p_0 on laser energy flux and optimum temperature T_0. 1 -- wedge nozzle, $h_* = 1.42$ mm, $\theta = 15^0$, $S/S_* = 15.5$; contoured nozzles with a point of inflection; 2 -- $h_* = 1.3$ mm, $\theta_0 = 33.5^0$, $S/S_* = 11.7$; 3 -- $h_* = 0.35$ mm, $\theta_0 = 41.6^0$, $S/S_* = 44.6$ [4.35] (θ_0 is the angle of inclination of the contour to the axis at the point of inflection)

and concentration ξ_{CO_2}. In CO_2-N_2-H_2O mixtures considerably higher power was obtained for $T_0 = 1400$ K, $p_0 = 17$ atm in a 7.5% CO_2 - 91.3% N_2 - 1.2% H_2O mixture [1.22]. For $\xi_{H_2O} = 0.5\%$ the peak of about P = 60 kW was registered for $T_0 = 1250$ K, when $\xi_{CO_2}:\xi_{N_2} = 12.4:87.1$, and for $\xi_{CO_2}:\xi_{N_2} = 7.4:92.1$ the output power linearly increases with temperature T_0 up to the values of P = 80 kW for $T_0 = 1200$ K [2.19].

The increase in the laser radiation power with pressure p_0 is observed for mixtures of different composition if the pressure is not high (not greater than several atmospheres, and according to some results, up to 15 to 20 atm [5.22,54,56,57]). If other parameters remain constant, then with a further growth of p_0 the value of P attains saturation and can even decrease (see Fig.5.17). On the other hand, by changing over from wedge nozzles to contoured ones and increasing the flow area ratio and initial temperature, we can achieve a further increase of laser power P with pressure (Fig. 5.17, and also [2.33, 5.57]). In this case the power extracted per unit cavity volume also increases considerably—up to 83.3 W/cm^3 for $p_0 = 115$ atm [4.35].

In CO_2 gasdynamic lasers both stable and unstable resonators are used. A considerable increase in power can be obtained by using multipass resonator schemes [1.32, 2.33, 5.52,56-58]. In [5.30,56] the optimization was performed with respect to the number of passages in the cavity. The highest power was obtained using a three-pass scheme (for high pressures—in a five-pass scheme [5.30]). For stable resonators the transmittance t of the output mirror was optimized; the most advantageous value according to [1.23, 5.57] is t = 0.1. The optimum value of t depends on the mixture composition and operating conditions of a GDL [5.54].

Let us go over to the *calculation of the output power* in a gasdynamic laser. First we shall simplify the resonator model as much as possible. We shall consider a plane-parallel Fabry-Perot resonator with rectangular mirrors, remaining within the framework of geometrical optics approximations. Suppose this resonator operates in a single-mode regime and radiation is generated on one transition, e.g., on P20 ($00^01 \rightarrow 10^00$).[5] To calculate laser power we shall use (2.3.30), assuming that radiation is generated at steady-state conditions. The relation for steady-state conditions can be written either in a local, (2.3.22-24), or in the integral, (2.3.26-29), form. The power P obtained on the assumption that the radiation intensity I along the flow in a cavity is constant (2.3.28) is only 20 to 30% higher than that obtained for $k_* = \text{const}$ (2.3.22). Within these accuracy limits any relation for steady-state conditions can be utilized.

The value of the saturated gain coefficient k_* is related to the active medium parameters (population of laser levels, pressure, gas temperature, etc.) through the same equation (5.1.6) that predicts the value of the unsaturated small-signal gain k_0. The difference is that now in the cavity the populations N_m and N_n and, hence, the vibrational temperatures T_2 and T_3 (or the energy e_2 and e_3), are determined not only by the collisional relaxation, but also by the radiation intensity I. To compute T_2 and T_3, one of the kinetics models described in Sect.5.1 should be used, e.g., (5.1.2). In the analysis it is natural to assume that there is a Boltzmann distribution of molecules over rotational levels. Typical operating conditions for a CO_2 gasdynamic laser as a rule provide for the Boltzmann distribution of molecules over the vibrational levels inherent to the given vibrational mode and also for the equality $T_1 = T_2$ due to the Fermi resonance in the CO_2 molecule (Sect.3.5). This makes it possible to retain the assumptions made in Sect.5.1 for the analysis of the small-signal gain k_0.

For a gas flow in the cavity the law of conservation of energy flux (4.1.6) taking account of (2.3.31) is valid,

5 In some experiments (e.g., [5.53]) it was in fact found that in the radiation spectrum of a CO_2 gasdynamic laser this transition is the most intensive. The neighboring lines (P18, P16, P22, etc.), however, can also make a considerable contribution to the total power. In many cases the radiation is of a multimode nature, which depends on the cavity design [1.22].

$$\frac{d}{dx}\left(h + \frac{u^2}{2}\right) = -\frac{k_* I}{\rho u} \quad . \tag{5.3.1}$$

If we neglect the dependence of the vibrational quantum on the rotational state, we find that, according to (2.3.31), the number of radiative transitions J between laser levels of the CO_2 molecule per unit time is

$$J = k_* IT/(\theta_3 - \theta_1)\xi_{CO_2} p \quad . \tag{5.3.2}$$

Then the relaxation equations for e_2 and e_3, which enter the system (5.1.2), for the cavity flow take the form

$$\frac{de_2}{dt} = \frac{(2e_2 + 1)^2}{6e_2^2 + 6e_2 + 1}\left[\frac{p}{\tau_2}(\bar{e}_2 - e_2) + \frac{3}{2}\varphi_{32} + J\right] \quad ,$$

$$\frac{de_3}{dt} = -\varphi_{32} - \xi_{N_2}\varphi_{34} - J \quad . \tag{5.3.3}$$

Here J is determined by (5.3.2), and all the other quantities have the same meaning as in Sect.5.1. These equations define the balance relations for laser level populations (2.5.8) for a CO_2 gasdynamic laser. Other equations describing the relaxing gas flow in a cavity do not change. The conditions for the gain at the cavity entrance (transition $k_0 \rightarrow k_*$) were discussed in Sect.2.5. The value of the radiation power (2.3.30) can be found from the solution of the system of equations for a one-dimensional steady-state inviscid gas flow (4.1.1,3,5) in the nozzle and (5.3.1) in the cavity, and also from the relaxation equations (5.1.2) replaced by (5.3.2,3) for the flow of a mixture in the cavity. This system is completed by the relation for steady-state radiation.

Figure 5.18 shows the results of power computation in the constant-intensity approximation according to (2.3.29) [2.19] in comparison with experimental results for two types of GDLs. Despite the difference between the real cavity and its simplified theoretical model and the different account of the oscillation spectrum, quite satisfactory agreement was observed. An example of analysis in the constant-gain approximation is given in Fig.5.19. At the cavity entrance T_2 and T_3 rapidly change, $k_0 \rightarrow k_*$, and the radiation intensity I grows [2.41, 5.43]. In the cavity flow the gasdynamic parameters vary insignificantly, but there is a noticeable decrease in the vibrational energy and increase in power P. The downstream increase of P is not proportional to the length of the mirrors, but it tends to saturation due to the drop of radiation intensity. These theoretical results

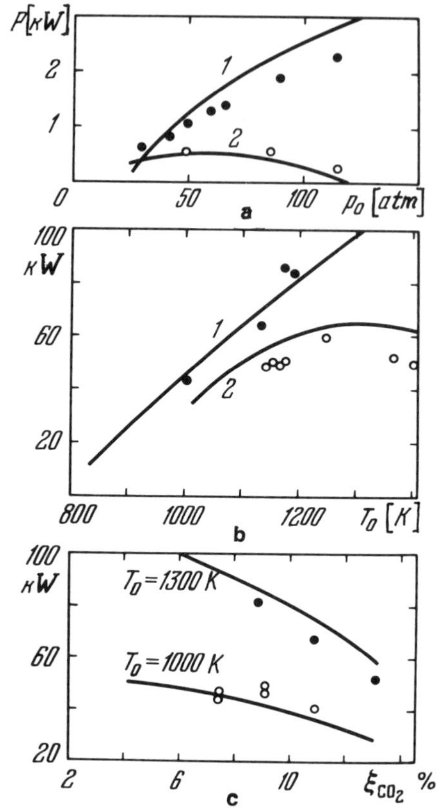

← Fig.5.18. Experimental (dots) and theoretical (lines) values of laser power in a shock tube (a) and a combustion-driven (b, c) GDL. Mixtures: (a) 1 -- 10% CO_2-40% N_2-50% He; 2 -- 10% CO_2-89% N_2-1% H_2O; (b) 1 -- 7.4% CO_2-92.1% N_2-0.5% H_2O; 2 -- 12.4% CO_2-87.1% N_2-0.5% H_2O; (c) 0.5% H_2O, other components -- CO_2-N_2. Loss coefficient α = 3%

Fig.5.19. Parameter distribution in a nozzle (left) and in a cavity (right). Solid lines: p_0 = 100 atm, T_0 = 3680 K, 5.3% CO_2-69.3% N_2-25.4% He, S/S_* = 50; dash lines: p_0 = 25 atm, T_0 = 3310 K, 8.1% CO_2-57.9% N_2-34% He, S/S_* = 33.3

↓

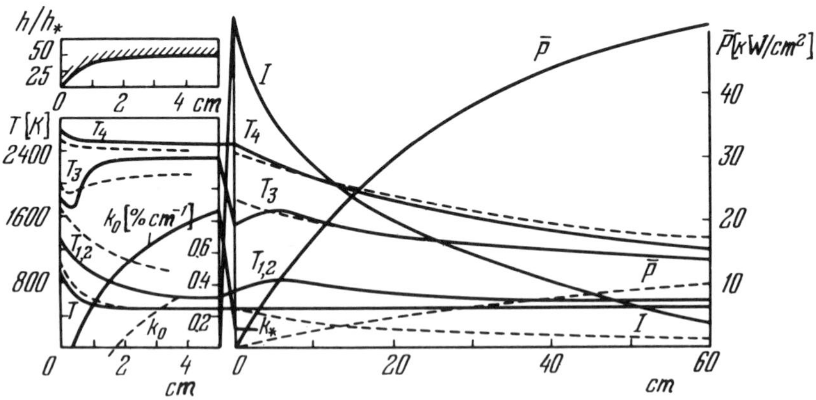

are within a factor of 2 agreement with experiment [4.35], despite some difference in the initial conditions, configuration of nozzles and cavity, and also the oscillation spectrum.

The loss factor α (Sect.2.3) used in these computations, as well as attenuation factors β and γ (Sect.2.5) are usually known with a low accuracy. The theoretical aspect of this problem was treated in Chap.2. In the experiments, α can vary even in the course of one experiment because of external effects on the mirrors (dirtying, deformation, etc.) and β and γ, because of impurities and pulsations. For this reason it is important to know the nature of the dependence of the output power on these quantities. The idea of this dependence is given by approximate expressions for P considered below. Sometimes the actual loss factor is determined from the comparison of theoretical and experimental results. Thus in [5.22b] $\alpha \approx 3$ to 5% was obtained; similar values were obtained in [2.19]. Another inaccuracy in the calculation of P is connected with the errors in the kinetic constants of exchange rate (or relaxation time) [1.2, 5.25]. According to [5.25a], the decisive role in determining radiation power is played by the V-T deactivation of nitrogen during collisions with water vapor molecules and by the V-V exchange $\nu_3 \rightleftharpoons \nu_1, \nu_2$ during collisions with N_2 and H_2O. The inaccuracy in P in this case can reach 20 to 30% [5.25a].

An important problem in the GDL theory is the selection of conditions providing higher useful power or, as is more convenient, relative power. The ratios $\bar{P} = P/S_*$ (S_* is the cross-sectional area of the nozzle throat) or $W = P/G$ ($G = \rho uS$ is the gas flow rate) can be used as characteristic parameters here. The first of these quantities characterizes the design features of a GDL, and the second, the weight factors; for this reason it is called the *specific power*. The optimization problem for the relative output power can be solved using the configuration method considered in Sect.5.2. In this case in addition to the parameters varied in Sect.5.2, ξ_j, T_0, α_i, and $\lambda = p_0L$, which determine the state of an active medium and a nozzle contour, we must take cavity parameters. For the simplest model of the Fabry-Perot resonator we shall limit ourselves here to only one varying parameter, i.e., transmittance t. Furthermore, the flow width and the corresponding cavity size will be d = 50 cm, the mirror length along the flow a = 60 cm, and the mirror height b equal to the height H of the nozzle exit, which is determined from the optimization of the geometrical characteristics of the nozzle. Cavity losses are included into the coefficient α and is taken to be not large (for two passages $\alpha' = 0.01$). The actual absolute values of $\bar{P}$ and $\bar{W}$ are apparently below the theoretical values for this level of losses. Since the loss factor α is not known with sufficient accuracy, main attention will be paid to the relative value of specific power.

The power optimization within the framework of the constant-gain model (2.3.23) in CO_2-N_2-He and CO_2-N_2-H_2O mixtures gives the following results [5.43]. In the optimum, after the fast nozzle flow expansion over the length of about 1 cm, it is expedient to have a constant cross-section channel about 4 to 5 cm long or somewhat longer (Fig.5.19; the sketch of the supersonic nozzle contour is shown in the left upper corner as the dependence of h/h_* on x for p_0 = 100 atm). It is clear that in the first region the nitrogen vibrational energy freezes (and to a certain extent, that of CO_2), while the second region is required for the energy transfer from N_2 to CO_2 (ν_3) and for the depopulation on the lower laser level, just as in the case of k_0 optimization. The results of optimization are summarized in Figs.5.20 and 5.21. For very large divergence angles not only is the one-dimensional nature of the flow violated, but the possibility of shock wave formation and separation of the flow from the wall appears. For this reason the quantity α_0 (the derivative of the function $\omega = h/h_*$ at the point x = 0) was limited in computations by the value of $\alpha_0/L = 100$, above which too small a value of h_* (below 0.1 to 0.2 mm) would be required. The optimum initial temperature T_0 and molar fraction of nitrogen increase with the initial pressure p_0, while the CO_2 and water vapor content decreases. In the CO_2-N_2-He mixture the limiting value of α_0/L is attained for $p_0 \geqq 30$ atm. The solid lines in Fig.5.21 correspond to the optimization results for the relative power $\bar{W}$ with respect to the mass rate (or specific power), and the dashed lines, to $\bar{P}$ (with respect to the throat area). The recalculation of the results of optimization of $\bar{P}$ for $\bar{W}$ (using the known values of ρ_* and u_*) leads to the values of $\bar{W}$ (dotted and dashed line) differing little from the results of direct optimization of $\bar{P}$. So we can optimize only one of the quantities, the other being calculated accordingly. In this way the values of $\bar{W}$ in Fig. 5.20 had been obtained. It is important that the relative power increases monotonically with initial pressure, if all the other parameters are optimized simultaneously. This was confirmed experimentally [1.23]. Specific power $\bar{W}$ rapidly increases with pressure only up to $p_0 \approx 30$ to 40 atm, above which it increases little. The output power for mixtures with helium and water vapor under optimal conditions differ insignificantly.

The optimum temperature T_0 is so high (in the mixtures containing helium it is higher than with water vapor) that we must take account of the chemical reactions. If this correction is made as described in Sect.5.2, we shall

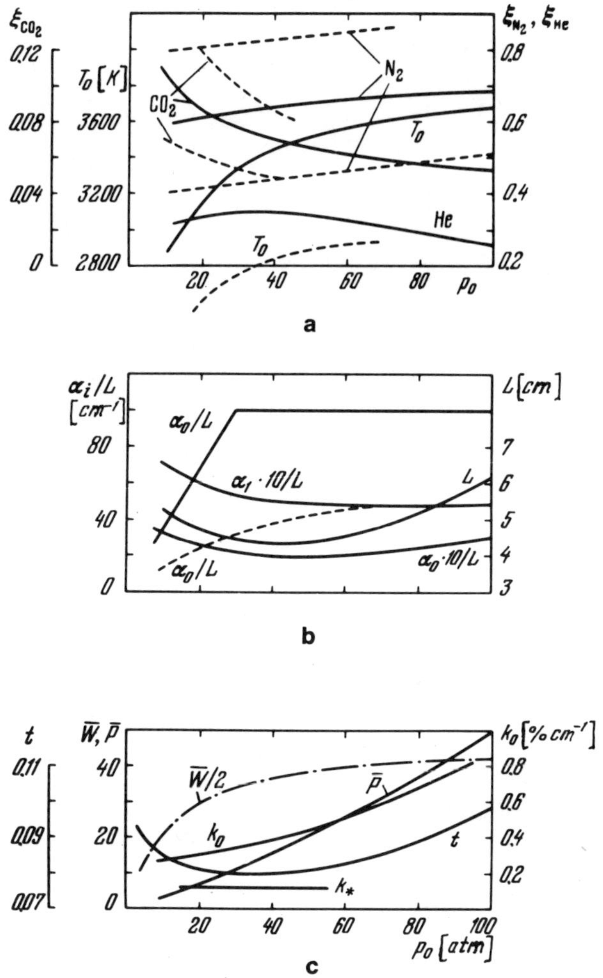

Fig.5.20. Optimum conditions for obtaining maximum relative power $\bar{P}$ (kW/cm^2) and $\bar{W}$ ($kW\ kg^{-1}s$) for a CO_2-N_2-He mixture

obtain the improved values of the initial concentration ξ_j' and of the ratio of the number of atoms N:O and H:O in the initial mixture for the optimum temperature T_0 (Fig.5.22).

Figure 5.23 gives an idea of the behavior of certain parameters near the optimum points for a CO_2-N_2-He mixture for p_0 = 25 atm. All the other parameters, besides those laid along the coordinate axes, have optimum values. Figure 5.23a shows the topography of $\bar{P}$ (solid lines, in percent of the peak value) and k_0 at the cavity entrance (dashed lines, in percent per cm); note that there is a difference in conditions for peak $\bar{P}$ and k_0 (the limiting lines $k_0 = k_*$ and $\bar{P} = 0$ naturally have the same contours). These results

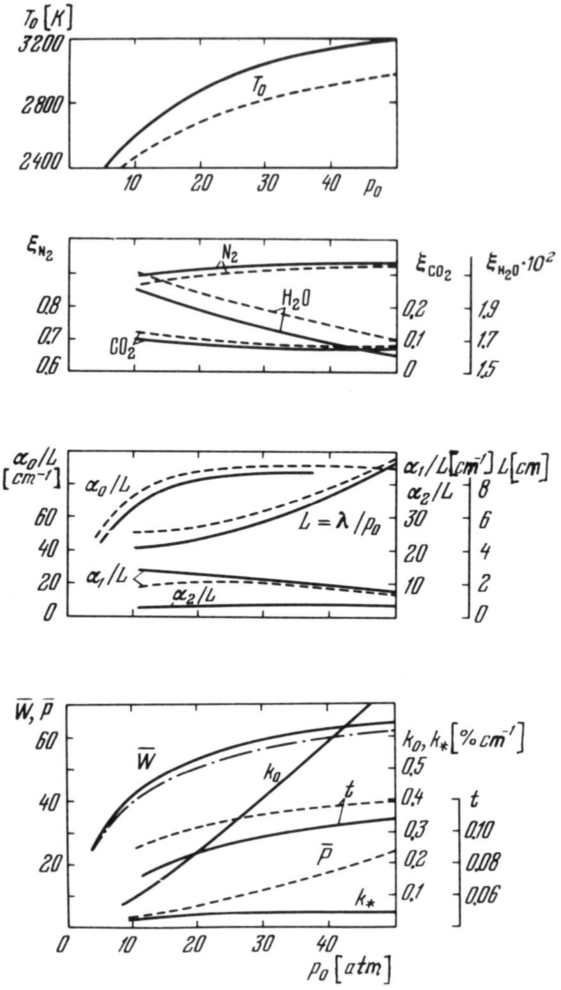

Fig.5.21. The same as Fig.5.20 for a CO_2-N_2-H_2O mixture

show in which direction and how far we can deviate from the optimum by simplifying the setting up of an experiment without a considerable loss in power. Dash lines in Fig.5.20 correspond to T_0, α_0/L, and other parameters providing 90% of the optimal value of $\bar{P}$, which makes it possible to carry out the experiments in essentially less severe conditions. The value of α_0/L in this case is noticeably below the limiting values used in calculations, and T_0 decreases approximately by 600 K. For considerable deviations from the optimum the level lines get close, which indicates the strong dependence of $\bar{P}$ on the parameters determining it.

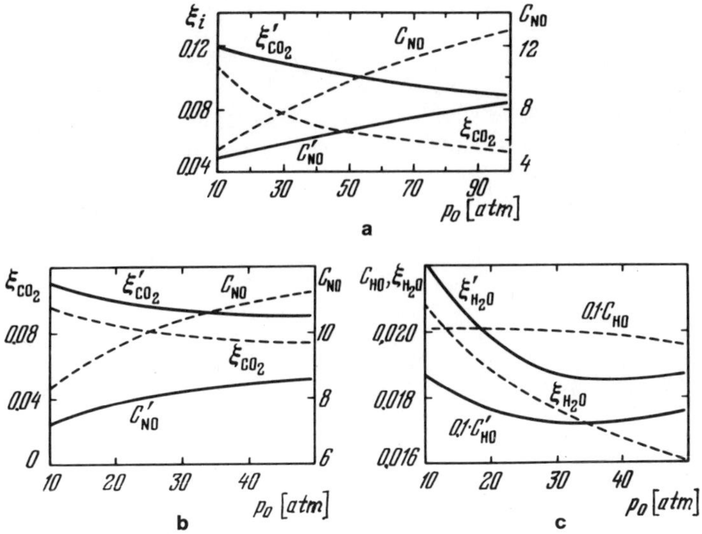

Fig.5.22. Correction for the variation of a chemical composition in optimum conditions (transition from dash lines to solid lines) for CO_2-N_2-He (a) and CO_2-N_2-H_2O mixtures (b, c). C_{NO} and C_{HO} are the ratios of numbers of atoms N:O and H:O, respectively

The nozzle contour obtained in these analyses in fact may not provide shock-free nozzle flow and sufficiently uniform flow at the exit, especially when the parameter α_0 is large, since the obtained nozzle is not contoured (in the sense of Sect.4.3).[6]

To assess the flow quality we must carry out the analysis of a two-dimensional nonequilibrium flow; the computational method must take into account the possible presence of shock waves. As was mentioned in Sect.4.3, such is the shock-capturing method developed in [4.39] and generalized for a nonequilibrium flow in [4.40]. The application of this method is illustrated by the results of [5.59] shown in Figs.5.24 and 5.25 for a two-dimensional flow of the CO_2-N_2-He mixture at p_0 = 10 atm for optimum values of all other parameters (except for α_0/L = 12.25, corresponding to the relative power $\bar{P}$ which is 10% below the optimum value; see Fig.5.20). To the same value of α_0 in a two-dimensional flow there correspond different fixed values of the divergence half angle θ and of the ratio $\delta = h_*/2L$, where L is the length of the supersonic nozzle [according to (5.2.1), $\alpha_0 = \tan \theta_0/\delta$]. Figure 5.24 shows

6 This difficulty can be avoided in solving the optimization problem by varying the shape of the already contoured nozzle.

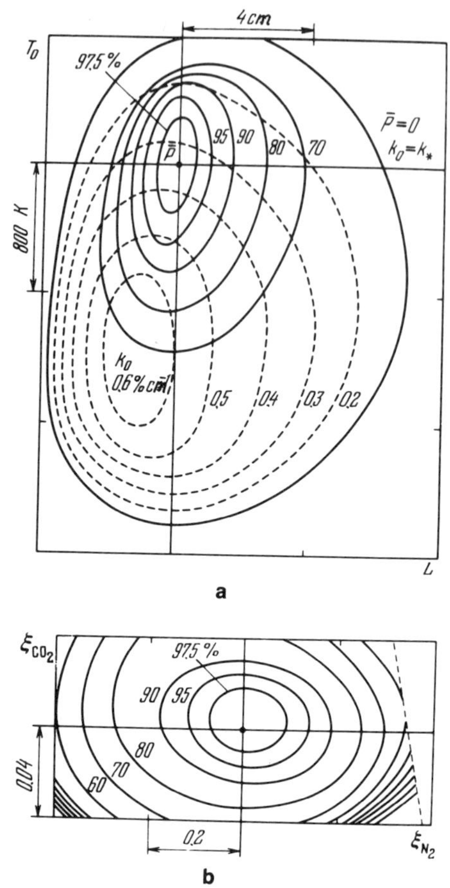

Fig.5.23. The relative power $\bar{P}$ (solid lines, in percent) and small-signal gain k_0 (at the cavity entrance, dashed lines, in % cm^{-1}) levels are plotted as function of coordinates T_0 and L(a) and as a function of ξ_{CO_2} and ξ_{N_2} (b)

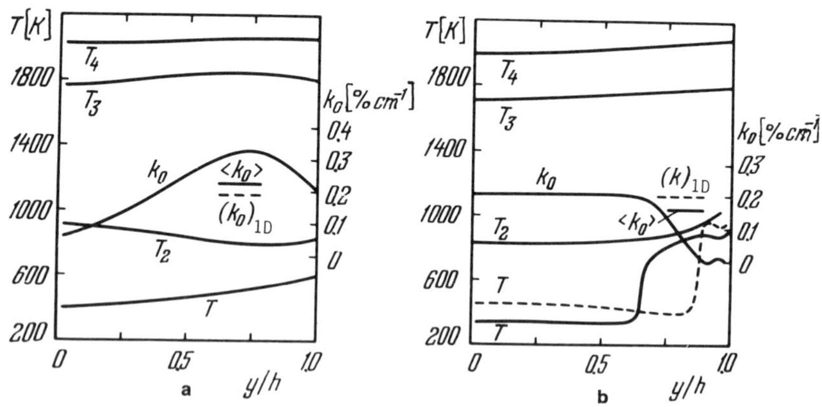

Fig 5.24a,b. Parameter distribution across the flow at the optimum nozzle exit. (a) $\theta_0 = 24^0$; (b) $\theta_0 = 36^0$

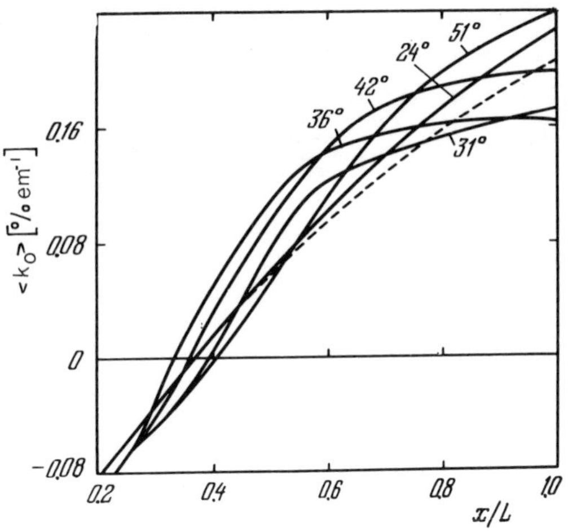

Fig.5.25. Small-signal gain averaged over the flow cross section $<k_0>$ for different angles θ_0 (numbers at the curves). The dashed line corresponds to a one-dimensional flow

the distribution of temperature T, T_i and the small-signal gain k_0 across the nozzle exit cross section (from the symmetry axis y = 0 to the nozzle wall y = 1) for $\theta_0 = 24^0$ (a) and 36^0 (b). The values of $<k_0>$, i.e., the small-signal gain averaged over the cross section, and $(k_0)_{1D}$ for a one-dimensional flow are plotted for comparison. From Fig.5.24b it is clear that a shock wave appears in the flow (dash line shows the distribution of T over the nozzle cross section at a distance x = 0.6L). A considerable increase of T and the growth of T_2 result in k_0 dropping to zero at the wall. The shock wave intensity increases with the initial angle θ_0, and the shock wave is "pressed" to the nozzle wall. The distribution of $<k_0>$ over the nozzle center line is shown in Fig.5.25. At an angle $\theta_0 \leqq 20^0$ shock waves in this nozzle are absent and $<k_0>$ is close to $(k_0)_{1D}$. As θ_0 increases, the difference between them becomes noticeable and for this nozzle flow it changes sign. This is due to the change in the configuration of the shock waves and of the regions where the gas is heated by these waves. In general, however, it should be noted that the shock waves appearing in this case in the flow do not deteriorate the operating conditions of a GDL (this is confirmed by experiments [4.29,30]).

The method of analysis of the output power considered above does not permit, however, evaluation of the power quickly and requires the numerical solution of the problem. For estimates it is sufficient to have approximate analytical expressions for power depending on the state of an active medium at the cavity entrance and cavity parameters. The maximum output power P_{max} can be obtained by determining the total vibrational energy ε_{II} of the second combined block of vibrational states CO_2 $(\nu_3)+N_2$ (see Sect.5.1) for the conditions at the cavity entrance, minus the part of the energy ε_{II} which corresponds to $k < 0$, where the inversion is absent and there is no lasing even for small losses. In this case only that part of the energy can be converted into radiation which is determined by the ratio η_q of the quantum of laser radiation to the vibrational quantum of the upper laser level (quantum efficiency); here $\eta_q = (\theta_3 - \theta_1)/\theta_3$, i.e., 0.41. According to Sect.5.1, $k = 0$ for $T_3 = \theta_3 T_1/\theta_1$, and P_{max} is obtained for $T_1 = T$. Thus

$$P_{max} = \eta_q G[\varepsilon_{II}(T_3) - \varepsilon_{II}(T_3')] \quad , \tag{5.3.4}$$

where G is the gas flow rate according to (4.1.1). The values of P_{max} are used to assess the upper limit of the output power of a CO_2 gasdynamic laser (Fig.5.26).

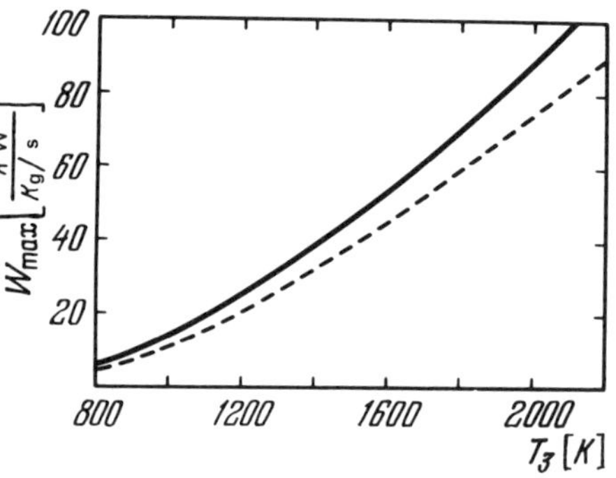

Fig.5.26. Effect of vibrational temperature $T_3 = T_4$ for T = 300 K on maximum specific power in 10% CO_2 - 90% N_2 mixture (solid line) and 10% CO_2 - 40% N_2 - 50% He mixture (dashed line)

The assumption about the constancy of all gasdynamic parameters in the cavity flow simplifies the determination of the power taking account of losses and the effect of saturation in the constant-gain approximation and in the approximation of constant radiation intensity along the cavity flow. The solution of this problem and analytic expressions for P are given in Appendix N. In the limiting case of a very fast energy supply from nitrogen molecules to the upper laser level of CO_2 according to (N.9,10), the power can be assessed by the formula

$$P = P_{max}\eta_r \quad , \tag{5.3.5}$$

where P_{max} is calculated from (5.3.4), and η_r is the resonator (or cavity) efficiency (2.3.34) or (2.3.35).

To extract the maximum power from the active medium there is no need for a too long (downstream) cavity: the maximum mirror length a_m is limited by the condition of existence of radiation determined by de-excitation and collisional deactivation; a_m is determined by the balance between the population of the upper level and the energy losses for radiation and collisional deactivation. In the constant-gain approximation a_m is determined from (N.5). The value of a_m can be estimated by the deactivation time for the upper laser level, τ_m, in the approximation of the model (5.1.13); then

$$a_m \approx \frac{u(\xi_{CO_2} + \xi_{N_2})}{p(\xi_{CO_2} k_{32}^0 + \xi_{N_2}/\tau_4)} \tag{5.3.6}$$

(notations were given in Sect.5.1). From (5.3.6) it follows that at the most typical conditions for a CO_2 gasdynamic laser a_m amounts to tens of centimeters.

The main part of the laser energy is contained in nitrogen, and the contribution of the CO_2 vibrational energy can often be neglected. In the approximation $k_* = \text{const}$ for a maximum-length cavity a_m (N.5) for the optimum transmittance (2.3.33), we have according to (N.8)

$$P = \eta_q G \varepsilon_4^0 (1 - \sqrt{\alpha/k_0 d})^2 \quad , \tag{5.3.7}$$

where $\varepsilon_4^0 = \xi_{N_2} R\theta_4 e_4^0/\mu$ is the vibrational energy of nitrogen (per gram) at the cavity entrance; the loss factor α is calculated for one pass. Relations (5.3.5,7) coincide with the results of the theory of the output power for gas lasers with nonmoving isotropic active media [2.21,22]. This theory was successfully employed to assess losses and saturation effect in gasdynamic lasers (for example, in [5.22b]). The power extraction analysis in the $k_* = \text{const}$ approximation according to (5.3.7) for CO_2-N_2-H_2O mixtures in a wide range of initial pressures and temperatures was carried out in [5.60]. Equations for P, accounting for vibrational exchange kinetics, are given in Appendix N.

To obtain effective transformation of energy into radiation in an unstable resonator, we can also limit ourselves to a certain maximum downstream mirror length, which does not exceed a_m (5.3.6) [2.30,42]. In this case it is important to properly select the ratio between the length a_1 of the "oscillator" part and a_2 of the "amplifier" part of the unstable resonator (see Fig.2.16). For the mixture of 15% CO_2 + 83.5% N_2 + 1.5% H_2O for u = 1500 m/s, p = 0.1 atm and T = 350 K, according to [2.42], the optimum length of the "amplifier" is a_2 = 15 cm. If we fix this value for a_2, the

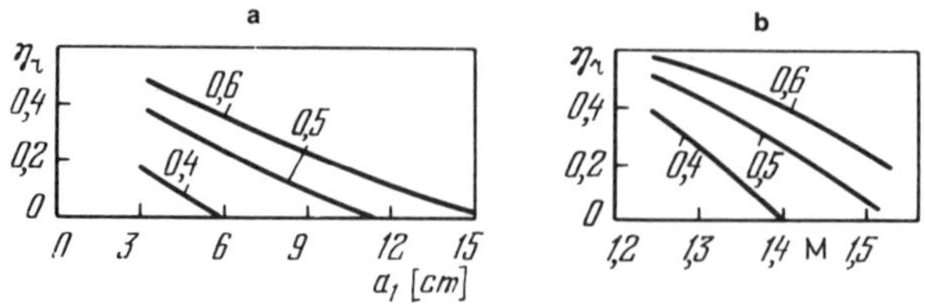

Fig.5.27. Effect of the "oscillator" part length a_1 (a) and magnification M (b) on the efficiency of an unstable telescopic resonator with a fixed length of "amplifier" part a_2 = 15 cm. Numbers at the curves give values of k_0d at the resonator entrance

efficiency of the entire unstable resonator drops with an increase of a_1 (Fig.5.27). The reduction of the length a_1 leads to deterioration of the beam quality, so that it is expedient to take a_1 at about one order of magnitude lower than a_m. In this case resonator losses should be such that k_0 exceeds twice (ln M - ln r)/d, r being the reflection coefficient and M the magnification of the resonator (see Sect.2.4). The nonuniformity in the intensity distribution along the resonator flow is more noticeable for spherical mirrors than for cylindrical ones [2.42]. To obtain a more uniform intensity distribution and thus to improve the beam quality, we should decrease the cavity efficiency.

5.4 Combustion-Driven Gasdynamic Laser. Effect of Impurities

The basic method of thermal excitation of a CO_2 gasdynamic laser, which provides a high laser output power in a continuous regime, is the combustion of substances yielding the required components of a laser mixture at a sufficiently high temperature and pressure. The highest power of laser radiation (above 200 kW [5.61]) with continuous operation had been obtained by this method. Carbon dioxide is one of the basic combustion products of many fuels. During the combustion, however, often too much water vapor and too little nitrogen are formed. For this reason in CO_2 gasdynamic lasers either the fuel mixture is diluted with nitrogen or use is made of oxidants, such as air or nitrous oxide [1.6, 5.62-66]. A good ratio of the mixture components was obtained upon combustion of the $CO-H_2$ mixture in O_2 with an admixture of N_2 or in air. Combustion of CO was utilized in gasdynamic lasers described in [1.22, 5.35,52,53,57,67]. A sufficient amount of water vapor in the combustion products can be obtained without hydrogen from the combustion of a small amount of methane [1.22, 5.52]. Laser power can be considerably increased if nitrogen added to the mixture is preheated (up to 1000 K [5.52]). Acetylene, ethylene, propane, benzene, natural gas [4.10, 5.35,41,56,64-67], and other substances (see [1.2, 5.63,68]) can be used as GDL fuels.

The selection of the initial fuel mixture is determined by many factors, among which, in addition to the best ratio of laser mixture components, T_0 and p_0, we must mention the engineering and economic expediency and low toxicity, which is extremely important in the development of continuously operating high-power lasers [1.2]. Let us consider the composition of mixtures obtained as a result of combustion of a fuel and used for creating an active medium. Both theoretical estimates and experimental results (see, for example, [5.56]) show that the state of the mixture obtained as a result of combustion in front of the nozzle entrance is often close to equilibrium. The analysis of the equilibrium composition for a given initial ratio of the number of atoms of different chemical elements in the initial mixture and for a certain temperature and pressure can be carried out by the method described in Appendix G. Of greatest interest here are systems containing C, N, O, and H atoms. The results of analysis for such a system are shown in Figs.5.28 to 5.30 [5.69]; in this analysis the possibility of formation of a condensed phase (carbon black) was taken into account. Figure 5.28 gives a ternary diagrams of the content of C, O, and H atoms in a mixture for the fixed fraction of N atoms equal to 0.6, for p_0 = 10 atm and T_0 = 2500 K. The set of points corresponding to similar concentrations of one of the components form the lines of equal concentrations. Figure 5.29 shows the lines of equal ratios $\xi_{CO_2}:\xi_{H_2O}$. Some of them correspond to the combustion products of certain fuels with respect to the ratio of the number of atoms C:O in the initial mixture. The dot-and-dashed line here is the line of the ternary mixture of CO_2-N_2-H_2O, in which there is a small amount of other admixtures. The results considered in previous sections were obtained precisely for the conditions along this line. For comparison the values of the gain coefficient k_0 determined in [5.53] are plotted on this line. The equilibrium composition at the points above this line are basically determined by the reaction $CO + H_2O \rightleftharpoons CO_2 + H_2$; CO and H_2 in this case are the main admixtures in the CO_2-N_2-H_2O mixture. Above these points equilibrium compositions already contain carbon black. It appears in noticeable amounts at the ratio C:O exceeding unity, which imposes limitations on the selection of the ratio between the fuel and the oxidant. Below the ternary mixture line O_2 prevails as an admixture. In addition to those mentioned above, equilibrium compositions contain some other admixtures; their molar fraction, however, is small. This is presented on the plots of Fig.5.30, where the temperature

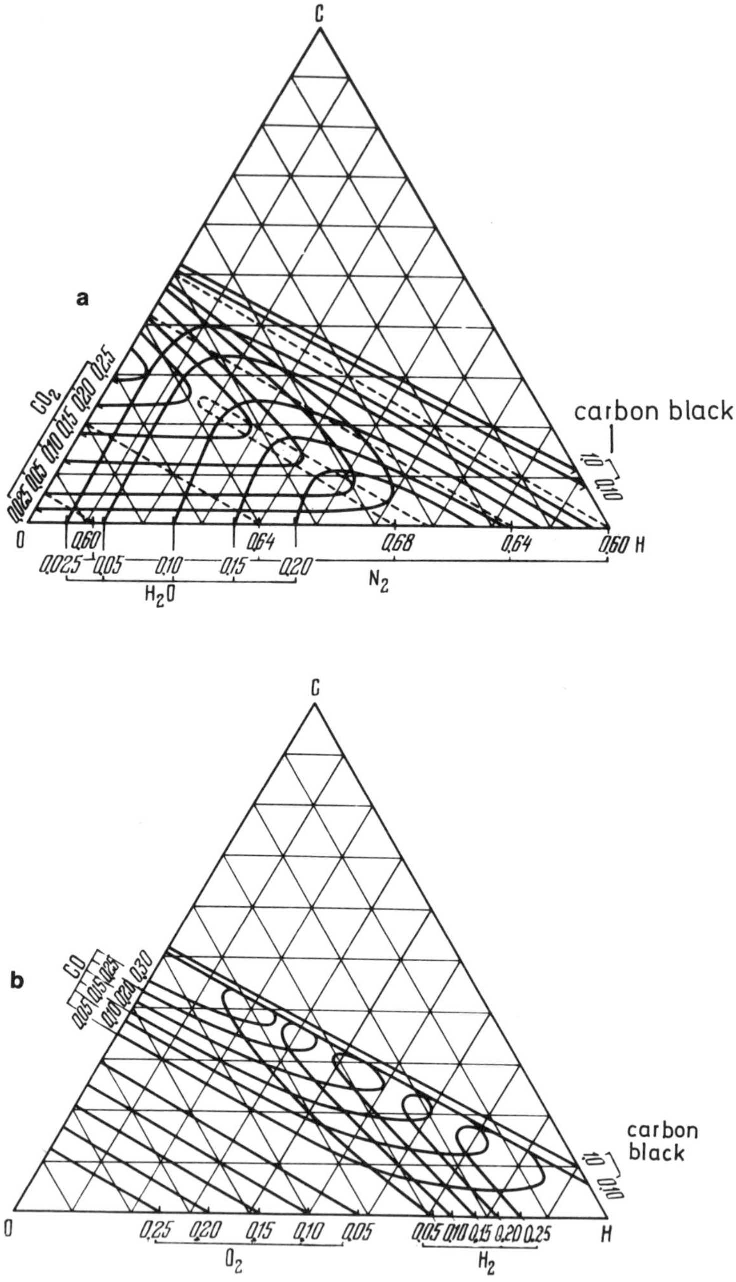

Fig.5.28a,b. Calculated N_2, CO_2, and H_2O (a) and CO, H_2, O_2 (b) concentrations in mole fractions and carbon black concentration (in g-atoms per kg mixture) for T_0 = 2500 K, p_0 = 10 atm

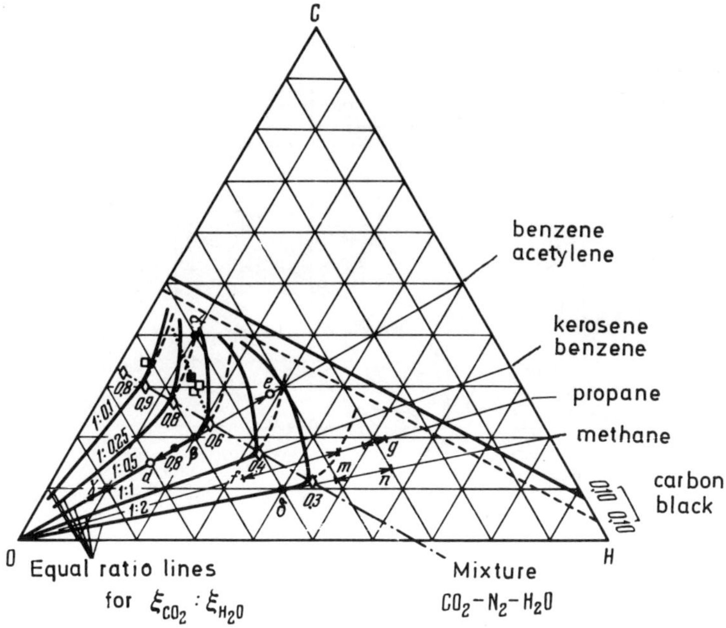

Fig.5.29. Molar concentration ratios CO_2:H_2O of triple mixtures CO_2-N_2-H_2O for T_0 = 2500 K, p_0 = 10 atm; for carbon black, the solid line corresponds to T_0 = 2500 K, dashed lines, to T_0 = 1500 K. Other notations are given in the text

dependence of the molar fractions of the components is shown in the temperature range from 1500 to 2500 K for the compositions indicated in Fig.5.29 by the points α, β, γ, δ. In this temperature range for pressure p_0 = 10 atm the content of the main components of the mixture changes little. These results reveal all the basic regularities, which are valid for other values of p_0, and also for a different nitrogen content in the mixture, since at these temperatures nitrogen is not appreciably reactive. Figure 5.29, in addition to the points obtained in [5.53], shows the regions of the diagram for which the values of k_0 acceptable for the GDL performance have been obtained. Thus, segment de had been analyzed in [5.41] with air-acetylene mixtures, and at point β the value of $k_0 \approx 0.6\%\ cm^{-1}$ was obtained. In [5.70] benzene-air mixtures were studied on the same line, and for T_0 = 1600 K it was found that $k_0 \approx 0.8\%\ cm^{-1}$. In [5.56] propane-air mixtures were investigated on segment fg ($k_0 \approx 0.4\%\ cm^{-1}$) and on mn, the mixtures containing methane ($k_0 \approx 0.3\%\ cm^{-1}$). The dotted line corresponds to the results of experiments [4.10], in which acetylene mixtures with CO, N_2, and O_2 had been studied. Squares in the figure indicate the laser radiation region observed in [5.71], which was ob-

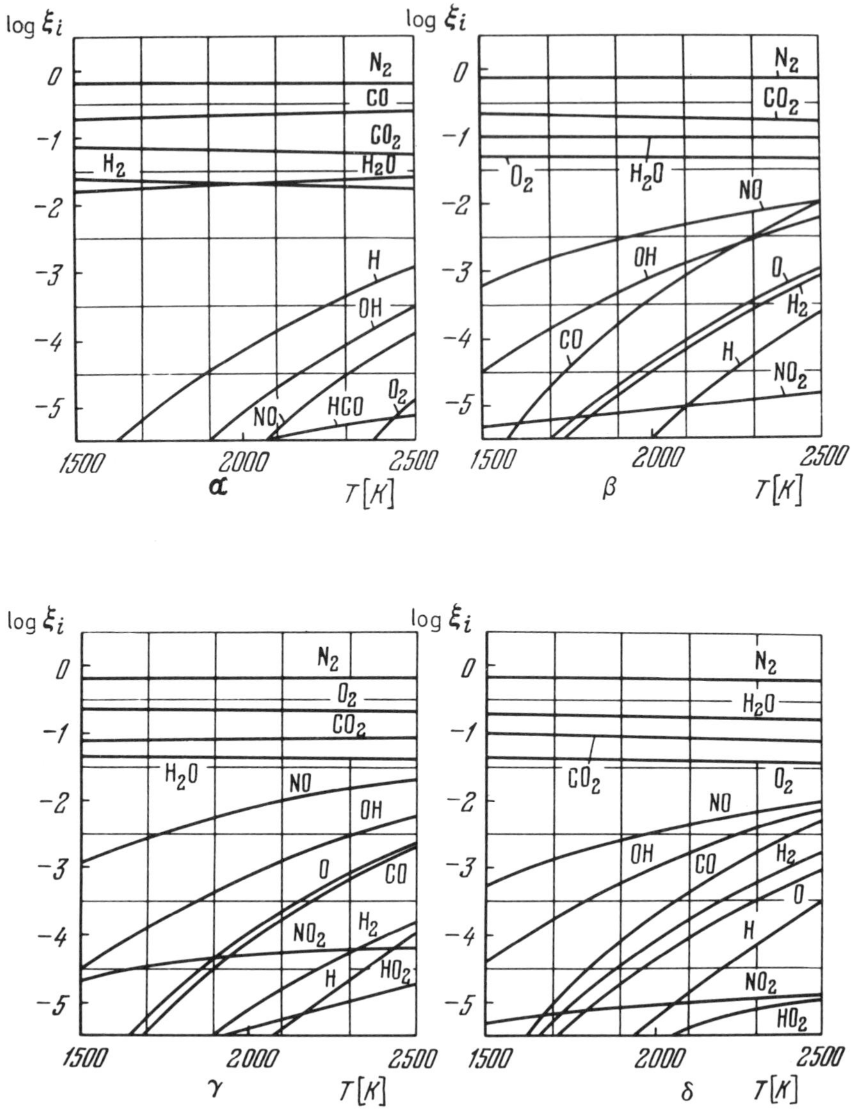

Fig.5.30. Effect of reservoir temperature on mixture composition for p_0 = 10 atm at points α, β, γ, δ in Fig.5.29

tained with combustion products of different solid nitrocompounds with air admixtures (light squares) and without them (black squares).

In the analysis of the flow for certain mixtures of combustion products in the nozzle and in the cavity similar diagrams can be plotted, in which the lines of equal levels of the small-signal gain k_0 (maximum along the flow) and of the specific output power $\bar{W}$ [J/g] can be plotted for the plane-parallel Fabry-Perot resonator with the distance between the mirrors $d = 2$ m and losses per pass $\alpha = 6\%$ (Fig.5.31 [5.72,73]. The calculations were carried out for a two-dimensional contoured minimum length nozzle with throat height $h_* = 0.2$ mm and $S/S_* = 30$. In the analysis it was assumed that the fraction of nitrogen atoms was 0.6, $p_0 = 20$ atm and $T_0 = 1600$ K. It is interesting that according to these calculations the highest $\bar{W}$ were obtained when in addition to CO_2, N_2, and H_2O, either O_2 or CO were present in these mixtures. It is clear from Figs.5.29,31 that methane and other fuels with high ratios H:C, which give large amounts of water vapor during their combustion, are less effective.

Too large amounts of water vapor are obtained during the explosion of solid substances; for example, the water vapor content in the detonation products of solid hydrazoic acid HN_4 in the mixture with CO_2, NO_2, and Xe, which had been used in the first GDL of the detonation type [1.25], reached 15 to 20%. The use of other explosives makes it possible to decrease this value [5.63,71,74]. Thus, by using an explosive on the basis of sodium azide NaN_3 with an admixture of other components [5.63], we can reduce ξ_{H_2O} to 1.6%, but in this case other components appear in the explosion products (and among them sodium), which can affect the state of the active medium. The estimates [1.2, 5.75] indicating a considerable decrease of the small-signal gain k_0 due to the presence of sodium in a mixture were, however, exaggerated, since they were based only on the theoretical values of the deactivation probability of N_2 by sodium atoms. In experiments [3.99] considerably lower effectiveness of this process (by a factor of 25) was observed [the deactivation rate constant for N_2 at $T = 300$ to 330 K was $k(Na) = 7 \times 10^5$ atm^{-1}s^{-1}]. The utilization of explosives and solid fuels (such as solid carbon) is a promising trend in the development of gasdynamic lasers (see Fig.5.32) [1.2, 5.63]. Another example of the source of hot nitrogen for a GDL is a burner in which atmospheric nitrogen is heated during oxidation of dispersed iron particles by oxygen, yielding F_3O_4 [5.63].

Figure 5.32 shows the effectiveness of some fuels and oxidants. It shows the results of computation of the maximum power P_{max} according to (5.3.4), which can be obtained in a resonator mounted at the nozzle exit [5.68]. The water vapor content here is taken at 1%, except for benzene, where $\xi_{H_2O} = 5\%$, which noticeably effects the value of P_{max}. The results of analysis for a solid fuel based on sodium azide are also presented here [5.63] with no account of the deactivating effect of the formed admixtures, Na, NaOH, CO,

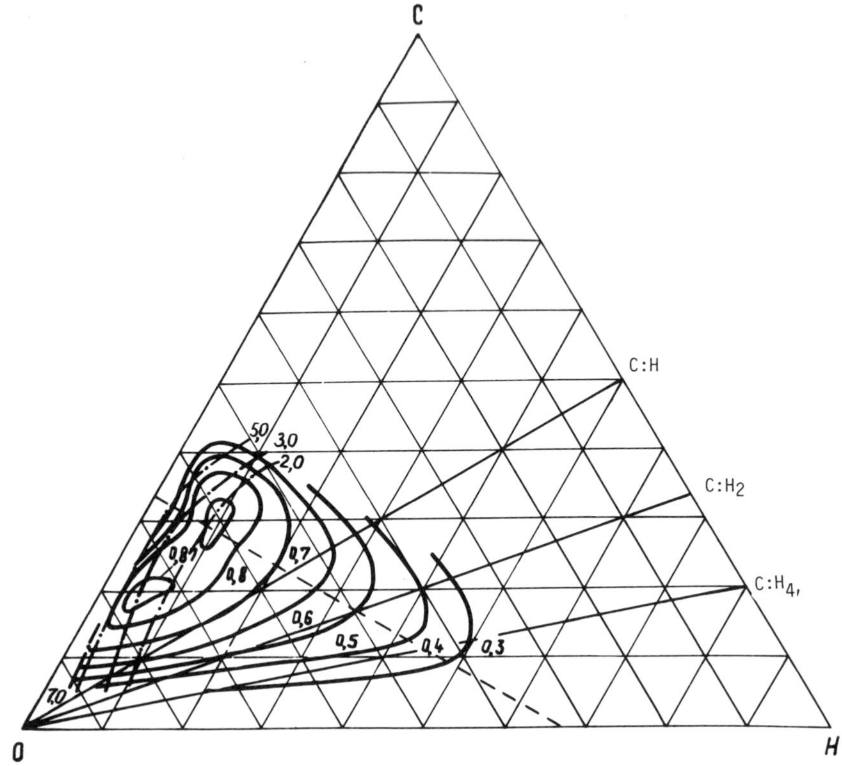

↑ ⟶

Fig.5.31. Small-signal gain k_0 (a) and specific power $\bar{W}$ (b) in a combustion-driven GDL using hydrocarbon fuels with a fixed ratio of C:H and a fraction of N atoms of 0.6; p_0 = 20 atm, T_0 = 1600 K. Dashed lines correspond to triple CO_2-N_2-H_2O mixture. Dashed-and-dotted lines in (a) are maximum k_0 downstream (in cm from the nozzle throat)

SO_2, SO, with a total content up to 18% (note that in experiments [5.76] it was found that SO_2 had a negative effect on the small-signal gain).

The major admixtures in the combustion products of hydrocarbon fuels, according to Figs.5.29,30 are carbon monoxide, oxygen, and hydrogen. Experiments [4.10] with a GDL driven by acetylene combustion with CO showed that the substitution of carbon monoxide for nitrogen results in a decrease of the small-signal gain. With complete substitution of CO for N_2, however, the gain remains positive but small. Similar results were obtained when carbon monoxide was added to mixtures of CO_2-N_2-He [5.36,76] and CO_2-N_2-H_2O [5.32]. In the latter case the effect of O_2 and H_2 admixtures to the 15% CO_2-83% N_2-2% H_2O mixture was studied. The results are shown in Fig.

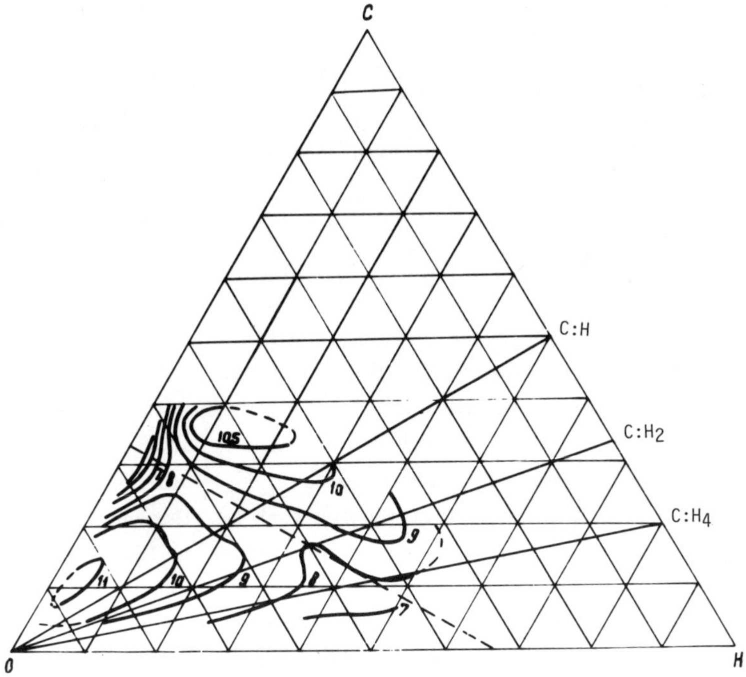

Fig.5.31b

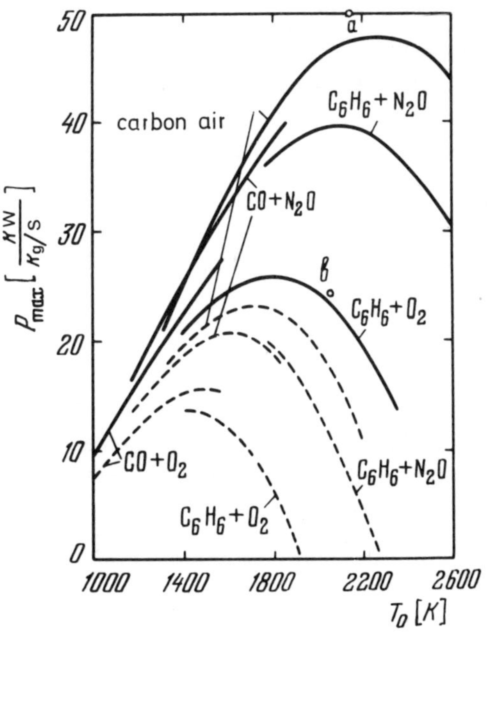

Fig.5.32. Effect of temperature at the nozzle entrance T_0 on maximum values of specific power which can be obtained usind indicated fuel mixtures and a solid fuel based on sodium azide (TT). Solid lines and point a correspond to a nozzle with $h_* = 0.3$ mm, $S/S_* = 50$ for $p_0 = 37.5$ atm; dashed lines and point b correspond to a nozzle with $h_* = 1$ mm, $S/S_* = 20$, $p_0 = 20$ atm

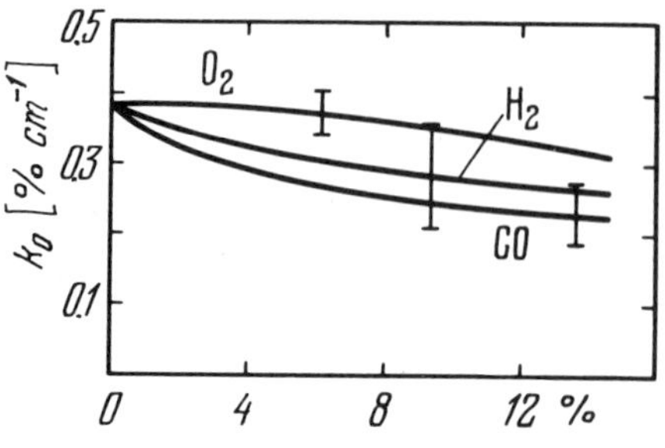

Fig.5.33. Effect of O_2, H_2, CO admixtures on small-signal gain k_0 in $CO_2:N_2:H_2O$ = 15:83:2 mixture in a wedge nozzle flow (h_* = 2 mm, $\theta = 15^0$) at a distance of 5.6 cm from the throat (T_0 = 1800 K, p_0 = 5.3 atm)

5.33 by the curves drawn through the experimental points using the least-squares method. It is clear that small admixtures of O_2 and H_2 have little effect on the small-signal gain. The replacement of half of the nitrogen by oxygen in the 10% CO_2-87% N_2-3% H_2O mixture decreases k_0 almost by half [5.77]. The complete replacement of N_2 by CO leads to k_0 = 0.02 to 0.1% cm^{-1}. and it is difficult to effect laser radiation for such k_0. The analysis of k_0 taking account of admixtures using the equations similar to (5.1.2) in [5.78,79], (5.1.11) in [4.9], and (5.1.13) in [5.36] made it possible to explain the experimental results by a decrease of pumping to the upper laser level of CO_2 and by an increase of the deactivation rate of N_2 due to the energy exchange with the added molecules.

Active medium characteristics in a CO_2 gasdynamic laser can be improved by using certain admixtures. Thus, the addition of chlorine to the combustion products of methane mixtures noticeably increases the small-signal gain [5.80]. When chlorine is introduced into the methane-oxygen-nitrogen mixture, water vapor molecules in the combustion products are replaced by hydrogen chloride molecules. According to calculations, the addition of 15% Cl to the mixture of $CH_3:O_2:N_2$ = 1:2:9.4 reduces ξ_{H_2O} in the combustion products from 17 to 2.5%. Experiments [5.80] confirmed the effectiveness of this admixture up to the value which is stoichiometric with respect to hydrogen. The utilization of chlorine, however, increases the toxicity of laser mixtures and makes them chemically aggressive.

Of great importance is the problem of the complete replacement of water vapor and helium by other gases, which could be as effective in creating the population inversion in a CO_2-N_2 mixture. One such possibility is the utilization of hydrogen, which actively affects the kinetics of population of the CO_2 vibrational levels; this was confirmed in experiments [5.29,32,62]. Experiments [5.81] proved the possibility of laser performance on CO_2-N_2-HCl mixtures. The effect of addition of H_2 and HCl on the small-signal gain and the optimum content of these gases in the mixture proved to be the same as

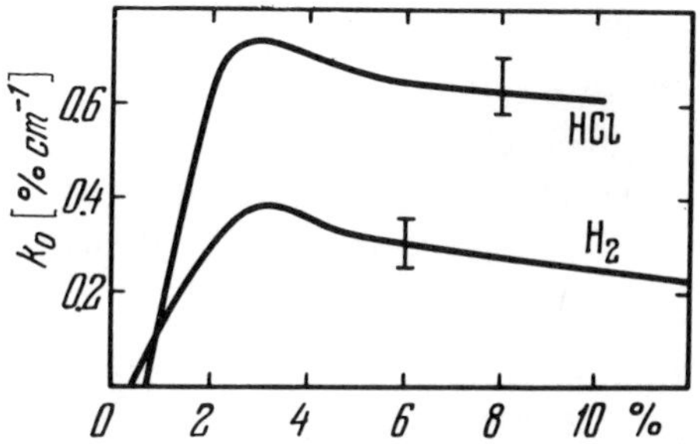

Fig.5.34. Variation of small-signal gain k_0 with addition of H_2 to CO_2:N_2 = 3:17 mixture (2 mm throat height, $S/S_* = 16$, $T_0 = 1800$ K, $p_0 = 5.3$ atm [5.32]) and of HCl to 10% CO_2 mixture with nitrogen (1 mm throat height, $S/S_* = 15$, $T_0 = 1500$ K, $p_0 = 10$ atm [5.76])

for water vapor (Fig.5.43), which can be explained by the same mechanism of admixture influence on the lower laser level. Amplification and laser radiation in a GDL working on a CO_2-D_2 mixture was observed in [3.35].

It was pointed out in Sects.2.5 and 3.7 that when a GDL operates, carbon black, water drops, and other condensed particles may occur in the active medium. According to [2.50], the heterogeneous relaxation in the nozzle of a CO_2 gasdynamic laser on the particles above 10^{-6} cm in diameter with a weight concentration of about 1% is small in comparison with the homogeneous relaxation, and in the cavity flow, when the rates of these processes may become equal, the role of collisional deactivation becomes insignificant. An analysis [3.140] showed that the effect of water condensation would be noticeable only in mixtures with a high water vapor content (above 5 to 10%), if the initial temperature is not high ($T_0 \leqq 1200$ K). The addition of water vapor to the mixture of CO_2-N_2-He (where the cooling rate is high) made it possible in experiments [5.82] to establish a considerable (to one-third) decrease of k_0 in the region where, according to the analysis [5.83], the intensive formation of condensate drops occurred. According to [3.140], losses due to absorption or dispersion of laser radiation by water aerosol particles in ordinary operating conditions of a GDL are insignificant. These losses may become noticeable only when the mixture pressure increases.

5.5 Supersonic Mixing in Gasdynamic Lasers

To increase the effectiveness of CO_2 gasdynamic lasers a scheme was suggested [1.17, 5.84] and realized for the first time [5.85] of separately heating the exciting gas, nitrogen, and then mixing it with cold radiating gas, CO_2, already in the supersonic nozzle flow. Later this scheme was developed considerably. Its advantages include the absence of energy expenditures for preheating CO_2 molecules (particularly, many levels of the symmetrical modes of the CO_2 vibrations, which determine the population of the lower laser level), and also molecules and atoms of admixtures (helium, water vapor), and the possibility of a considerable increase of the nitrogen temperature without losses for CO_2 dissociation. It is also important that the vibrational state of undiluted nitrogen is easily frozen during cooling in the supersonic nozzle.

The main problem in the realization of this scheme is to provide fast mixing of nitrogen with carbon dioxide for the transfer of the vibrational energy from N_2 molecules to the upper laser level of CO_2. The characteristic length of mixing, ℓ, is close to [5.86]

$$\ell = \kappa uF/\mathcal{D} \quad , \tag{5.5.1}$$

where F is the cross-sectional area of a mixer, u is the velocity of the flow, $\mathcal{D}$ is the diffusion coefficient, and κ is the constant determined by the degree of mixing. In a laminar flow $\mathcal{D}$ is equal to the molar diffusion coefficient D and $\kappa = 0.1$ for a 10% nonuniformity of the relative concentration of the averaged components $\Delta\bar{\xi} = (\xi_{max} - \xi_{min})/\xi_m$. Since D is very small, for a laminar flow, according to (5.5.1), mixing proceeds very slowly. To increase the rate of this process, the flow must be made turbulent; in practice for this purpose bluff bodies are introduced into the flow or gas streams are injected across the flow. To evaluate ℓ for a turbulent mixing (without liberation of heat) (5.5.1) can be used, but for $\mathcal{D}$ we should take the turbulent diffusion coefficient D_t, which depends in a complicated way on the nature of the flow, and in determining κ account should be taken of the "instantaneous" nonuniformity of concentration ξ' due to turbulent pulsations. According to [5.86], $\kappa = 50$ for $\Delta\xi_\Sigma = 10\%$, $\kappa = 10$ for $\Delta\xi_\Sigma = 32\%$, and $\kappa = 1$ for $\Delta\xi_\Sigma = 350\%$, where $\Delta\xi_\Sigma = \Delta\xi^2 + 16\xi'^2$, and D_t can be determined from the empirical formula

$$(D_t/u\sqrt{F}) = 4.2\times10^{-2}\sqrt{c_f\varphi} + 1.5\times10^{-2}\sqrt{\rho_{in}u_{in}^2S_{in}/\rho u^2F} \quad , \tag{5.5.2}$$

where c_f is the friction coefficient for a body introduced into the flow to make it turbulent, φ is the extent of the blockage of the flow by this body, ρ_{in} is the density of the injected gas, S_{in} is the cross-sectional area of the source opening, and u_{in} is the velocity component of injection, perpendicular to the main flow. The first term on the right-hand side of (5.5.2) describes the contribution to D_t from bluff bodies. The friction coefficient c_f for such bodies varies from 0.5 (for a sphere) to 1.2 to 1.5 (for a plate placed across the flow), and for a cylinder $c_f = 1$. The extent of the blockage φ equals the ratio S_m/F, where S_m is the maximum cross-sectional area of a body in the flow (midsection area). The second term in (5.5.2) is determined by the streamline flow of injection (along the component perpendicular to the main flow). As the velocity of the main flow increases, the effectiveness of the turbulent mixing drops, so that in supersonic flows, according to (5.5.1,2), good mixing is usually obtained only over the length of the order of (50 to 100) h. The decrease of height h and of extent of nonuniformity during injection promotes the reduction of this length, which is important to provide for the optical homogeneity of the cavity flow. Note that the shock waves appearing during injection may result in the noticeable increase of temperature and deactivation of nitrogen molecules. Numerical modelling of mixing in a supersonic flow of a GDL in [5.87,88] confirmed the correctness of the estimate according to (5.5.1).

Figure 5.35 shows different schemes of mixing CO_2 into a hot nitrogen supersonic flow. The system in which mixing proceeds in almost parallel supersonic flows introduces the smallest perturbations into the supersonic flow (Fig.5.35a, [5.89,90]); the mixing length in this case, however, is large. In a gasdynamic laser operating according to this scheme [5.89], nitrogen mixed with argon was heated by an electric-arc discharge up to a temperature of about 1700 K and then supplied to the nozzle array at a pressure of about 9 atm. Cold carbon dioxide was simultaneously accelerated through another array (p_0 = 11 atm), and then the flows were mixed. The resulting mixture had the following composition: 27% CO_2-63% N_2-10% Ar. Laser radiation was extracted from two resonators arranged at a distance of 6.2 and 13.5 cm from the exit of the nozzle arrays, and the total specific power was $\bar{W}$ = 1.15 kW $kg^{-1}s$. In [5.85,91] a cold CO_2-He mixture was injected along the normal to the center line into a completely formed supersonic flow of vibrationally excited nitrogen (Fig.5.35b). With the electric-arc heating of nitrogen ($T_0 \approx 2000$ K, $p_0 \approx 1$ atm) the specific power in the 20% CO_2-40% N_2-40% He mixture amounted to 0.83 kW $kg^{-1}s$ [5.85]. Experiments [1.32,33, 5.38,92,93] showed that the most effective are the systems in which the additional expansion of the flow occurs after injection. For this purpose a cold gas is supplied into the flow near the throat either from the additional central pilon (Fig.5.35c) or from the wall (Fig.5.35d). In [1.32] the tubes with holes were inserted across the nozzle throat. In these schemes the small-signal gain k_0 can be considerably increased and it can reach record values for CO_2 gasdynamic lasers (up to 2.6% cm^{-1} for nitrogen heated by a shock wave [5.38,39] and above 3% cm^{-1} for an electric-arc heating [1.33]). The small-signal gain distribution across the flow is determined by the type

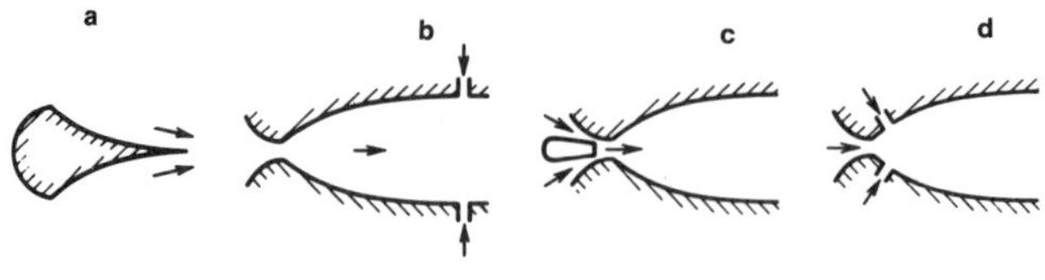

Fig.5.35a-d. CO_2 admixing systems

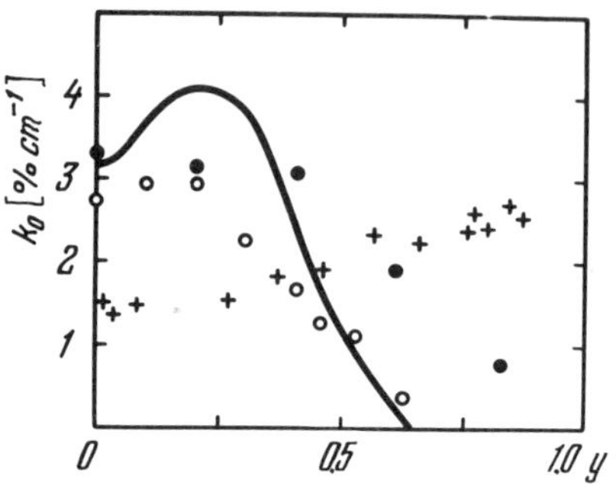

Fig.5.36. Measured small-signal gain distribution across the flow during injection (dots -- [1.33], crosses -- [5.38,39]); solid line -- theoretical curve

of injection system: with the injection through the wall k_0 is maximum near the wall, and vice versa (Fig.5.36). With the injection at the nozzle center line k_0 can be increased at the periphery of the flow (black circles in Fig.5.36) if the injection openings are oriented at an angle to the center line. The increase of pressure of the injected mixture also promotes this effect. In [5.94] mixing occurred during supersonic CO_2 injection perpendicular to the main flow at the edge of the slit through which nitrogen heated in a high-power induction discharge flowed. Laser radiation was obtained when CO_2 was injected into a D_2-He mixture [5.95].

If a flow has large dimensions, then mixing occurs only at a considerable distance downstream, and the uniformity of the flow is greatly distorted. By decreasing the characteristic dimensions of the main and the injected flows, we can achieve more rapid mixing and improve the homogeneity of the active medium. For this purpose in [5.96] a mixing gasdynamic laser on the basis of a screen nozzle array [4.58] has been developed. The apparatus used there combined two such screen nozzle arrays: through one of them the main gas, nitrogen, flowed, and a CO_2-H_2O mixture was supplied through the other (Fig.5.37). Nitrogen was heated in a shock tube. The maximum specific power attained in these experiments amounted to 26 J/g; the authors believed, however, that this value could be considerably increased if larger setups were created. It is very important that all gas density perturbations in the flow behind the screen nozzle array are so small that their effect on the

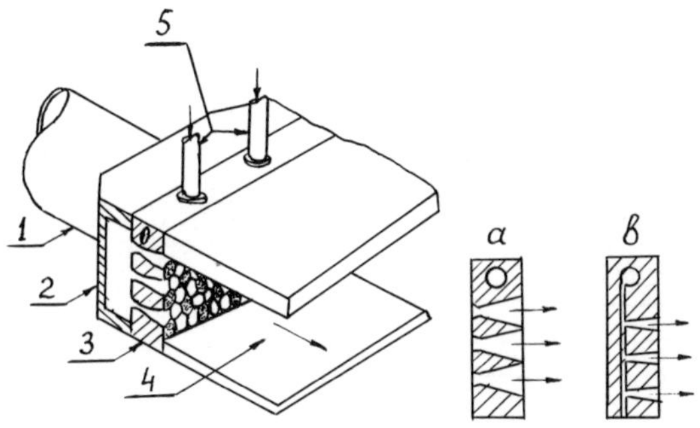

Fig.5.37. Screen mixing nozzle array. 1 -- shock tube; 2 -- transition section; 3 -- nozzle block; 4 -- laser cavity; 5 -- manifolds of CO_2-H_2O; a -- plane of N_2 nozzles; b -- plane of CO_2-H_2O nozzles

optical parameters of the active medium in a resonator can be neglected. The advantages of this scheme were convincingly demonstrated in [5.98].

To determine the optimum operating conditions of a mixing gasdynamic laser, we must vary the composition of the injected mixture and analyze the effect of different admixtures. In [5.91] it was found that when a CO_2-He mixture was injected into nitrogen, the laser power was maximum at the ratio of N_2:He = 1:1. The optimum composition of the injected mixture depends on temperature [5.99]: in low-temperature mixing gasdynamic lasers high small-signal gains were obtained when a mixture with a low helium content was injected. With increasing temperature we must increase the helium content, since this promotes quicker cooling of the mixture being formed. Laser radiation was also observed when nitrogen and carbon dioxide were mixed in the absence of helium; the level of power in this case, however, is several times lower [5.38,39,91]. Laser power can be increased with increase of the fraction of the injected gas in the total flow rate [5.91]. The optimum in this case has not yet been established. The replacement of nitrogen by air reduces the small-signal gain after mixing only by 20 to 30% (for T_0 = 1500 to 6000 K) [5.38,39]. The effect of admixtures of CO, H_2, SO, and other gases into nitrogen in a mixing GDL was also studied in [5.93,96,100,101].

The theoretical analysis of the vibrational exchange during mixing of N_2 and CO_2 and calculation of the small-signal gain can be carried out both for a laminar [5.88] and a turbulent [5.81,97] flow and makes it possible to determine the fields of temperature, the concentrations of the components being mixed, and the small-signal gain (Fig.5.38). In the analysis [5.81] good agreement with experiments [1.33] was obtained for the injection into the central part of a transonic nozzle flow (solid line in Fig.5.36).

The component mixing in a supersonic flow increases the effectiveness of a gasdynamic laser: along with the extracted specific energy (with respect to the nitrogen flow rate) of 20 to 25 J/g [1.32,33, 5.96,98] values up to

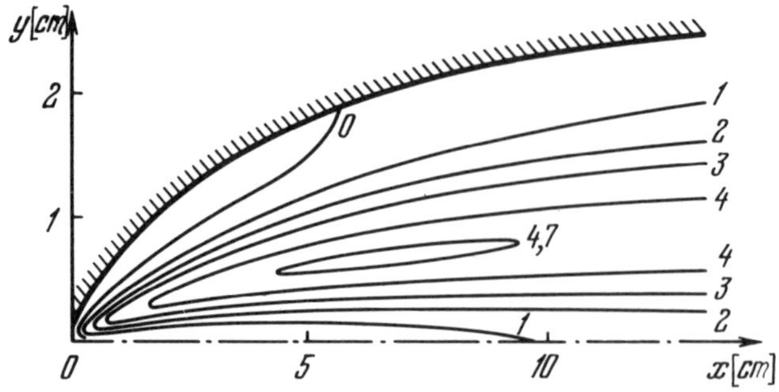

Fig.5.38. Calculated small-signal gain levels (numbers at the curves are in % cm^{-1}) for mixing in the throat region according to [5.81] using computer techniques of [5.97]. Flow proportion is 44% N_2, 39% He, 17% CO_2. Stagnation conditions for N_2 are T_0 = 2500 K, p_0 = 15 atm

50 J/g and record values of the small-signal gain of about 5% cm^{-1} [1.34] were obtained. Further improvement of mixing systems makes this trend of the development of gasdynamic lasers very promising. Particularly, among the new methods of mixing, the injection of the solid CO_2 aerosol into vibrationally excited nitrogen flow [3.138, 5.102] should be mentioned. Two-phase mixing proceeds much quicker than homogeneous mixing; the heat of sublimation of aerosol particles provides additional decrease of the gas temperature. According to [5.102], the specific energy extracted in this case can amount to 120 J/g. Continuous wave CO_2 gasdynamic lasers driven by aftermixing jets through orifice arrays have been analyzed also in [5.103].

5.6 Losses and Efficiency

The efficiency of a gasdynamic laser cannot be high, since the major part of the energy of a compressed and heated gas in the nozzle flow is transformed into kinetic energy. This circumstance makes us investigate more thoroughly all the sources of losses influencing the effectiveness of transformation of the remaining portion of energy into radiation. These losses are mainly due to the vibrational deactivation in the nozzle, and also due to light absorption in the mirrors and incomplete realization of the population inversion on laser levels in the cavity, i.e., they are determined by the design and characteristics of the nozzle and the cavity. The total efficiency of a GDL depends on the efficiency of the heater (η_h), efficiency of thermal excitation (η_t), the nozzle efficiency (η_n) the resonator (cavity) efficiency (η_r), and, finally, on the quantum efficiency (η_q).

The efficiency of a heater η_h is determined by the type of the heater and inevitable energy losses including the losses due to dissociation. In a combustion-driven laser the fuel combustion can be incomplete. Electric-arc heaters in gasdynamic lasers have efficiency η_h exceeding 50% (e.g., in [5.89] η_h = 56%). The question of the heater efficiency is beyond the scope of this book; therefore, we shall compute efficiency assuming that $\eta_h = 1$.

The effectiveness of thermal excitation η_t indicates the fraction of the useful vibrational energy in the total enthalpy of a gas in front of the nozzle, calculated from the initial temperature T_{in} of the heated mixture. Therefore, for mixtures under consideration we obtain, disregarding the vibrational energy of water vapor molecules,

$$\eta_t = (\xi_{CO_2}\theta_3 e_3 + \xi_{N_2}\theta_4 e_4)/[(T_0 - T_{in})(2.5 + 1.5\xi_{H_2O} + \xi_{N_2} + \xi_{CO_2}) + \xi_{CO_2}(\theta_1 e_1 + 2\theta_2 e_2) + \xi_{N_2}\theta_4 e_4] \quad , \tag{5.6.1}$$

where all the values of e_i are taken at the temperature T_0. Figure 5.39 shows the dependence of η_t on temperature T_0 for the most typical mixtures and for pure nitrogen.

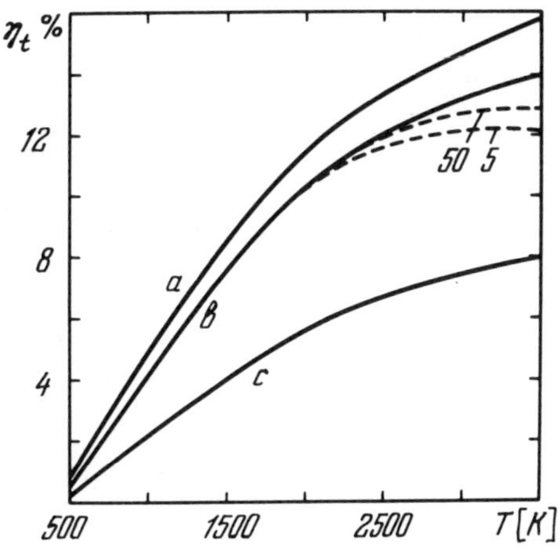

Fig.5.39. Effect of increasing temperature on a working gas efficiency η_t with respect to vibrational energy store at thermal excitation. a -- undiluted nitrogen; b -- 15% CO_2-83% N_2-2% H_2O mixture; c -- 10% CO_2-40% N_2-50% He mixture. Solid lines -- with no account for dissociation, dashed lines -- with the account of dissociation for b (numbers at the curves give pressure in atm)

Nozzle efficiency η_n is determined by the useful vibrational energy ε_{II}, which is conserved in the nozzle gas flow,

$$\eta_n = \frac{\xi_{CO_2}\theta_3 e_3^0 + \xi_{N_2}\theta_4 e_4^0}{\xi_{CO_2}\theta_3 e_3(T_0) + \xi_{N_2}\theta_4 e_4(T_0)} \quad , \tag{5.6.2}$$

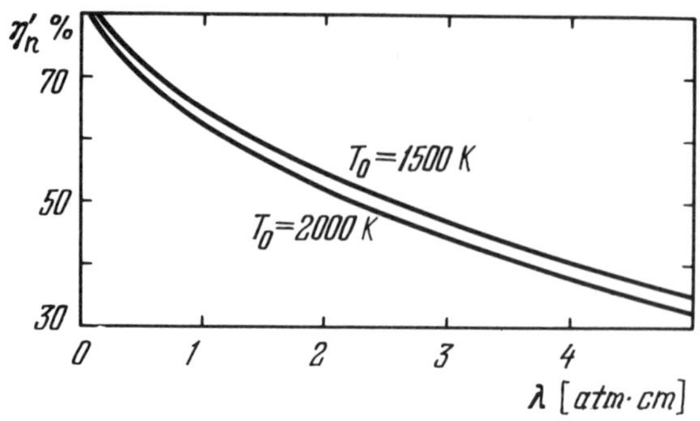

Fig.5.40. Two-dimensional wedge nozzle efficiency η'_n for a shockless inviscid flow of 10% CO_2-40% N_2-50% He as a function of parameter $\lambda = p_0 h_*/(2 \tan\theta)$ for $S/S_* = 30$

where the subscript 0 indicates the conditions in front of the cavity entrance. The fact that η_n differs from unity is connected with the deactivation during molecular collisions and depends, firstly, on the rate of freezing of the vibrational energy level in the nozzle and, secondly, on the losses in boundary layers, shock waves, and also the losses during heterogeneous relaxation. The decrease of the parameter $\lambda = p_0 h_*/(2 \tan\theta)$ increases the rate of freezing in the supersonic nozzle; for low pressures freezing in the subsonic flow is also possible. The value of η'_n for a shock-free inviscid flow of laser mixture can be obtained by solving the problem described in Sect. 5.1; Figure 5.40 gives an idea of the results of this solution. In a viscous gas a part of the useful vibrational energy Δ_e is lost during deactivation in the boundary layer, so that $\eta_n = (1 - \Delta_e)\eta'_n$. The value of Δ_e is determined by

$$\Delta_e = \frac{1}{H}\int_0^H \left[1 - \frac{\xi_{CO_2}\theta_3 e_3 + \xi_{N_2}\theta_4 e_4}{\xi_{CO_2}\theta_3 (e_3)_e + \xi_{N_2}\theta_4 (e_4)_e}\right] \frac{\rho u}{\rho_e u_e}\, dy \quad , \tag{5.6.3}$$

where H is the nozzle exit height and the subscript e corresponds to the inviscid core of the flow.[7] The losses in the viscous wake behind the nozzle array edge can be taken into account in a similar way. The values of Δ_e are obtained from the solution of the problem for a viscous flow of a relaxing gas in the laminar boundary layer and in the turbulent wake on the basis of the equations given in Sect.4.4. The results of such a solution [4.52] are shown in Fig.5.41. In this figure we can also see the relative decrease of the small-signal gain Δ_k due to effect of viscosity,

7 The effect of viscosity, heat conduction, and wall conditions involving accommodation of vibrational energy on the flow field and small-signal gain are numerically investigated in [5.51].

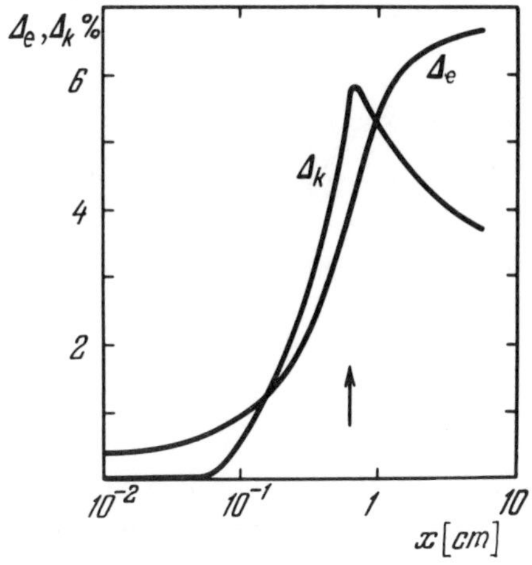

Fig.5.41. Relative part of vibrational energy Δ_e and of gain coefficient Δ_k lost as a result of deactivation in laminar boundary layers and in turbulent wakes behind a nozzle array as a function of a distance from the nozzle throat ($h_* = 0.2$ mm). The arrow indicates the nozzle exit ($S/S_* = 10.8$). 5% CO_2-93% N_2-2% H_2O mixture, $T_0 = 1400$ K, $p_0 = 25$ atm. Nozzle wall temperature $T_w = 300$ K

$$k = \frac{1}{H}\int_0^H \left[1 - \frac{k_0}{(k_0)_e}\right] dy \quad . \tag{5.6.4}$$

In the viscous flow the losses of vibrational energy steadily grow, but the drop of k_0 at the nozzle exit ceases, and in the wake k_0 increases as a result of the temperature smoothing and the absence of the deactivation effect of the walls. As the temperature of the nozzle walls increases, the drop of k_0 becomes considerable (in the example in Fig.5.41 for $T_w = 700$ K, Δ_k reaches 21% at a distance of 3 cm from the nozzle throat). The losses grow with the increase of Reynolds number [4.52]. The shock waves appearing in the flow at $p \geq 0.02$ atm, when the collisional broadening of lines prevails, result in a decrease of the small-signal gain [5.47,48]. As the pressure drops, the small-signal gain in a shock wave may increase. This occurs in oblique shock waves: for $p < 0.02$ atm at $M_n \leq 2$, where $M_n = M \sin \alpha$, α is the inclination angle of a shock wave to the flow, and in a wider range of M_n, if the pressure is considerably lower. In any case, however, shock waves result in an additional deactivation of the vibrational energy. According to [2.50], the part of heterogeneous deactivation in a CO_2 gasdynamic laser is small.

The fraction of laser energy extracted from the cavity determines its efficiency (Sects.2.3 and 5.3). For the simplest plane-parallel Fabry-Perot resonator with maximum length a_m along the flow in the constant small-signal gain approximation the value of η_r is directly obtained from (N.7). Then, taking into account, according to Sect.2.5, the decrease of radiation intensity due to pulsations in the flow (attenuation factor β per cm) and due to the presence of condensate particles (coefficient γ), we arrive at

$$\eta_r = \frac{t}{2k_0 d}\left[\frac{k_0 d}{\alpha + (\beta + \gamma)d + t/2} - 1\right] , \qquad (5.6.5)$$

where α is the loss factor per one pass and t is the transmission coefficient. For the optimum transmission coefficient t_0 (2.3.33) instead of (5.6.5) we obtain (2.5.20). Figure 5.27 gives an idea of the efficiency of an unstable resonator. Usually η_r for multipass resonators amounts to tens of percentage points, e.g., in [1.32], about 25%; in [1.22], 26%; in [5.56], about 30%. According to [5.104a], the highest experimentally obtained resonator efficiency is $\eta_r = 0.8$. We remind our reader that quantum efficiency η_q for CO_2 is 41%.

The total efficiency of a CO_2 gasdynamic laser is computed as a product: $\eta = \eta_q\eta_r\eta_n\eta_t$. In experiments with a combustion-driven GDL this value was found to be: in [1.22] $\eta = 0.41 \times 0.26 \times 0.49 \times 0.076 = 0.4\%$; in [5.57], up to 0.2%, and in [5.56], 0.6%. Similar values were obtained in the analysis [5.104b]. The considerable increase of η, however, cannot be achieved even with optimization, since it is limited by the small value of η_t.

Selective heating of undiluted nitrogen increases η_t and promotes the growth of the nozzle efficiency η_n due to fast freezing of nitrogen. This explains the higher efficiency of mixing gasdynamic lasers, which utilize the injection of cold CO_2 into the supersonic flow. Indeed, in experiments [1.33] carried out with such a scheme the total efficiency was increased to $\eta = 1.6\%$ (in [5.104b] the possibility of the increase of efficiency up to 2 to 3% was indicated). Nevertheless, even in this case the efficiency of gasdynamic lasers remains considerably lower than the efficiency of lasers with electrical pumping.

From the viewpoint of the prospects of the efficiency increase, closed-cycle gasdynamic lasers are very promising. Rapid-flow electric-discharge lasers with a closed cycle have sufficiently good characteristics. In a closed-cycle GDL, after expansion in the supersonic nozzle (or nozzle array) and light radiation in the cavity flow, the gas is decelerated in a diffuser, where it is compressed and heated. Because of inevitable losses of pressure the gas must then be supplied to a compressor for the pressure recovery to the initial value of p_0. The compressor performance, however, will be effective only when the gas supplied to it is cooled in a heat exchanger [5.105]. If 40% of this thermal energy is transformed into mechanical work of the compressor, then with an ideal cavity ($\eta_r = 1$) and complete pressure recovery in the diffuser the efficiency of the closed-cycle GDL will be close to 20% (for $T_0 = 1700$ K, frozen vibrational temperature 1417 K and the Mach number at the cavity entrance $\underset{\sim}{M} = 4$). This limit is basically connected with the decrease of the drag pressure as a result of radiation. The account of the diffuser losses (and they are really high!) reduces η to 5%, if the diffuser recovery coefficient is close to 0.5. Nevertheless, the development of closed-cycle systems is promising.

In the search for ways of increasing GDL efficiency many other ideas have been put forward. Their fruitfulness has to be verified experimentally, since in the theoretical analysis considerable simplifications are often introduced or only approximate estimates are possible because of the lack of the required data. Among the suggestions of greatest interest are those which consider the possibility of combined pumping of molecules to the upper laser level. In this case, in addition to a gasdynamic method of creating the population inversion, the employment of optical, chemical, and electrical methods of excitation is suggested. We shall discuss here only the principal features of these methods simultaneously with the gasdynamic method of pumping on the example of the CO_2 gasdynamic laser.

Optical excitation of CO_2 molecules by the radiation from the CO flame made it possible to obtain radiation in stationary conditions at a wavelength of 10.6 μm [5.106]. According to [5.106], during burning of CO or hydrocarbons, more than 20% of the heat effect of the reaction can be transformed into infrared radiation, a large portion of which is the radiation on the 001 → 000 transition of CO_2; absorption of this radiation by carbon dioxide results in the population inversion. If the laser operates using the combustion of the indicated fuel, the laser mixture is irradiated at the entrance of the GDL cavity with the aid of a specially designed optical scheme. In this case the efficiency should be increased when a considerable part of the radiation exceeds the equilibrium level. In another variant of optical pumping it is suggested that an obstacle should be introduced into the supersonic flow behind the cavity. Then the radiation of the CO_2 molecules from the shock wave thus formed penetrates the cavity and provides the additional pumping [5.107]. The main difficulty in the realization of these suggestions is connected with the difference in rotational temperatures in radiating and absorbing gases, which hampers this pumping and makes it ineffective. The small-signal gain of a CO_2 gasdynamic laser can also be increased by irradiation the flow at the throat by some part of laser radiation [5.108].

Chemical excitation, in essence, is connected with the formation, as a result of chemical reactions, of the excess amount of vibrationally excited molecules. An example of such a reaction is the decomposition of hydrazoic acid, which supplies a considerable amount of excited nitrogen molecules [5.109,110]; after this the fast cooling of the mixture in the GDL (before thermal equilibrium sets in) should be effected. In [3.19] the possibility of pumping the molecules to the upper laser level as a result of the recombination $CO + O + M \to CO_2 + M$ was investigated, which proceeded during cooling of a highly heated and dissociated mixture in an expanding flow. Calculations were based on (3.6.11) and on the assumption that the dissociation energy D, after recombination, is completely transformed into vibrational energy of the formed molecule and is uniformly distributed over all vibrational degrees of freedom of CO_2. Then in (3.6.11) $E_1^* = 0.25$ D, $E_2^* = 0.5$ D (taking account of the ν_2 mode), and $E_3^* = 0.25$ D. The results of analysis showed that during the expansion of a mixture stoichiometric with respect to composition of dissociated CO_2, the pumping of energy to the 00^01 state is not effective. This is mainly due to the rapid chemical "freezing" and cessation recombination.

Experiments [5.111,112] revealed the possibility of effective additional pumping at the expense of the chemical reaction $N_2O + CO \to CO_2 + N_2$ with the formation of the superequilibrium concentration of vibrationally excited nitrogen (see also [5.113]). This reaction proceeds in front of the GDL nozzle entrance and facilitates the increase of T_3 in comparison with the values obtained at preliminary prepared mixtures of the reaction products (in all the cases the mixtures under investigation contained helium).

The combination of gasdynamic and electrical methods of creating population inversion is very promising. Fast flow of the medium in electric-discharge CO_2 lasers provided high radiation power in continuously or periodically operating lasers. These are two results of the fast gas motion: (1) the active medium is rapidly substituted in the cavity and, hence, a large mass of the gas participates in lasing, and (2) this substitution prevents the active medium heating, destroying the inversion. The velocity of the mixture flow reaches supersonic values. In [5.114] laser power of 20 kW was obtained with an efficiency $\eta = 24\%$ with respect to the electrical energy expenditures. In these schemes, however, the gas flow does not create the population inversion. Along with this, according to [5.115], there is a possibility to develop gasdynamic lasers with additional electrical pumping. To do this, at the supersonic nozzle exit in a constant cross section channel a distributed electric discharge is passed through the active medium. The arc discharge does not appear because the power supplied to the discharge and absorbed in a unit volume of the discharge space is limited. The initial concentration of electrons in the flow can be created by adding to the mixture

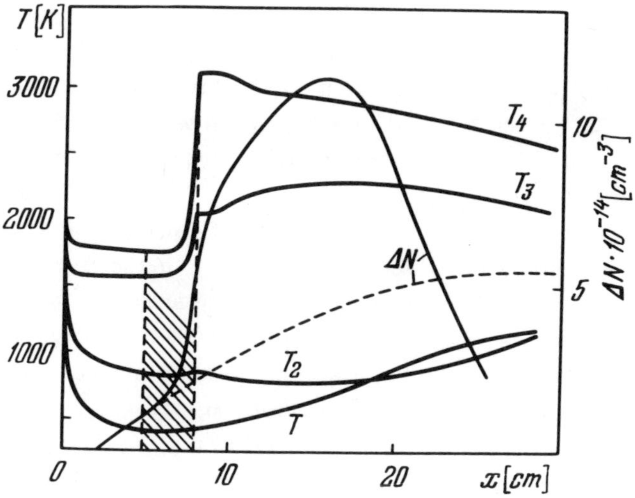

Fig.5.42. Nozzle flow with additional electrical pumping for CO_2:N_2:He = 15:30:50 mixture with 5% ionized admixture, T_0 = 2000 K, p_0 = 10 atm

rapidly ionized admixtures (or using electron beams or ultraviolet radiation). When the gas is cooled in the nozzle, the electron concentration rapidly freezes. Figure 5.42 gives the results of analysis [5.115] of the distribution of temperature, degree of ionization, and inversion with additional electrical pumping (solid lines) and without it (dash line). The discharge zone is hatched in the figure. It was assumed that the field intensity in the discharge zone provided for a constant temperature of electrons, T_e = 1.2 eV, which corresponds to the maximum probability of excitation of molecular vibrations by electron impact. Thus electron pumping increases T_4 by more than one thousand degrees, and the absolute value of population inversion by almost a factor of 3. When working with this scheme, according to [5.115] the efficiency can be increased by a factor of 3 to 5 compared to an ordinary gasdynamic laser. In this case, however, the efficiency of obtaining electrical energy was not taken into account.

The examples given in this section show only a few possibilities of improving CO_2 gasdynamic lasers and the ways of increasing their efficiency. Many of the above considerations can be applied to gasdynamic lasers of other types. But as was mentioned above, one of the advantages of the CO_2 gasdynamic laser is that it can operate using the combustion products of common fuels, which makes it very promising.

5.7 Twin-Mode CO_2 Laser

The possibility of obtaining population inversion in the levels of the twin symmetrical modes of the CO_2 molecule (see Fig.3.2), predicted in [3.54], served as the basis for the development of the GDL employing the transitions between the levels of these twin modes. According to Sects.3.4 and 3.5, the formation of this inversion is connected with anharmonicity and nonequidistancy of lower levels of the symmetrical modes of CO_2 owing to the Fermi resonance. If the vibrational energy of these modes is high and translational temperature of the gas is low, then as a result of the fast V-V exchange in the absence of a noticeable V-T exchange the quasistationary distribution of CO_2 molecules over the levels of these modes is formed, which creates the inversion on the transitions $03^10 \to 10^00$ and $04^00 \to 11^10$ (see Fig.3.2). This mechanism is similar to the mechanism of inversion formation in the Treanor distribution (see Sect.3.4) and thus differs in principle from the way of creating the population inversion of CO_2 molecules on the transitions $00^01 \to 10^00$, which was described in Sect.5.1. In fact this mechanism, connected with the effect of anharmonicity, is close to the mechanism of inversion formation in GDL using diatomic molecules, e.g., CO (see Sect.6.2).

Proceeding from (3.5.2), we find that for creation of inversion on the transition between two levels E_m and E_n, which belong to adjacent multiplet systems in the twin modes (for quantum numbers v + 1 and v, where $v = 2v_1 + v_2$, see Sect.2.1), it is required that the inequality [3.51]

$$\frac{T_2}{T} > \frac{1}{1 - \frac{E_m - E_n}{\theta_2}} \tag{5.7.1}$$

should hold, where T_2 is the vibrational temperature of the twin modes ($T_1 = T_2$). For the 10^00 and 03^10 levels this occurs at $T_2/T > 5$.

Thus, to create a twin-mode laser we must rapidly and considerably cool CO_2 to obtain the effective freezing of the vibrational energy of ν_1 and ν_2 modes. This can be done by diluting CO_2 with argon and using the nozzle with a high area ratio. Figure 5.43 shows the results of analysis [5.116] of the absolute inversion ΔN and the small-signal gain k_0 on the transition P15 of the band $03^10 \to 10^00$ in a free jet of the mixture of CO_2:Ar = 1:1 behind a slit nozzle. These values correspond to a fixed gas temperature in a jet T = 100 K (at lower temperatures the condensation of CO_2 begins to produce an effect). This transition corresponds to a wavelength of about

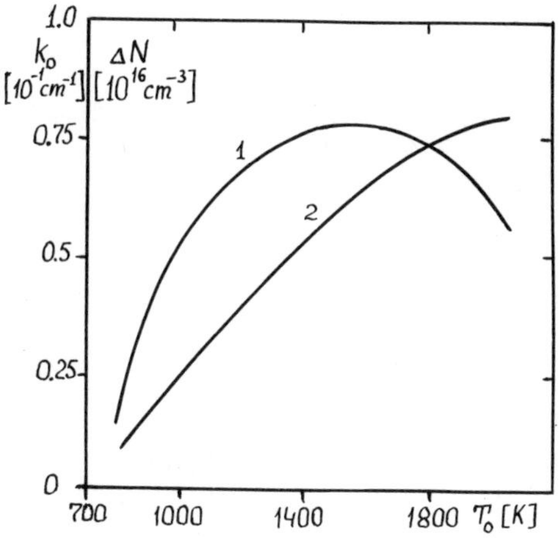

Fig.5.43. Effect of stagnation temperature on population inversion ΔN (1) on the transition $03^10 \rightarrow 10^00$ and on small-signal gain k_0 (2) for CO_2:Ar = 1:1 mixture with initial density $\rho_0 = 7.2 \times 10^{19}$ cm^{-3}

19 μm. The tendency of the population inversion and the small-signal gain to drop at high initial temperatures, observed in Fig.5.43, is connected with the increase of the rate of the V-T relaxation of the energy of symmetrical modes. Analysis [5.116] pointed out the possibility of obtaining $k_0 \sim 0.1$ cm^{-1} on the $02^00 \rightarrow 01^10$ transition in the range of $\lambda \approx 16$ μm with the cooling of this mixture in a jet so that $T_2/T > 12$.

Experimental studies of the possibility of developing a GDL on transitions between the levels of the CO_2 twin modes proved to be successful [5.117]. Laser radiation at a wavelength of 18.38 μm (transition Q7 $03^10 \rightarrow 10^00$) during cooling the mixture of CO_2:Ar = 1:9 at the slit nozzle had been observed. The mixture was heated by a shock wave reflected from the shock tube end face. In these experiments, however, radiation from other transitions was not observed, which indicates the possibility of deterioration of quasistationary conditions and occurrence of processes not accounted for in the analysis.

The efficiency of a twin-mode CO_2 gasdynamic laser can be much higher (about 10%) than that of the laser on the transitions $00^01 \rightarrow 10^00$ and $00^01 \rightarrow 02^00$, which is connected with a high quantum efficiency and heat capacity of the symmetrical vibrations of CO_2 [5.116,117].

In conclusion it should be noted that laser radiation on the transition between the levels $10^00 \rightarrow 01^10$ and $02^00 \rightarrow 01^10$ (wavelengths of 14 and 16 μm, respectively) in a GDL can be obtained if the levels 10^00 and 02^00 are overpopulated as a result of radiation on the transitions $00^01 \rightarrow 10^00$ and $00^01 \rightarrow 02^00$ in an ordinary scheme or in a scheme of component mixing [5.118-120]. Such a GDL works in two stages: first it radiates at a wavelength of

10.6 μm (or 9.4 μm) and then at 14 μm (or 16 μm). The addition into the mixture of a small amount of hydrogen, which effectively deactivates the lower laser level 01^10, improves the conditions for the secondary laser radiation [5.119].

5.8 Creation of an Active Medium in Expansion Waves and Shock Waves

Along with the development and improvement of gasdynamic lasers in which the population inversion is obtained during acceleration of a laser mixture in the nozzle, attempts are being made to create a laser operating in other gasdynamic conditions. For this purpose the cooling in an expansion wave, rapid heating in a shock wave, and other methods are suggested.

Attempts have been made to obtain population inversion in a one-dimensional expansion wave formed during unsteady expansion of a heated CO_2-N_2-He mixture after the removal of the diaphragm at the end face of a shock tube [5.121], but these attempts have not been successful. The gas cooling in a multidimensional expansion wave is more promising. Thus, to create inversion in [5.122] it had been suggested that use should be made of the divergence of products from the detonation of a cylindrical charge of a fuel mixture, e.g., acetylene with air. In a similar scheme [5.123,124], after the explosive liberation of energy on a certain straight line, the cylindrical shock wave heats the prepared mixture of CO_2-N_2-He (or H_2O). Immediately behind the shock front the gas pressure is high, and therefore equilibrium sets in quickly. Then, as the shock wave spreads and the gas volume expands, adiabatic cooling in the cylindrical shock wave occurs. For the analysis of the gas state in the expansion wave behind the cylindrical shock wave the equations of nonstationary gas dynamics are used,

$$\frac{\partial}{\partial t}(r\rho) + \frac{\partial}{\partial r}(r\rho u) = 0 \quad , \quad \frac{\partial}{\partial t}(r\rho u) + \frac{\partial}{\partial r}[r(p + \rho u^2)] = p \quad ,$$

$$\frac{\partial}{\partial t}\left[r\left(\frac{1}{2}\rho u^2 + \rho h - p\right)\right] + \frac{\partial}{\partial r}\left[r\rho u\left(h + \frac{u^2}{2}\right)\right] = 0 \quad , \qquad (5.8.1)$$

where r is the distance from the explosion axis and h is the specific enthalpy. These equations are supplemented with equations of state (3.1.6,10) and the equations of vibrational relaxation kinetics for the CO_2-N_2-He (or H_2O) mixture in the form (5.1.2). The conditions on the shock wave are used as the boundary conditions. When determining the initial conditions it is assumed that at t = 0 the liberation of energy E_* per unit length of an infinitely long cylinder of radius r_* occurred, so that the gas temperature in this region is T_*, and pressure is p_*, so that $T_* \gg T_0$, $p_* \gg p_0$, where T_0 and p_0 are initial temperature and pressure in an unperturbed gas. The resulting system of equations is solved by the numerical method described in [4.12].

The creation of the population inversion of the CO_2 levels 00^01 and 10^00 during adiabatic cooling of the mixture in the expansion wave proceeds in the same way as in a nozzle flow. The small-signal gain k_0 in this scheme will be maximal at a certain optimum distance from the explosion axis. In fact, the gas entrained by the shock wave at an early stage of its development is characterized by too high values of temperature and pressure (the Mach number of the shock wave $\underset{\sim}{M}_s$ is high: $\underset{\sim}{M}_s \gg 1$). Therefore, despite the subsequent cooling, population inversion is not created. In time the temperature drops, but simultaneously the rate of cooling decreases. On the contrary, far from the explosion axis the velocity of propagation of the shock wave decreases to such an extent ($\underset{\sim}{M}_s \to 1$) that the gas temperature in it increases insignificantly and does not provide the required pumping of molecules to the upper laser level. There is also an optimum in the dependence of k_0 on the parameter $\pi = \sqrt{p_0 E_*}$: for $\pi \to 0$ the energy of the charge is too low for pumping, and for $\pi \to \infty$ the vibrational and translational temperatures become too close, and inversion disappears. The results of the solution of this problem [5.123] for the mixture of 7.5% CO_2-38% N_2-54.5% He are shown in Fig.5.44. As was shown in [5.125], for real explosives the liberation of energy is much higher, so that to obtain the desired effect we must cool the gas still more. According to [5.125], this can be achieved by expanding the heated gas into a vacuum. To do this the laser mixture is placed into a thin-wall cylinder coaxial with the cylindrical charge of explosive; outside this cylinder a vacuum is created.

The appearance of the population inversion of the levels 00^01 and 10^00 of CO_2 can be expected also in other expanding flows, e.g., in the Prandtl-Meyer flow and in the flow over side surfaces or tail parts of blunted bodies.

It is known that the nonequilibrium state of a gas in certain conditions is observed also behind the shock front [1.7,8]; therefore we can speak about obtaining the population inversion in this case as well. Thus, behind the front of a normal shock wave propagating in the CO_2-N_2-He mixture with initial pressure p_0 = 0.1 atm, according to calculations [1.2] using the simplified scheme of vibrational exchange kinetics (5.1.13), we can expect

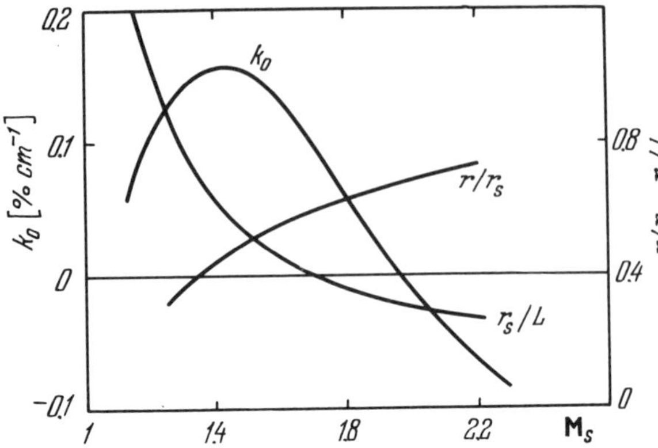

Fig.5.44. Small-signal gain k_0 at a distance r from the shock axis as a function of a cylindrical shock wave Mach number M_s; r_s--distance from the wave front to the axis, $L = \sqrt{E_*/p_0}$, p_0=0.2 atm

the formation of the population inversion between the levels 040—001 (transitions in the range of 50 μm) and 200—001 (22 μm) in the zone of the width of several millimeters. This is attributed to the fact that behind the shock front, symmetrical vibrations are excited more rapidly than antisymmetrical ones. The analysis based on a more precise kinetics [(5.1.2) with the additional account of the different between T_1 and T_2[8] for the CO_2-N_2-H_2O mixture] showed, however, that the zone of inversion for these transitions behind the shock front is considerably shorter (Fig.5.45). In fact, the vibrational temperature T_3 and even more so T_4 relaxe behind the shock front more slowly than do T_1 and T_2. The gas temperature T increases on the wave front jumpwise and then somewhat decreases due to expenditures of energy for the vibration excitation. In contrast to the cooling flow, in this case the lower laser level is the 001 state, and nitrogen facilitates its depopulation. For this reason here there is no necessity of introducing into the CO_2-N_2 mixture helium or water vapor. As the velocity of the shock wave increases, the inversion passes through the peak, and the inversion zone considerably reduces [5.126]. The inversion and laser radiation in such thin layers will be difficult to observe due to diffraction.[9] A wider zone of population inversion for the same stagnation parameters is obtained in oblique shock waves because of the supersonic nature of the flow.

---

8 In the approximation which accounts for the inequality $T_1 \neq T_2$, the concept of vibrational temperatures T_1 and T_2 loses its meaning [3.51] (see Sect.3.5). All the same, in the analysis the values of T_1 and T_2 turned out to be practically equal.

9 In [5.127] laser radiation of HF molecules behind the front of an overcompressed detonation wave propagating in F_2O-H_2-Ar mixture had been observed.

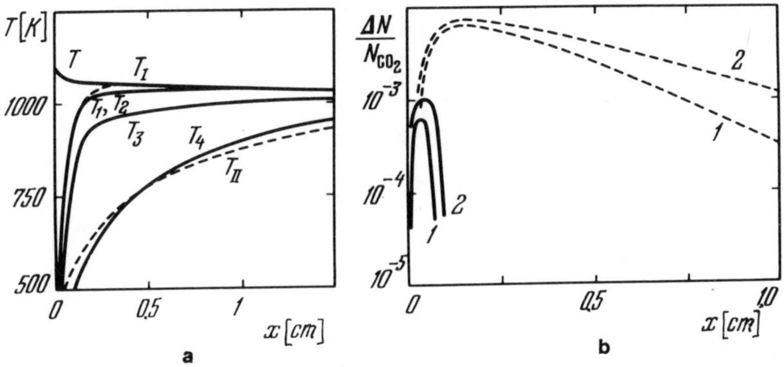

Fig.5.45. Distribution of temperature (a) and relative inversion (b) on the transitions 040 → 001 (1) and 200 → 001 (2) behind the shock front at M = 4 in 10% CO_2-88% N_2-2% H_2O mixture (initial pressure p_0 = 6 torr). Dashed lines correspond to a simplified kinetic scheme (5.1.13)

A similar mechanism of creating population inversion of vibrational levels behind the shock front is possible for shock waves propagating in nitrous oxide [5.128] and water vapors [5.129].

6. Gasdynamic Lasers with Other Active Medium

6.1 N_2O and CS_2 Gasdynamic Lasers

The considerable progress in the development of CO_2 gasdynamic lasers stimulated the search for other active media for GDLs. In this connection two problems were solved: the obtaining of higher values of extracted power in continuous operating conditions and the possibility of operation at other wavelengths and in other spectral regions. To date carbon monoxide, nitrous oxide, and carbon sulphide gas dynamic lasers have been created. These lasers, like CO_2 lasers, generate infrared radiation, that is, they operate on transitions between vibrational energy levels of molecules. The mechanisms of inversion formation for lasers on diatomic and triatomic molecules, however, differ considerably. We shall first consider the N_2O and CS_2 gas dynamic lasers, since the nature of the processes occurring in them is in fact close to that of the CO_2 gasdynamic laser, which has been considered in the previous chapters.

N_2O electric-discharge lasers are inferior to CO_2 lasers in respect to the extracted power due to the dissociation of N_2O during collisions with electrons in the discharge. The effect of dissociation, however, can be considerably decreased with thermal excitation and the subsequent cooling in a gasdynamic flow [1.17,18]. This promoted progress in creating the N_2O gasdynamic laser [1.27,28, 6.1a,b], whose performance was studied in detail in [6.1-5].

The N_2O molecule is in many respects similar to the CO_2 molecule (the scheme of the lower vibrational levels for these molecules is given in Fig. 6.1), but the nuclei in it are arranged asymmetrically (NNO). Like the CO_2 molecule, it is a linear molecule possessing three vibrational modes: symmetrical valence $\nu_1(\theta_1 = 1850$ K), bending ν_2 ($\theta_2 = 848$ K), and antisymmetrical ν_3 ($\theta_3 = 3202$ K). The second mode is doubly degenerate and is in Fermi resonance with the first mode. All the levels of N_2O molecules are somewhat

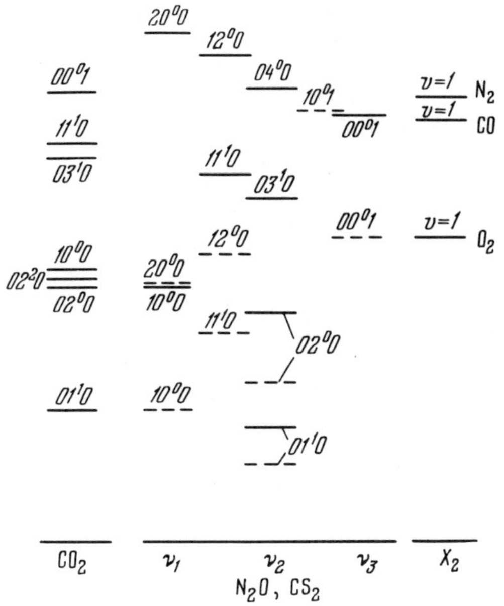

Fig.6.1. Lower vibrational levels for CO_2, N_2O, CS_2 (dashed lines), N_2, CO, and O_2

lower than the respective levels of CO_2 molecules; for this reason the energy defect between the first nitrogen vibrational level and the 00^01 state is higher than that between the levels of CO_2 and N_2 (see Fig.6.1). The pumping to the upper 00^01 laser level in the N_2O gasdynamic laser occurs during collisions with N_2 (or CO), and the inversion is created on the transition $00^01 \rightarrow 10^00$ (at a wavelength of about 10.9 μm). The gain coefficient k_0 was measured during cooling the mixtures of N_2O with nitrogen, helium, carbon monoxide, and water vapor for the flow from a two-dimensional slit nozzle of a certain configuration, which was fixed at the end face of a shock tube [1.28, 5.26, 6.1,2,6]. In contrast to a CO_2 laser, in this case a sufficiently high gain level in a binary N_2O-N_2 mixture was reliably established (Fig. 6.2) (the optimum ratio of the components was 1:4). The addition of helium to this mixture increases k_0 (see Fig.6.3), while water vapor added in the amount of 1 to 3% effects the gain little (Fig.6.4). The gain was also observed in a binary N_2O-He mixture. Generally, the observed values of the small-signal gain k_0 differ little from the results for CO_2, but in this case lower temperatures are required, which is very important to provide the continuous operation of a GDL.

The scheme of the vibrational energy exchange in these mixtures is similar to that discussed in Sect.3.5 for the mixtures containing CO_2. As before, every mode of vibration can be approximated by the harmonic oscillator model

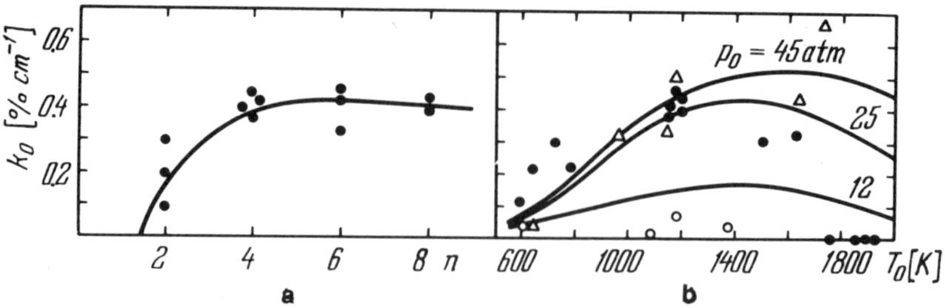

Fig.6.2. Small-signal gain k_0 as a function of $n = \xi_{N_2}/\xi_{N_2O}$ for an N_2O-N_2 mixture [$(p_0)_{N_2O}$ = 4.4 atm, T_0 = 1200 K] (a) and of temperature T_0 (b) for n = 4. Hollow circles — p_0 = 12 ± 2 atm, black circles — p_0 = 25 ± 5 atm, triangles — p_0 = 45 ± 5 atm. Solid curves — theory [6.16]

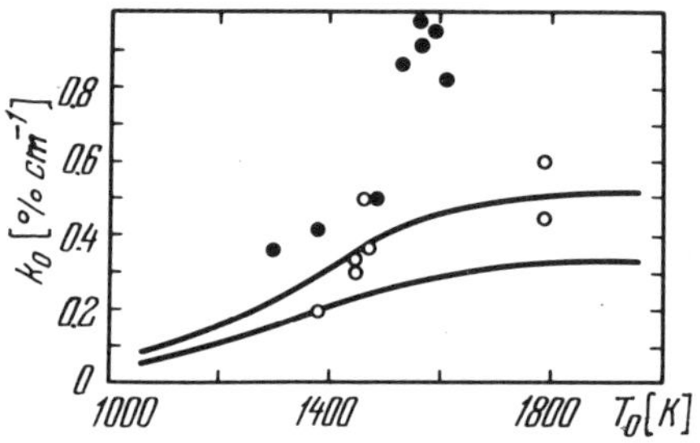

Fig.6.3. Temperature dependence of small-signal gain k_0 for N_2O:N_2:He = 1:2:7 mixture. Hollow circles — p_0 = 14.5 ± 2 atm, black circles — p_0 = 45 ± 6 atm. Solid curves — theory [6.16]

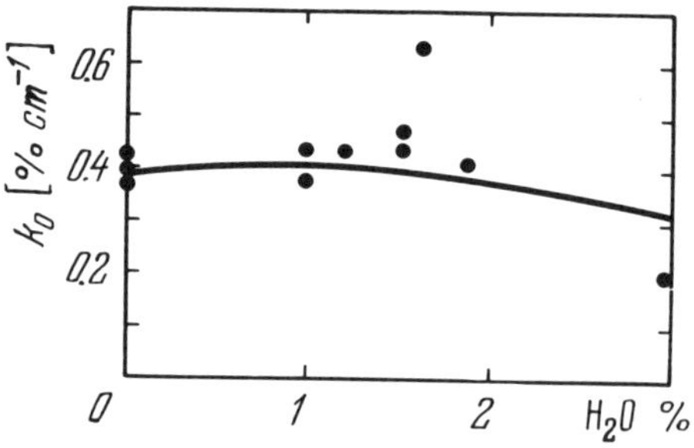

Fig.6.4. Effect of H_2O concentration on small-signal gain k_0 for N_2O:N_2 = 1:4 mixture. Experimental points for p_0 = 20 ± 4 atm, T_0 = 1100 ± 100 K, solid line — theoretical results [6.2] for p_0 = 20 atm, T_0 = 1100 K

and thus we can assume that there is a Boltzmann distribution of molecules over the levels, thus characterizing the vibrational energy in each mode by a certain vibrational temperature. From the arrangement of energy levels of N_2O (Fig.6.1) it is clear that we must account for the transitions involving a larger number of states of the ν_1, ν_2 modes: (11^10), (12^00), (03^10), and (04^00). These states participate in the V-V exchange with other molecules [6.16]. Thus on the right-hand side of the relaxation equation (3.4.21) the terms of the third and the fourth order will appear (according to the number of quanta of the ν_2 mode of the N_2O vibrations). The account of these additional exchange channels decreases the population inversion. The values of the vibrational exchange probabilities involving the N_2O molecules (and, hence, the relaxation time and exchange rate constants) can be found in

[6.1b-3]. The gain coefficient, as before, is computed by (5.1.3-6); the Einstein coefficient A_{mn} and the collisional width of the spectral line for N_2O were obtained in [6.7] and given in [6.1b-3].

The results of computation of k_0 for the P20 ($00^01 \rightarrow 10^00$) transition are in good agreement with experiments in the wide range of variation of parameters (Figs.6.2-4). In contrast to the CO_2-N_2 mixture, in this case in the process of cooling there is a moment when T_3 becomes higher than T_4 (the subscript 4, as before, is related to nitrogen). This is connected with the transition of the V-V' exchange to the quasistationary conditions, when $\theta_3 < \theta_4$ and the gas temperature T is sufficiently low [see (3.4.23)]. The optimization of the conditions to obtain the peak population inversion ΔN in the nitrous oxide GDL for a given nozzle contour showed for a mixture of N_2O:N_2:He = 1:2:7 that $\Delta N \approx 10^{16}$ cm^{-3} can be achieved at p_0 = 60 atm, T_0 = 1100 K, while for similar conditions in the CO_2 gasdynamic laser a temperature of about 2500 K would be required. The analysis of a two-dimensional flow in the N_2O gasdynamic laser was carried out in [5.49b].

Of the energy liberated during the decomposition of molecules in the heated nitrous oxide a considerable part is concentrated in vibrations of the nitrogen molecules. This effect, as was pointed out in Sect.5.6, can promote the additional chemical pumping of molecules in gasdynamic lasers. In N_2O-CO-He mixtures during heating the vibrationally excited CO_2 molecules are formed as a result of reactions $N_2O + M \rightleftharpoons N_2 + O + M$, $O + CO + M \rightleftharpoons CO_2 + M$, $N_2O + CO \rightleftharpoons N_2 + CO_2$ [5.113, 6.8]. When this mixture is used in a GDL, the gain and laser radiation are observed simultaneously for two molecules — N_2O and CO_2 [6.4-6]. The addition of CO to the N_2O-N_2-He mixture does not improve the active medium characteristics.

Continuous operating conditions for the N_2O gasdynamic laser have been realized in [6.9,10]. In this case the scheme of mixing nitrogen with nitrous oxide in coupled supersonic flows has been used. The replacement of N_2O by CO_2 in this laser made it possible to carry out the direct comparison of the effectiveness of these molecules. It turned out that for the same initial parameters the CO_2 laser is twice as powerful as the N_2O laser. This, however, does not decrease the importance of nitrous oxide lasers, since in some respects they are more advantageous. Thus, during energy transfer over large distances in the earth's atmosphere the radiation from a CO_2 laser is absorbed to a much greater extent than that from N_2O laser. It is interesting that the replacement of nitrogen by air in the mixing N_2O laser results in an increase of the output power. This can be explained by reactions occurring with oxygen [6.10]. In [5.128] the possibility of creating population in-

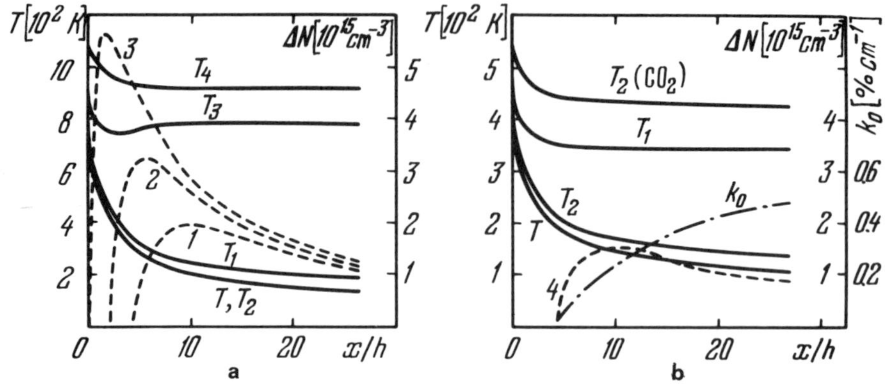

Fig.6.5a,b. Theoretical curves for slit nozzle flows of mixtures containing carbon sulphide as a function of x/h (x — distance from the slit, h — slit height of 0.8 mm). Dashed lines — population inversion on the levels 00 $00^01 \to 10^00$ for CS_2 (1), $00^01 \to 20^00$ (2), $00^01 \to 12^00$ (3), $10^00 \to 01^10$ (4); dotted and dashed line — small-signal gain k_0 on transition $10^00 \to 01^10$. Mixtures $CS_2:CO_2:He = 1:5:1.4$, $T_0 = 1400$ K, $p_0 = 20$ atm (a) and $CS_2:CO_2 = 1:19$, $T_0 = 600$ K, $p_0 = 15$ atm (b)

version between certain vibrational energy levels of N_2O in a narrow zone behind the shock front in the N_2O-N_2-He mixture was indicated.

Already in [1.17] it has been suggested that carbon sulphide GDLs can be created. Lasing on the $00^01 \to 10^00$ transition of CS_2 at a wavelength of 11.4 μm (see Fig.6.1) was obtained during cooling the CS_2-CO-He mixture in a nozzle [1.30] and in a free jet [1.29] behind a slit at the end face of a shock tube. The maximum power of 2 W has been registered in [1.30] for the flow of this mixture (the component ratio was 3:57:40 and 2:68:30) in a two-dimensional contoured nozzle ($h_* = 0.8$ mm, $S/S_* = 25$) for $T_0 = 900$ K, $p_0 = 9$ atm. The content of CS_2 in the mixture in these experiments was varied from 1 to 10%; the ratio He:CS_2, from 5 to 25; the temperature, from 600 to 1800 K; and pressure, from 2 to 20 atm. The nonresonant V-V' exchange with the CO molecule facilitates the pumping of molecules to the upper laser level in quasi-equilibrium conditions [1.30] which set in during gas cooling (see Sects.3.4 and 4.2).

Lasing in a CS_2 gasdynamic laser is possible on other transitions as well (see Fig.6.1) in CS_2-O_2 and CS_2-CO_2 mixtures. When creating inversion on the transition ($10^00 \to 01^10$) of CS_2 at the wavelength of 38.3 μm, we can expect a considerable increase of almost 20% in the GDL efficiency — for $T_0 = 340$ K — if the losses due to collisional deactivation and incomplete energy conversion in the cavity are not taken into account [1.29]. The results of an analysis of the flow through a slit for mixtures containing CS_2 (Fig.6.5a,b) made it possible to estimate the effect of deactivation and confirmed the possibility of creating such gasdynamic lasers. The data on probabilities of collisional and radiative transitions in CS_2, required for the analysis, can be found in [1.29,30].

These examples illustrate possibilities in the development of GDLs with triatomic molecules, especially for those with the appropriate low vibrational levels (COS, HCN, etc.). The difficulties connected with the high toxicity of these gases can be overcome by using closed-cycle gas systems.

6.2 Carbon Monoxide Gasdynamic Laser

While nitrous oxide and carbon sulphide gasdynamic lasers are just at the beginning of their development, the utilization of diatomic molecules as an active medium began immediately after the first works on CO_2 gasdynamic lasers had appeared. In these lasers carbon monoxide producing laser radiation at a wavelength in the region of 5 μm is used [1.26, 6.11]. Heated mixtures of CO with argon [6.11] or nitrogen [1.26] admixture are cooled in large axiosymmetrical nozzles (of about 1 m length) with the area ratio S/S_* exceeding 500, i.e., with a deep gas cooling. Subsequently laser radiation was obtained in shorter two-dimensional nozzles and slits for mixtures with inert gases [6.12,13]. Figures 6.6 to 6.8 show the dependence of the laser specific power on the gas composition, initial temperature T_0, and pressure p_0. Figure 6.6 gives the power expressed as the relative laser efficiency $\eta' = \bar{W}/\varkappa h_0$, where $\bar{W}$ is the registered specific power per unit mass rate of the mixture, h_0 is the specific enthalpy of the gas at the nozzle entrance, and $\varkappa$ is a correction factor equal to the ratio of the resonator area to the area of the entire flow at the same cross section. For the axiosymmetrical nozzle this correction differs considerably from unity, which reduces the actual efficiency of a gasdynamic laser. Figure 6.7 gives laser power in relative units and Fig.6.8, the intensity of the output beam. The addition of argon to the CO-N_2 mixture leads to a noticeable increase of power; this laser operates well even without nitrogen. The output power increases with the decrease of the molecular mass in the mixture of CO with inert gases, since in this case the rate of cooling grows and the gas temperature drops. Under the conditions studied the optimum values of T_0 are close to 2100 K, and useful power grows with pressure. For $p_0 \geq 80$ atm the dependence becomes linear [6.14]. The analysis of the spectral composition

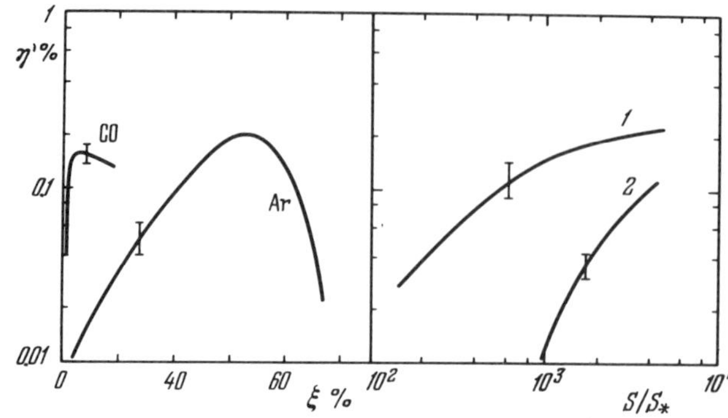

(a) (b)

Fig.6.6a,b. Conditional efficiency of a carbon monoxide GDL as a function (a) of concentration of CO added to 80% Ar mixture instead of nitrogen and Ar concentration (in $CO:N_2$ = 1:3 mixture), (b) of expansion area ratio S/S_*, 1 — 5% CO - 15% N_2-80% Ar; 2 — 25% CO-75% N_2 [6.14]

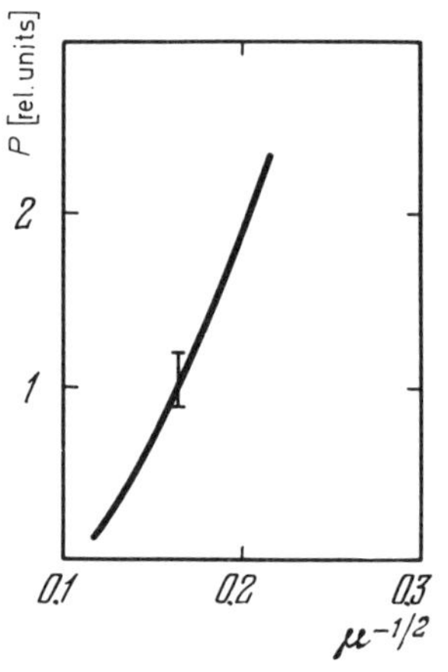

Fig.6.7. Laser power as a function of molecular mass μ in a CO mixture with inert gases [6.13]

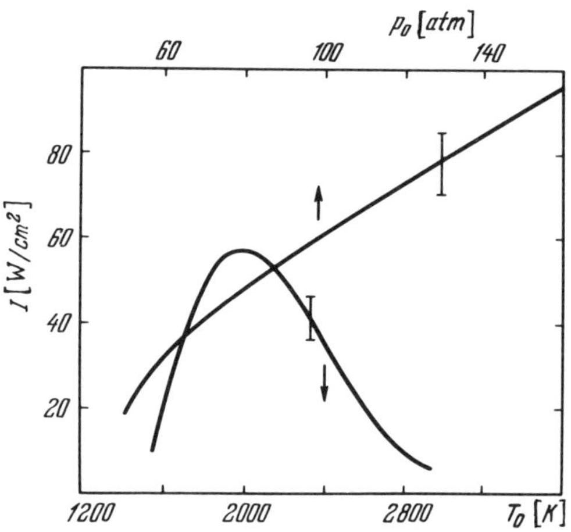

Fig.6.8. Effect of increasing initial temperature T_0 (for p_0 = 90 atm) and pressure p_0 (for T_0 = 2100 K) on laser intensity for 5% CO-15% N_2-80% Ar mixture (d_* = 1.27 cm, S/S_* = 577) [6.14]

of laser radiation [6.14] showed that for T_0 = 2100 K, p_0 = 85 atm in the 5%CO-15% N_2-80% Ar mixture at the exit from the nozzle with S/S_* = 2730 lasing occurs on the following transitions of the CO molecule $n \rightarrow n-1$ (and corresponding wavelengths λ):

$n \to n-1$	$2 \to 1$	$3 \to 2$	$4 \to 3$	$5 \to 4$	$6 \to 5$	$7 \to 6$	$8 \to 7$	$9 \to 8$
λ in μm	4.78	4.84	4,90	4.96	5.02	5.07	5.14	5.26

The line intensity decreases with an increase of n.

The observed laser radiation is connected with the anharmonicity, which in the quasi-equilibrium V-V exchange disturbs the Boltzmann distribution over the levels in a peculiar way (see Fig.3.1). To carry out the analysis of the processes in a GDL in this case we must consider the system of balance equations for the population of vibrational levels of the CO molecule simultaneously with gasdynamic equations. The population x_n of a certain vibrational level is determined by the V-T and V-V exchange during collisions and also by radiative transitions,

$$\frac{dx_n}{dt} = \left(\frac{dx_n}{dt}\right)_{VT} + \left(\frac{dx_n}{dt}\right)_{VV} + \left(\frac{dx_n}{dt}\right)_r \quad . \tag{6.2.1}$$

For mixtures of CO with an inert gas, taking account of only single-quantum transitions, the first two terms on the right-hand side are of the form of (3.4.1) and (3.4.26), respectively.[1] The probability of multiquantum radiative transitions is also considerably lower than that for single-quantum transitions [6.17,18]. Therefore, the last term in (6.2.1) can be written as

$$\left(\frac{dx_n}{dt}\right)_r = A_{n+1,n}x_{n+1} - A_{n,n-1}x_n \quad , \tag{6.2.2}$$

where $A_{n+1,n}$, $A_{n,n-1}$ are the probabilities of spontaneous emission (Einstein's coefficients). The values of these coefficients for $1 \leq n \leq 14$ are given in Table 6.1 [6.17]. The relative values of the Einstein coefficients are determined in [6.18].

Table 6.1.

n	1	2	3	4	5	6	7
$A_{n,n-1}$ (s^{-1})	33.4	64,5	92,9	118.0	142.0	164.3	182.0
n	8	9	10	11	12	13	14
$A_{n,n-1}$ (s^{-1})	200.1	214.1	227.1	238.8	248.5	256.3	262.5

1 The case of a two-component system of anharmonic oscillators is more complicated (see, for example, [3.39, 6,15,16]).

Equations (3.4.1,26) include the probabilities of the V-T transitions ($P_{n,n-1}$) and V-V transitions ($Q_{n,n-1}^{m,m+1}$). As before, to determine these probabilities it is expedient to make use of the results of experimental studies. Knowing P_{10}, we can obtain the values of $P_{n,n-1}$, required for calculations, using (3.3.13) if n is not too high [6.19] (see also [6.20,21]). For CO + M collisions P_{10} has been determined in a wide temperature range and for many partners M (see references in [6.20], and also [6.22,23]). For CO + CO collisions, according to [6.20], these results can be described using the equation for the rate constant $k_{10} = zP_{10}$ [z is the frequency of collisions—see (3.2.9)][2]

$$k_{10}(CO) = 2.662 \times 10^{12} T \exp(-193.7T^{-1/3} + 1410T^{-1}) \text{ cm}^3/\text{mole s} \quad . \quad (6.2.3)$$

From (6.2.3) it follows that the probability of deactivation of the lower vibrational levels of the CO molecule for collisions with CO is very small. For the collisions with He atoms this probability is much higher (Fig.6.9) [6.18].[3] The probability of the V-T deactivation of CO during collisions with water vapor molecules is high—$k_{10}(H_2O) = 1.2 \times 10^{11}$ cm^3/mole s for T = 297 K [6.25], which is more than 10^6 times as high as $k_{10}(CO)$. For CO + Ar collisions at $T \leqq 300$ K in [6.20] it was assumed that $k_{10}(Ar) = 0.2k_{10}(CO)$.

The probability of the V-V exchange between CO molecules was experimentally studied in [6.17,26,27]; in these works there are also references concerning the previous studies and comparisons with theoretical data. Proceeding from these data, for the rate constant $k_{n,n-1}^{m,m+1} = zQ_{n,n-1}^{m,m+1}$ for T = 100 to 500 K, we can recommend the relation [6.27]

$$k_{n,n-1}^{m,m+1}(CO) = 1 \times 10^{13} n(m+1)(0.919T^{-1/2} + 61.9T^{-3/2}) \times \exp\{-[24.63(n-m-1)^2/T]\} \text{ cm}^3/\text{mole s} \quad . \quad (6.2.4)$$

This relation is valid for $0 \leq (n - m - 1) \leq 0.4T^{1/2}$, and the values of $k_{n,n-1}^{m,m+1}$ for m + 1 > n should be determined from (6.2.4), formally changing the notation ($m + 1 \rightarrow n'$, $n - 1 \rightarrow m'$) and using the principle of detailed

2 To pass over in the rate constant from cm^3/mole s, adopted here, to s^{1} atm^{-1}, we must multiply by 1/82 T, where T is the gas temperature.

3 Figure 6.9 gives the theoretical values of $k_{n+\Delta n,n}(He)$ according to [6.24], which for n = 0, Δn = 1 differ little from experimental results (see references in [6.24]).

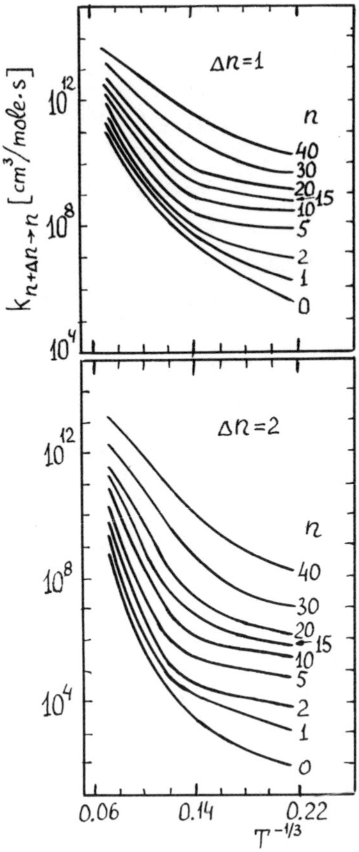

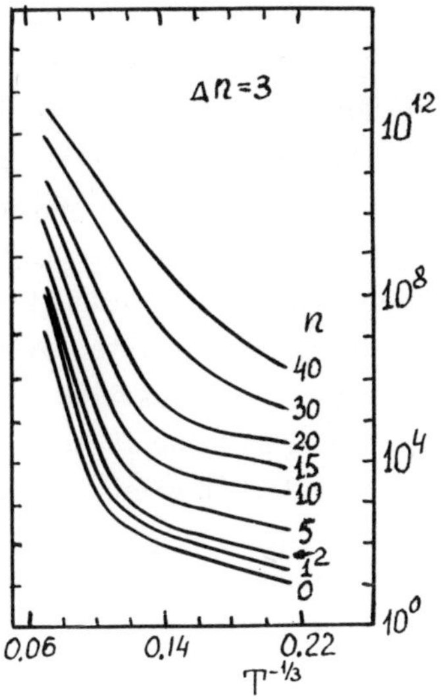

Fig.6.9. Transition rate constants $k_{n+\Delta n \to n}$ for the V-T exchange during collisions CO + He as a function of $T^{-1/3}$ ($100 \leq T \leq 3000$ K) according to [6.24]

balancing (3.4.25). Comparing (6.2.4) with (3.3.17) and (3.3.18), we see that the V-V exchange theory with the account of the long-range attraction forces between the molecules describes the experimental results well. The theoretical dependences for $Q_{n,n-1}^{m,m+1}$ obtained with account of the short-range repulsion forces were used in analyses [3.37, 6.14,20,21]. To improve the agreement with experiment, corrections have been introduced into these dependences.

For the analysis of the V-V' exchange between CO and other molecules, of great importance is the information on the probabilities of the V-V' exchange $CO \rightleftharpoons N_2$ [6.17,23,25,28]. In the temperature range from 100 to 650 K these results for the rate constant of the V-V' exchange in the exothermal direction can be described by

$$k_{n-1,n}^{m,m-1}(N_2) = 1.075 \times 10^8 mn\sqrt{T}(1 + 3.86 \times 10^3 T^{-1} + 1.98 \times 10^4 T^{-2})\exp(-\delta^2/T) \text{ cm}^3/\text{mole s} \quad , \tag{6.2.5}$$

where $\delta = 18.911 + 2.586n - 3.085m$ ($\delta > 0$), n numbers the levels of CO, and m, the levels of N_2. For $\delta > 0$ (6.2.5) should be transformed similarly to (6.2.4), taking into account the difference in vibrational quanta of N_2 and CO. For higher temperatures the probability of this process, according to measurements with shock tubes (see references in [6.23]), rapidly increases and does not obey (6.2.5) any longer. This change in the temperature dependence of the vibrational exchange probability can be explained by the change of the mechanism of intermolecular interaction, which determines the effectiveness of this exchange (the transition from the long-range attraction forces to the short-range repulsion forces, Sect.3.3).

The solution of the system of balance equations (6.2.1) together with gasdynamic equations and equations of state for the carbon monoxide flow (in the general case in the mixture with other gases) gives the following results. During gas cooling in the nozzle first the intensive V-T deactivation occurs, which is completed by the freezing of a certain stock of vibrational energy (Sect.4.2). In contoured nozzles with a point of inflection this freezing occurs near the nozzle throat, and in the nozzles with a small divergence angle the freezing occurs later. After the V-V deactivation of lower levels has stopped, the distribution of molecules differs both from the Boltzmann distribution—due to the nonresonance V-V exchange which continues here—and from the Treanor distribution—as a result of the V-T deactivation of the upper levels (Fig.6.10). With further cooling a quasistationary state sets in (the *Gordiets distribution*, see Fig.3.1). From Fig.6.10 it is clear that the population inversion for all the vibrational levels is practically absent. As the gas temperature lowers, however, the redistribution of molecules among the rotational levels occurs because of rotational relaxation, which results in the appearance of a noticeable inversion on the separate rotational levels belonging to neighboring vibrational states (*partial*, or *local inversion*). This permits one to obtain the positive gain coefficient k_0 for small values of the rotational quantum number even if there is no population inversion for the vibrational level as a whole (Fig.6.11). To compute the small-signal gain the relations (2.2.8-10) are used, in which the populations of separate rotational levels are determined by (D.3,4). The Einstein coefficients A_{mn} depending on the number of the vibrational level are given in Table 6.1, and collisional line width can be found in [6.29]. Since the gas pressure is low, often it can be assumed that the line broadening is of the Doppler type. The values of k_0 (maximum with respect to the rotational quantum number J), obtained in the analysis [6.20] for the 20% CO-80% Ar mixture, are given in Fig.6.12. In this figure the dashed

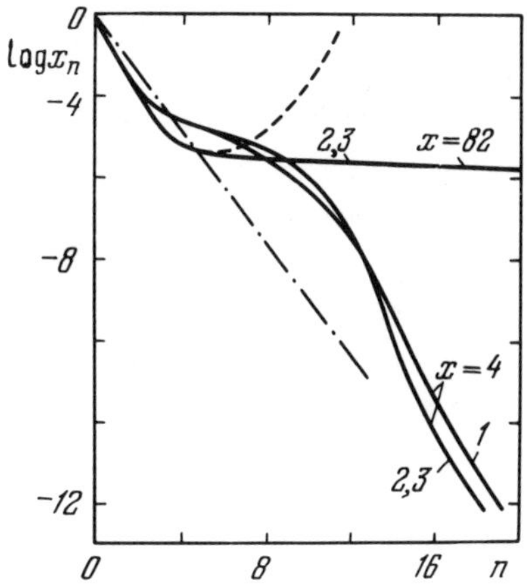

Fig.6.10.

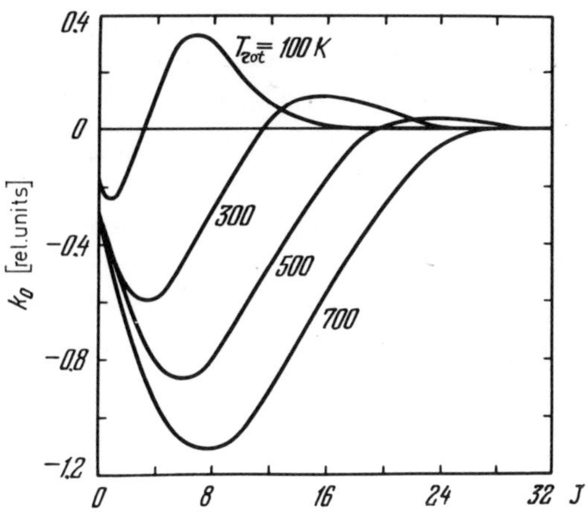

Fig.6.11.

Fig.6.10. Population x_n of CO levels as a function of level number n for the flow of a 20% CO-80% Ar mixture (T_* = 1800 K, p_* = 100 atm) in three types of nozzles. 1 — two-dimensional wedge nozzle, $\theta = 26^0$; 2 — two-dimensional contoured nozzle $S(x)/S_* = 1 + 100\ [1 - \exp(-x/2)]$; 3 — $S(x)/S_* = 1 + 100\ [1 - \exp(-x)]$; h_* = 0.4 mm for all the nozzles, x — distance from the nozzle throat in cm. Dashed line — Treanor distribution for flow parameters at point x = 82 cm for nozzle 2. Dotted and dashed line — Boltzmann distribution for T_* = 1800 K [6.20]

Fig.6.11. Small-signal gain as a function of rotational quantum number of CO molecule for x_n/x_{n-1} = 0.8 and different rotational temperatures T_{rot}

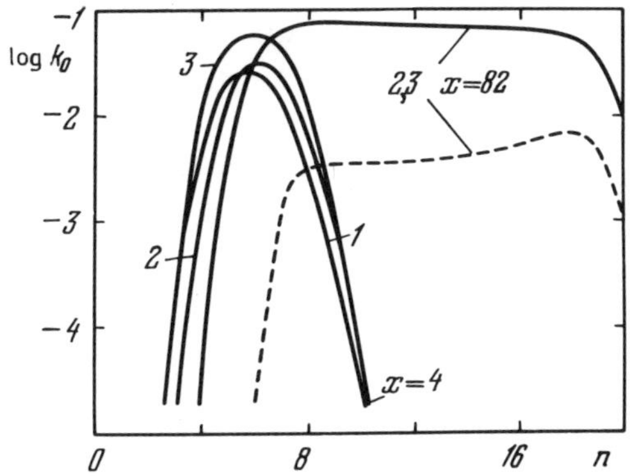

Fig.6.12. Small-signal gain (in cm^{-1}) maximum with respect to J as a function of level number n in the same nozzle conditions for a 20% CO-80% Ar mixture as in Fig.6.10. The dashed line gives k_0 for two-quantum transitions $n \to n - 2$

line corresponds to k_0 for two-quantum transitions in the wave length range 2.3-2.7 μm. The correspondence of theoretical results to the experiment has been established by the cessation of laser radiation with an increase of the CO content in the mixture (for $\xi_{CO} \geq 52\%$ lasing is absent both in experiments and according to the theory) [6.20]. The small-signal gain determined in [6.30] for a carbon monoxide GDL amounted to 3% cm^{-1}; theoretical results are close to this value. A good agreement between the theoretically determined distribution of the population of the CO levels and experimental results was observed in [6.31] for nozzle flows of CO-N_2-Ar mixtures.

The peak laser power P_{max} for the carbon monoxide GDL is close to

$$P_{max} = \eta_q G(\varepsilon_{max} - \varepsilon') \quad , \qquad (6.2.6)$$

where G is the gas flow rate, ε_{max} is the maximum vibrational energy per gram of a mixture, ε' is the residual vibrational energy at the cavity exit, and η_q is the quantum efficiency. For carbon monoxide η_q is close to unity because of the cascade mechanism of radiation, when the same molecules consecutively participate in radiation during the transitions from the upper levels to the lower ones. The value of P_{max} can be estimated using the principle of instantaneous freezing (4.2.3) and assuming that the distribution of CO molecules over the levels is quasistationary (Sect.3.4). The maximum efficiency of the carbon monoxide GDL (with no account of the losses due to V-T deactivation and absorption of energy in the cavity mirrors) can reach 10%. The specific power extracted from the carbon monoxide GDL in optimum conditions exceeds the maximum specific power of the CO_2 gasdynamic laser [6.32]. The combined scheme of electrical excitation and pumping of the CO molecules with subsequent cooling in the supersonic flow [6.33-35] is also effective. All this enhances the further development of the carbon monoxide GDL.

6.3 Gasdynamic Lasers for the Far Infrared Spectral Region

Let us consider some projects of other types of gasdynamic lasers. Of great interest is the development of GDLs for the far infrared, visible, and ultraviolet regions of the spectrum.

In the infrared and submillimeter wavelength range so far only low-power lasers on the vibration-rotational and purely rotational transitions in molecules without gasdynamic pumping are in operation (see references in [6.36,37]). The main difficulty in creating high-power lasers in this range is connected with very fast pumping, which is required because of high rates of the rotational relaxation and V-R transitions in polyatomic molecules. In [6.36,37], suggestions are made about a gasdynamic laser on rotational transitions in the hydrogen chloride molecule in a HCl-He mixture. This mixture should be heated to a temperature of several hundreds of kelvins (and according to estimates [6.37], even to T_0 = 2300 K) and then rapidly cooled in a supersonic nozzle. The probability of single-quantum rotational transitions decreases with an increase of the number J (as the temperature decreases, these transitions are dominating in comparison with multiquantum transitions). When the thermal equilibrium is distorted, the rotational relaxation occurs first of all in lower levels, and the population of the upper levels can be frozen, which leads to inversion. For this it is required that the number of the R-T transitions β_J ($J \rightarrow J + 1$), occurring during the characteristic time of expansion t^* in the nozzle, be such that $\beta_J \gg \beta_{J+1}$ [6.36]. As a result of numerical solution of the kinetic equations for the population of rotational levels together with gasdynamic equations for the supersonic nozzle flow of the HCl-He mixture for T_0 = 600 K in [6.36], a possibility of inversion on the levels with J = 13 to 20 was found, if the gas temperature in the nozzle drops to T < 200 K. These transitions correspond to the wavelength range from 20 to 40 μm. The laser power depends on the initial conditions, nozzle shape, and the parameters of intermolecular interaction used in calculations, which determine the probabilities of rotational transitions. In these estimates, the power reached 175 W per meter across the flow. The efficiency of such a laser amounts to about 0.1%.

There are suggestions concerning the creation of gasdynamic lasers in which the population inversion of vibrational levels is formed as a result of the interaction of polyatomic molecules with aerosol particles in the flow (during nonequilibrium adsorption at a surface, the lower vibrational levels are deactivated more intensively) [6.38] or as a result of nonequilibrium condensation of a supersaturated vapor [6.39]. Owing to the liberation of the association energy, the intermolecular types of vibrations of the complexes formed during condensation turn out to be excited, e.g., dimers A_2^* in the process $A_1 + A_2 \rightarrow A_2^*$, which subsequently are deactivated during collisions ($A_2^* + A_1 \rightarrow A_2 + A_1$). With a considerable supersaturation the unexcited dimers A_2 irreversibly turn into trimers during condensation, so that in a certain region of the jump of condensation the concentration of A_2^* can exceed the concentration of A_2. According to estimates [6.39] such a gasdynamic condensation-laser operating on the vibrational levels of dimers of water molecules, formed in a rapidly cooling mixture of water vapor and air, can provide for a laser power of about 1 kW per 1 kg of the mass in the spectral region from 25 to 100 μm.

6.4 Recombination Lasers on Electronic Transitions

For the solution of many scientific and technological problems high-power sources of laser radiation operating in the visible and ultraviolet spectral regions are required, in which lasing occurs on the transitions between electronic levels of atoms and molecules. For this purpose gasdynamic lasers may be developed in which an active medium is formed as a result of the chemical or electron-ion recombination proceeding in a rapidly cooled gas or plasma (in the latter case such lasers are called *plasmadynamic*, see Sect.6.5). Presently this scientific trend is being intensively developed, but so far these lasers have not been practically realized. We shall limit ourselves here and in Sect.6.5 only to a brief review of the basic ideas and suggested estimates of various versions of these lasers.

The first of the devices mentioned here—*recombination laser on electronic transitions*—is essentially a chemical laser, but the suggested way of creating inversion due to cooling (and thus stimulating the recombination) in a flow makes it gasdynamic in the sense adopted here. If at the nozzle entrance the temperature is so high that the XY molecules are dissociated into atoms or radicals X, Y, then during cooling in the nozzle intensive recombination occurs, which is accompanied by the formation of molecules in an excited (XY^*) and the ground (XY) electronic state as a result of mechanical (3.6.14) or optical (3.6.13) stabilization, i.e., during triple or double collisions,

$$X + Y + M \rightarrow XY^* + M \quad , \tag{6.4.1}$$

$$X + Y + M \rightarrow XY + M \quad , \tag{6.4.2}$$

$$X + Y \rightarrow XY^* \rightarrow XY + h\nu \quad . \tag{6.4.3}$$

The XY^* state is usually close to the dissociation limit (see Fig.6.13); in principle this can be weakly coupled XY' complex of two colliding particles X and Y [6.40,41]. The recombination reaction (6.4.1) with a subsequent fluorescence $XY^* \rightarrow XY + h\nu$ and the process (6.4.3) itself determine the recombinational chemiluminescence (photorecombination). The population inversion of the levels belonging to the states XY^* and XY can be provided when the rate of recombination into the XY^* state (6.4.1) is higher than the rate

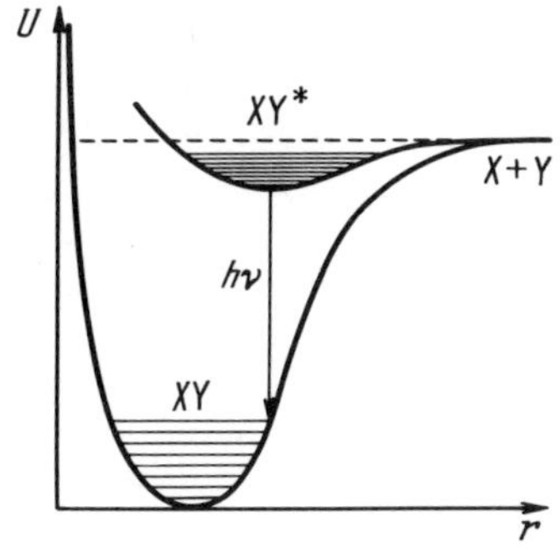

Fig.6.13. Potential curves of interaction between X and Y atoms resulting in molecules XY^* and XY. hν is the energy of optical transition $XY^* \rightarrow XY$

of recombination into the ground state (6.4.2), or when simultaneously with the formation of XY according to (6.4.2) another sufficiently powerful mechanism is acting (e.g., other chemical reactions), which removes the XY molecules. On the other hand, the number of the excited XY^* molecules decreases as a result of quenching, and we have

$$XY^* + M \rightarrow XY + M \quad . \tag{6.4.4}$$

The combination of processes (6.4.1-4) determines the possibility of creating inversion between XY^* and XY. The part of the optical stabilization (6.4.3) in these processes is usually small, since to obtain a noticeable gain a sufficiently high pressure is required, where the most essential is the three-particle recombination (see Sect.3.6). Partial inversion between separate vibrational levels of XY^* and XY, e.g., between lower vibrational levels of XY^* and upper vibrational levels of XY (Fig.6.13), can be created most easily. In this case the creation of inversion is facilitated by the rapid vibrational deactivation of XY^* and XY during gas cooling. The closer this system is to the vibrational equilibrium, in the conditions when the role of quenching processes (6.4.4) is small, the more molecules are on the upper level and the less on the lower ones. The value of the population inversion can be found by solving the system of kinetic equations of the type (3.6.12) for the processes (6.4.1-4) (and for other chemical reactions occurring here) simultaneously with vibrational transitions of molecules together with gasdynamic equations (Sect.4.1). The main difficulty in the solution of this problem is the lack of information about the reaction rate constants and transition probabilities. Kinetics of these processes is considered in [3.2, 6.42,43].

The prospects of the development of recombination lasers on electronic transitions can be established without solving this difficult kinetic problem just by simple estimates of the upper boundary of the possible values of the inversion and small-signal gain. For this purpose, according to [6.40,44], we shall assume that among free atoms (radicals) X,Y and the molecules in the XY state in the reaction (6.4.1) equilibrium rapidly sets in, and the distribution of molecules over vibrational levels is of the Boltzmann type with the temperature equal to the translational temperature of the gas. The first of these assumptions makes it possible to determine the number of molecules XY^* if we know the concentration of X and Y. By the law of mass action (Appendix G) we have

$$N_X N_Y / N_{XY^*} = Q_X Q_Y / Q_{XY^*} \quad , \tag{6.4.5}$$

where Q_j is the partition function for the jth component. According to the Boltzmann equation (3.1.1), the number of molecules on the upper (m) and lower (n) laser levels is, respectively,

$$N_m = (N_{XY^*}/Q_{XY^*}) g_m \exp(-E_m/kT) \quad ,$$

$$N_n = (N_{XY}/Q_{XY}) g_n \exp(-E_n/kT) \quad . \tag{6.4.6}$$

Substituting (6.4.6) into the inversion condition (2.2.6), using (6.4.5), and expressing the concentrations in terms of partial pressures, we obtain

$$p_{XY} < p_X p_Y \frac{\exp(-h\nu/kT)}{K_{eq}(T)} \quad , \tag{6.4.7}$$

where ν is the frequency of the electronic transition under consideration (Fig.6.13) and $K_{eq}(T)$ is the equilibrium constant for the reaction (6.4.2). Thus, most easily the inversion is created with the frozen recombination (6.4.2) at not very high temperatures and for long-wave spectral region. The upper limit of the small-signal gain k_0^+ (2.2.12) can be assessed assuming that the lower laser level is not populated. The value of k_0^+ is proportional to the population of the upper level. Proceeding from (2.2.4), we obtain, taking account of (2.1.11), (6.4.6) and (6.4.5),

$$k_0^+ = (c^2/8\pi\nu^2) a^*(\nu) N_X N_Y \quad . \tag{6.4.8}$$

The proportionality factor $a^*(\nu)$, which depends on the probability of spontaneous emission at a frequency ν, can be easily found from the experiments on recombination chemiluminescence (see [6.42,45,46]). The value of $a^*(\nu)$ for a monochromatic radiation is connected with the measured rate constant k_r^* of the recombination chemiluminescence in the entire spectral region through the relation

$$\frac{a^*(\nu)}{k_r^*} = \frac{I(\nu)}{\int I(\nu)d\nu} , \qquad (6.4.9)$$

where $I(\nu)$ is the intensity of monochromatic radiation. Passing over to the wavelength scale and expressing k_r in s^{-1} atm^{-2}, we obtain from (6.4.8)

$$k_0^+ = \frac{\lambda^4}{8\pi c} \cdot \frac{I(\lambda)}{\int I(\lambda)d\lambda} k_r^* p_X p_Y \quad . \qquad (6.4.10)$$

Note that the factor λ^4 in any case leads to a considerable shift in the spectral distribution of the small-signal gain $k_0^+(\lambda)$ to the red with respect to $I(\lambda)$. In particular, the peak of k_0^+ is always in the region of longer waves than the peak of $I(\lambda)$. Thus, to estimate k_0 we must know the spectral distribution of intensity of the recombination chemiluminescence $I(\lambda)$ (in relative units), concentrations of the recombining particles, and the rate constant k_r^*, which depends on temperature. This formula is used when the chemiluminescence spectrum is continuous. To compute k_0 for the line spectrum we must use relations (2.2.8-10).

The concrete estimates of the small-signal gain, which can be expected in the recombination laser on electronic transitions in molecular oxygen, were made in [6.47-49], in diatomic sulphur, in [6.49], and also in dioxides of nitrogen, carbon and sulphur in [6.45,47,50-58]. These estimates for real conditions give very low values of the small-signal gain, as a rule from 10^{-4} to 10^{-5} cm^{-1} and below. In certain cases these estimates proved to be exaggerated, for example, for the Herzberg bands of O_2 in [6.48]. The input data for these estimates are not always reliable, which also makes the results questionable (in particular, the estimate of k_0^+ for the long-wave region of the CO_2 spectrum in [6.49,55] is evidently overstated). The maximum value of k_0^+ (cm^{-1}) according to these calculations is

$$k_0^+ = b(1000/T)^s \exp(\theta/T) N_X N_Y \quad , \qquad (6.4.11)$$

where for different pairs of atoms and radicals X and Y at a wavelength λ we have the values given in Table 6.2 [6.45,54,55].

The low values of the small-signal gain are due to a large width of the chemiluminescence spectrum, reaching hundreds and thousands of angstroms, and also due to a low probability of the majority of optical transitions under consideration. It is clear from the preceding chapters that for a practical GDL realization it is required that the gain coefficient $k_0 \geq 1 \times 10^{-3}$ cm^{-1}.

Table 6.2.

X + Y	O + O		Cl + Cl
λ, Å	4800 (Herzberg bands)	5000 (Schuman-Runge bands)	11700
$b \times 10^{42}$, cm^5	3.4×10^{-5}	300	1
s	1.5	1.5	1.5
θ, K	8470	-14800	3300

X + Y	Br + Br		S + S	NO + O	SO + O
λ, Å	10 200	13 000	5460	10 500	4500
$b \times 10^{42}$, cm^5	0.8	0.1	700	7.5	15
s	1.5	1.5	0.5	2	2
θ, K	0	2900	2900	0	0

Let us turn to the estimates of operating conditions of the recombination laser. Applying the principle of instantaneous freezing (4.2.9), we obtain for a flow of a dissociated diatomic gas in a two-dimensional wedge nozzle the values given in Table 6.3 [6.49] (for the Herzberg bands of O_2 the gain coefficient is very small: $k_0^+ \approx 10^{-6}$ to 10^{-7} cm^{-1}). The most optimistic estimates were obtained for sulphur [6.59]. Free sulphur atoms are formed after heating triatomic gases — hydrogen sulphide H_2S, carbon sulphide CS_2, carbon oxisulphide COS, sulphur dioxide SO_2, etc. The upper limit of k_0 can be obtained assuming that in the nozzle flow of the heated gas mixture the chemical composition is frozen at the throat. The required values of temperature and pressure of the initial gas at the nozzle entrance can be determined from Fig.6.14. These estimates give the values of the small-signal gain $\sim 10^{-3}$ cm^{-1}.[4] It is more difficult to operate on the transitions of the re-

4 Laser radiation on the electronic transitions B → X of the S_2 molecules was obtained in [6.60] with optical pumping, which confirmed the possibility of using in lasers S_2 molecules. The estimates of the small-signal gain for the recombination of the S atoms were made also in [6.61].

Table 6.3.

XY	p_0[atm]	T_0[K]	h_*[mm]	k_0^+[cm^{-1}]
Br_2	18	3250	0.5	(0.5 to 1) × 10^{-4}
Cl_2	56	4100	0.5	(6 to 10) × 10^{-4}
O_2 (Schuman-Runge bands)	325	8000	0.1	1 × 10^{-4}

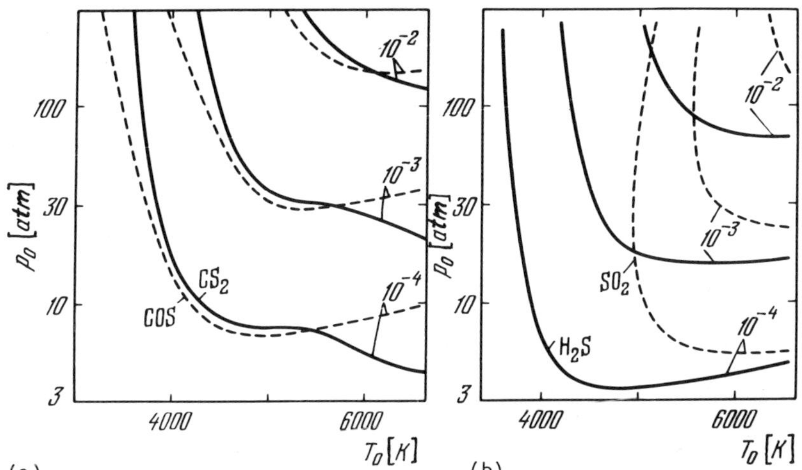

Fig.6.14. Effect of initial conditions on maximum values of small-signal gain (numbers at the curves, cm^{-1}, for S + S recombination (λ = 5460 Å) for flows of heated gases COS, CS_2 (a), H_2S, SO_2 (b)

combining triatomic gas NO + O and SO + O (Fig.6.15) [6.59]. The most severe conditions for NO_2 are due to the fact that during the dissociation of nitrogen dioxide a large number of O_2 and N_2 molecules is formed, while the concentration of NO and O is not very high. An example of the numerical solution for the problem of two-dimensional nozzle flow of a nonequilibrium recombining mixture formed during the dissociation of SO_2, taking account of other chemical reactions, is given in Fig.6.16.

The attempts to create recombination GDLs on electronic transitions were not successful despite the use of the mirrors with high reflectance. In the experiments with Cl_2, NO_2, SO_2 [6.57,58] and O_2 [6.62,63], lasing has not been registered. Even in the experiments with a 50-cm wide nozzle laser

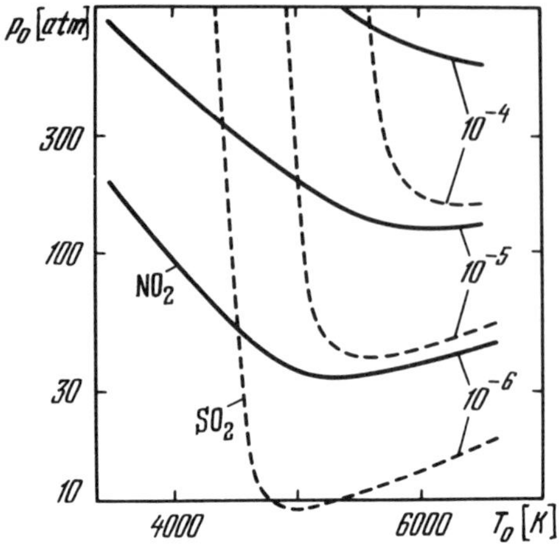

← Fig.6.15. The same as Fig.6.14 for XO + O recombination in NO_2 molecules (X ≡ N, λ = 10 500 Å) and SO_2 (X ≡ S, λ = 4500 Å)

Fig.6.16. Variation of temperature T and concentration ξ_j during recombination of SO_2 in a two-dimensional wedge nozzle flow ($h_* = 0.1$ mm, $\theta = 15^0$, $T_0 = 4000$ K, $p_0 = 20$ atm). Dashed lines correspond to equilibrium flow

↓

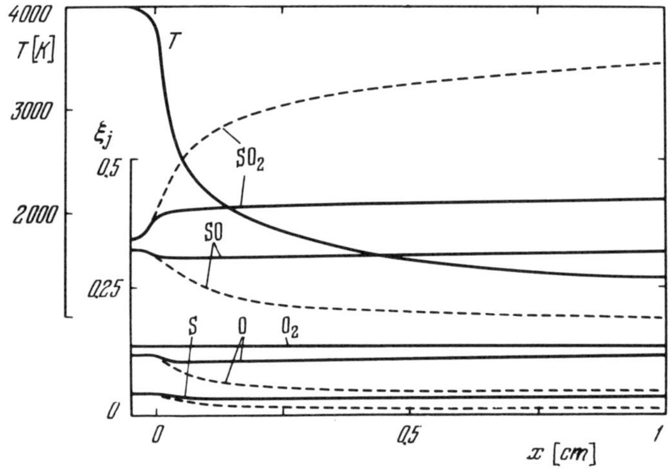

radiation in O_2 has not been observed. Nevertheless, the problem of creating these lasers continues to attract attention. This is connected with a prospect of obtaining high output power, since as a result of recombination highly excited electronic states are formed with energy close to the energy of dissociation. According to estimates [6.54,59], the maximum radiation power extracted from a Br recombination laser should make up about 0.5 kW per centimeter of the flow width with a maximum efficiency of about 20%. During the recombination S + S in the dissociated COS flow ($T_0 = 5000$ K, $p_0 = 30$ atm) through a 20-cm wide nozzle with a throat height $h_* = 0.5$ mm the power output at a wavelength λ = 5460 Å can reach 200 kW.

Further research in the development of recombination lasers is connected with the selection of new working media and the elaboration of different schemes of these lasers. A principal disadvantage of this laser is the adiabatic cooling, resulting in a considerable decrease of the flow density. It can be eliminated by using a rapid nonadiabatic cooling of a recombining gas, e.g., by delivering heat to a cooled array mounted in the flow [6.55]. In this method it is convenient to use reactions with a gain considerably increasing with a decrease of temperature, e.g., O + O, Br + Br, Cl + Cl [6.64a]. The possibility of accumulating energy in metastable electronic states with a large lifetime and of creating inversion as a result of the transfer of this energy to admixtures should be pointed out (for instance, from the nitrogen molecules to sodium atoms). The utilization of schemes involving component mixing in the supersonic flow, as was done with the CO_2 infrared lasers (Sect.5.5), may turn out to be promising. The mixing of carbon sulphide and nitrogen oxide with atomic oxygen at the nozzle exit is not effective {the estimate values of the small-signal gain on the transitions in SO_2 and NO_2 molecules are about (2.5 to 5) $\times 10^{-5}$ cm^{-1} [6.47]}. According to (5.5.1,2), the mixing in the throat region considerably increases with the formation of electronically excited states NO_2^* and HNO^* during mixing atomic oxygen with molecular complexes (clusters) of nitrogen oxides $(NO)_n$ [6.45,50,51]. For example, the molecular complex N_2O_3 ($NO_2 \cdot NO$) can be used as a working medium in the recombination laser [6.64b]. Large clusters-$(NO)_n$ drops — can also participate in the reactions with atomic oxygen at considerable cooling. In this case, with a heterophase mixing, according to the analyses [6.65,66] we can expect to obtain a small-signal gain $k_0 \approx 10^{-3}$ to 10^{-4} cm^{-1}. Turbulent mixing and the presence in a flow of condensate particles, however, considerably increase the level of cavity losses. These losses grow with a decrease of the wavelength (Sect.2.5).

One more possibility of creating the recombination laser in high-density gasdynamic flows consists in utilization of rapid heating of an exothermal gas mixture behind a shock front. According to [6.67], this heating, which causes the dissociation of unstable compounds, can provide high concentrations of atoms and radicals at nonequilibrium recombination. The principal difficulties of this method are associated with the necessity of using thin (of the order of tens of the wavelengths of light being amplified) active layers. In [6.67] it was shown that the zone of exothermic reaction behind the shock front may serve as a gas dielectric waveguide for radiation propagating across the flow. This waveguide may localize radiation in the region of the flow where the population inversion is realized. In [6.68], as an example, the gain and radiation at $\lambda = 0.6$ μm as a result of recombination $O + O \rightarrow O_2 + h\nu$ in a layer of the width of 5 μm behind the shock front in the mixture of O_3:Ar = 1:5 was considered (the Mach number for the shock wave was $\underset{\sim}{M} = 6.4$, initial pressure at the front $p_0 = 17$ atm, and $T_0 = 300$ K). In this case the small-signal gain can amount to 1.5×10^{-3} cm^{-1}, the specific energy, 41 J/g. These conditions can be realized either in pulses (for example, in a shock tube) or in a steady-state supersonic flow, where the position of the nonmoving shock front is determined by the flow geometry.

6.5 Plasmadynamic Lasers

During adiabatic expansion of plasma its temperature drops, and in the plasma the processes of electron-ion recombination develop. If the rate of plasma cooling is high, the ionization equilibrium is distorted, and the freezing of the concentration of electrons, N_e, occurs [1.7, 6.69,70]. As a result of recombination the upper levels of the formed atoms are predominantly populated. The population inversion in the cooling plasma is created, when the rate of deactivation of lower levels is higher than the deactivation rate of the upper levels. This deactivation occurs as a result of radiation or due to collisions of atoms with other particles (radiative or collisional mechanism of inversion formation). The first case is the most typical for a not very dense plasma, when the radiation can leave the volume occupied by plasma without reabsorption (optically thin plasma). In the optically thick plasma the radiation is again absorbed in the same volume, and deactivation is determined by collisional processes.

To create the nonequilibrium population of the levels in a recombining plasma, it is required that

$$\tau_r \gg t^* \quad , \tag{6.5.1}$$

where τ_r is the characteristic time of recombination and t^* is the gasdynamic scale of time (4.2.1) determined by the rate of plasma cooling. At a sufficiently high rate of relaxation of the atomic distribution over the excited levels the quasistationary state sets in (the constant sink flow approximation, see [6.70,71]; then we have

$$\tau_r = |(1/N_+)(dN_+/dt)|^{-1} = (\beta N_e)^{-1} \quad , \tag{6.5.2}$$

where N_+ is ionic concentration (for multiply charged ions $N_+ \neq N_e$) and β is the coefficient of the electron-ion recombination. In the simplest case of not very high temperature (e.g., for hydrogen below 3000 K) $\beta = 5.2 \times 10^{-22} \times z^3 N_e T_e^{-9/2}$, where z is the multiplicity of the recombining ion and T_e is the electron gas temperature (electronic temperature), expressed here in thousands of degrees kelvin [1.7]. Suppose that we managed to create inversion on the m and n levels,

$$\delta_{nm} = (g_m N_n / g_n N_m) < 1 \quad . \qquad (6.5.3)$$

The peak value of the laser radiation power, P_{max}, extracted from the unit volume, can be obtained if we assume that all the recombining atoms get to the level m, and the downward transitions from this level are only due to the energy escape because of laser radiation, so that $N_+ \tau_r^{-1} = B_{mn} \rho N_m$, where B_{mn} is the Einstein coefficient and ρ is the density of laser radiation (2.1.9). Then, proceeding from (2.3.31), we arrive at

$$P_{max} = h\nu N_+ (1 - \delta_{nm})/\tau_r \quad . \qquad (6.5.4)$$

Thus, the power in a plasmadynamic laser is proportional to $N_+ N_e^2$ and it considerably increases with a decrease of the electronic temperature $T_e (P_{max} \propto T_e^{-11/2})$. For very high electron concentrations it is difficult to comply with condition (6.5.1) and create inversion (6.5.3). Thus, there is an optimum value of N_e, which usually is at 1/2 to 1/3 of the value for which inversion disappears. Consequently, to realize a plasmadynamic laser we must cool the electronic component at the optimal value of its concentration. This cooling is mainly determined by the collisions of electrons with ions and is most effective in a plasma of light atoms (hydrogen, helium, etc.) [6.72].

To determine the population inversion, small-signal gain, and radiation power for a plasmadynamic laser, we must solve the system of kinetic balance equations for the population of atomic levels together with the equations of the radiation transfer and gasdynamic equations, in which we must account for the losses due to radiation and also energy exchange of electrons with ions and atoms. The form of these equations is given, for example, in [6.70, 71] and also in [1.7, 6.37,73]. In the simplest case of cooling the hydrogen plasma and plasma of hydrogenlike ions the radiative mechanism of inversion formation is realized. The analysis in the quasistationary approximation of a constant sink flow with no account for reabsorption [6.74] indicates the possibility of obtaining the maximum values of $k_0 \sim 10^4$ cm^{-1} and $P_{max} \approx 500$ kW/cm^3 for $N_e = 3 \times 10^{15}$ cm^{-3} and $T_e = 1000$ K on the transition

$3 \rightarrow 2$ ($\lambda = 6563$ Å) in hydrogen. In this case to comply with conditions (6.5.1), we must provide for cooling of the plasma during $t^* < 10^{-9}$ s, which is a difficult problem. If we limit ourselves to the values of $t^* \sim 10^{-6}$ s, then P_{max} will be about 100 W/cm^3 for $N_e = 1 \times 10^{15}$ cm^{-3} and $T_e = 3000$ K. To do this we must cool this plasma in a flow with a velocity about 10 km/s. The rejection of the quasistationary nature and complete numerical solution of the system of kinetic balance equations gives considerably lower values of the small-signal gain and the values of the cooling rate which are difficult to achieve with the adiabatic expansion of plasma [6.75]. The account of reabsorption practically leads to the disappearance of the population inversion on the $3 \rightarrow 2$ transition in hydrogen because of the absorption of radiation on the $1 \rightarrow 2$ transition and population of level 2 [6.72,76]. The inversion and gain coefficient on the transitions $4 \rightarrow 3$ ($\lambda = 1.88$ μm) and $5 \rightarrow 4$ ($\lambda = 4.05$ μm) in a freely expanding jet of hydrogen plasma are in principle sufficient for a plasmadynamic laser performance [6.72] (the analysis was carried out in the quasistationary approximation taking account of absorption), but the maximum power P_Σ extracted from the unit area of the resonator mirror and the efficiency η of this laser are, according to [6.72], small: $P_\Sigma = 0.35$ and 7.5×10^{-3} W/cm^3, $\eta = 0.03\%$ and 0.007% for the transitions $4 \rightarrow 3$ and $5 \rightarrow 4$, respectively. To decrease the effect of reabsorption it is expedient to use two-dimensional nozzles, slits, and nozzle arrays.

Unlike hydrogen, in more complex atoms and ions the collisional mechanism of inversion formation can be realized, when the redistribution of particles among different systems of levels occurs [6.37]. According to [4.8], the cooling and recombination of nitrogen plasma in a nozzle flow must lead to the excessive population of the upper level of nitrogen atoms for the transition at a wavelength of 1745 Å. The study of the kinetics of the xenon plasma expanding in the nozzle [6.77] revealed the possibility of creating inversion between the levels 7p and 5d, 4f and 5d of the Xe atom, which corresponds to the transition in the wavelength range from 0.9 to 1.46 μm.

The population inversion between the electronic levels in an expanding plasma was observed for the first time in [1.31] (see [6.78]) for the levels $n = 5$ and 4 of hydrogen in the supersonic argon-hydrogen flow. In experiments [6.79] the inversion in argon between low levels of the 5p group with respect to the levels of the 4p group and in helium between the levels $n = 4$ and 3 was observed. The existence of inversion in neon was established in [6.80], and the possibility of obtaining inversion in hydrogen was experimentally confirmed in [6.81,82], but the absolute value of the inversion ΔN in these experiments was not high. In experiments [6.83] the conditions for creating the population inversion of levels n = 3 and 4 in a recombining helium flow was studied in detail, and it was shown that the inversion is formed due to the difference in the rates of collisional deactivation of these levels. The computed values of the small-signal gain at a wavelength $\lambda \approx 1.7$ μm, according to the results of these experiments, were at 2 to 5% cm^{-1}. One of the possibilities for the successful solution of the problem is the addition of different admixtures to the plasma jet. In [6.84] a considerable population inversion (up to $\Delta N \sim 10^{14}$ cm^{-3}) on the transition $6S_{1/2} \rightarrow 5P_{3/2}$ of the carbon atom in the ultraviolet region (λ = 2478 Å) during the recombination of the argon-helium plasma with the admixture of methane into the jet behind the supersonic nozzle was reported. Unfortunately, this was not confirmed in later studies.

To increase the cooling rate of a plasma, use can be made of electromagnetic accelerators. In the argon jet flowing out of such an accelerator at a velocity of above 10 km/s, according to [6.85], there is a considerable population inversion of the argon ionic levels ($\Delta N \sim 10^{11}$ cm^{-3}) for the transition 4p → 4s (λ = 4880 Å). A quasistationary plasmadynamic laser operating according to this scheme was realized in [6.86]. The gain on the transitions of the same group of argon levels, but now at λ = 5145 Å, was registered in experiments [6.87], in which laser properties of ionized argon in the supersonic flow had been studied. The excitation of argon in these experiments occurred in the electric arc rotated by a magnetic field in the direction perpendicular to the gas flow. In these devices the combined method of pumping is realized, in which the excitation by the electronic impact is combined with the plasma recombination in the supersonic flow.

Appendixes

A. Derivation of the Relations Between Einstein's Coefficients

In thermal equilibrium, according to Boltzmann's formula, we have

$$\bar{N}_m/\bar{N}_n = (g_m/g_n)\exp(-h\nu/kT) \quad , \tag{A.1}$$

where k is Boltzmann's constant and T is the gas temperature. Substituting (A.1) into (2.1.12), we get

$$\bar{\rho}_\nu = \frac{a_{mn}(\nu)/b_{mn}(\nu)}{[g_n b_{nm}(\nu)/g_m b_{mn}(\nu)]\exp(h\nu/kT) - 1} \quad . \tag{A.2}$$

On the other hand, the spectral density of radiation for the systems in thermal equilibrium, according to Planck, is

$$\bar{\rho}_\nu = \frac{8\pi h\nu^3}{c^3}\,\frac{1}{\exp(h\nu/kT) - 1} \quad , \tag{A.3}$$

where c is the velocity of light. Comparing (A.2) and (A.3), we arrive at (2.1.11). If in the region under consideration the frequency varies insignificantly, so that the quantity $\bar{\rho}_\nu$ can be considered constant, then, according to (2.1.10), we obtain similar relations for the integral Einstein coefficients,

$$g_n B_{nm} = g_m B_{mn} \quad , \quad A_{mn} = (8\pi h\nu^3/c^3) B_{mn} \quad . \tag{A.4}$$

B. Approximate Analytical Representation of the Voigt Functions

Let us introduce

$$b = \Delta_c/2 + \sqrt{(\Delta\nu_c/2)^2 + \Delta\nu_D} \quad .$$

Then the Voigt contour width of the spectral line is approximately equal to [B.1]

$$\Delta\nu_V = b + 0.1(1 - \Delta\nu_c/b) \quad . \qquad \text{(B.1)}$$

Let us denote $\xi = \Delta\nu_c/\Delta\nu_V$ $\eta = 2(\nu - \nu_0)/\Delta\nu_V$. According to [B.1],

$$H(\xi,\eta) = \frac{\Delta\nu_D}{\Delta\nu_V}\left[(1 - \xi)\exp(-\eta^2 \ln 2) + \frac{1}{\sqrt{\pi \ln 2}}\,\frac{\xi}{1 + \eta^2}\right] \quad . \qquad \text{(B.2)}$$

This expression for $\eta < 1.6$ gives exaggerated values of the Voigt functions with the error less than 3%, and for $\eta > 1.6$, understated values with the maximum error of about 25% (for $\xi \approx 0.1$, $\eta \approx 3$), which decreases with a further increase of η. A more precise relation (with the error below 3% everywhere) can be found in [B.1].

The asymptotic expression for the Voigt function as a function of parameters a and ω (2.1.20,21) in the form of an infinite series is given, for example, in [B.2]. In the limiting case of a purely Doppler broadening ($a \to 0$) we have

$$H(0,\omega) = \exp(-\omega^2) \quad . \qquad \text{(B.3)}$$

On the contrary, under the suppressing effect of collisional broadening ($a \to \infty$) the Voigt function has the form

$$H(\infty,\omega) = a/\sqrt{\pi}(a^2 + \omega^2) \quad . \qquad \text{(B.4)}$$

At the center of the line

$$H(a,0) = \exp(a^2)[1 - \mathrm{erf}(a)] \quad , \quad \text{where}$$

$$\mathrm{erf}(a) = (2/\sqrt{\pi}) \int_0^a \exp(-t^2)dt \quad . \qquad \text{(B.5)}$$

For $a < 0.2$ we can use the expansion

$$H(a,0) = 1 - 1.12838\,a + a^2 - 0.752\,a^3 + 0.5\,a^4 + \ldots \quad ;$$

and for $a > 1.4$,

$$H(a,0) = (1/\sqrt{\pi}a)(1 - 1/2a^2 + 3/4a^4 + \ldots) \quad .$$

C. Small-Signal Gain as a Function of Intensity

Let τ_m and τ_n determine the lifetimes of the levels m and n as a result of collisions of particles and let S_m be the rate of pumping of atoms (molecules) to level m. The number of optical transitions $m \rightleftharpoons n$ per unit cavity volume per unit time, according to (2.3.31), is $kI/h\nu$, where k is the gain coefficient and I is the radiation intensity. Then, taking into account (2.2.12), we can write the balance equation for the population of the levels under consideration in the form

$$\frac{dN_m}{dt} = -\frac{N_m}{\tau_m} - \sigma_{mn}N_m\frac{I}{h\nu} + \sigma_{nm}N_n\frac{I}{h\nu} + S_m \quad ,$$

$$\frac{dN_n}{dt} = -\frac{N_n}{\tau_n} + \sigma_{mn}N_m\frac{I}{h\nu} - \sigma_{nm}N_n\frac{I}{h\nu} \quad . \tag{C.1}$$

In the stationary state $(dN_m/dt) = (dN_n/dt) = 0$, and

$$N_m/\tau_m + N_n/\tau_n = S_m \quad , \tag{C.2}$$

i.e., the rate of pumping equals the deactivation rate. Let us write for this state the system of equations including one of the equations (C.1) and (C.2). Solving this system for constant values of S_m, τ_m, τ_n, we obtain

$$N_m = \frac{S_m\tau_m(1 + \tau_n\sigma_{nm}I/h\nu)}{1 + (\tau_n\sigma_{nm} + \tau_m\sigma_{mn})I/h\nu} \quad , \tag{C.3}$$

$$N_n = \frac{S_m\tau_m\tau_n\sigma_{mn}I/h\nu}{1 + (\tau_n\sigma_{nm} + \tau_m\sigma_{mn})I/h\nu} \quad . \tag{C.4}$$

Substituting these values into (2.2.12), we get

$$k = \frac{k_0}{1 + I/\bar{I}} \quad , \tag{C.5}$$

where k_0 is the value of k for I = 0 and $\bar{I}$ is the saturation parameter,

$$\bar{I} = h\nu/(\tau_n\sigma_{nm} + \tau_m\sigma_{mn}) \quad . \tag{C.6}$$

When deriving (C.5), we assumed that the nature of interaction of every particle in the state m or n with the field of radiation in the cavity does not depend on the direction of motion of these particles; therefore, the change of the gain coefficient for a given frequency is proportional to the total number of induced transitions, which is characteristic of the homogeneous line broadening [2.21]. Then for the center of the line $\sigma_{mn} = \lambda^2 A_{mn}/8\pi\Delta\nu_c$. For a nonhomogeneous line broadening the balance equations similar to (C.1) should be written for the particles with fixed values of the projection of the velocity of thermal motion on the direction of radiation propagation. In the limiting case of pure Doppler broadening $k = k_0/\sqrt{1 + I/\bar{I}}$ [2.21]. The dependence k(I) for the line with the Voigt contour is more complicated.

D. Simplest Models of Molecular Motion

D.1 Translational Motion

When describing the translational motion of molecules, it is simplest to assume that molecules are rigid spheres, which elastically interact during collisions (rigid spheres model [3.1,3]). The solution of the problem of the motion of molecules in a closed volume V shows that the difference between translational energy levels is extremely small—much smaller than the mean thermal energy of the system, kT (this problem is similar to considering electromagnetic vibrations in a resonator in Sect.2.3, since the motion of molecules can be reduced to the propagation of the de Broglie waves). The partition function for translational degrees of freedom Q_t is calculated from the results of this solution and is expressed (see [D.1,2] or some other textbook on statistical thermodynamics) as

$$Q_t = (2\pi mkT/h^2)^{3/2}V \quad , \tag{D.1}$$

where m is the mass of molecules and h, k are the Planck and the Boltzmann constants, respectively. Owing to the small difference between energy levels, the translational energy e_t can be conveniently determined from the classical law of the energy distribution over the degrees of freedom (kT/2 per each degree per a molecule). Then

$$e_t = (3/2)kT \quad . \tag{D.2}$$

D.2 Rotation of Molecules

For a rigid rotor model, energy levels are distributed according to (2.1.1); the population of the Jth level is (3.1.1)

$$N_J = (N/Q_{rot})g_J \exp[-J(J + 1)h^2/8\pi^2 IkT_{rot}] \quad . \tag{D.3}$$

where I is the moment of inertia of a molecule, T_{rot} is the rotational temperature, g_J is the statistical weight of the Jth rotational level, $g_J = 2J + 1$, Q_{rot} is the partition function for rotational states, and N is the total number of molecules. In spectroscopy the quantity $h/8\pi^2 cI$ (c is the velocity of light) is called the rotational constant B (cm^{-1}). The quantity $\theta_{rot} = hcB/k$ has the dimension of temperature, is determined only by the molecular constants, and is called the characteristic rotational, temperature. The value of θ_{rot} for the simplest molecules is 85.4 K (H_2), 2.86 K (N_2), and 0.57 K (CO_2), so that in ordinary conditions θ_{rot} is considerably lower than the gas temperature. We can rewrite (D.3) with the account of θ_{rot} in the following form,

$$N_J = (N/Q_{rot})g_J \exp[-J(J + 1)\theta_{rot}/T_{rot}] \quad . \qquad (D.4)$$

The partition function Q_{rot} should be computed by (3.1.2) where due to the closeness of rotational energy levels, the sum can be replaced by an integral with respect to J. Then (see [D.1,2]) we obtain

$$Q_{rot} = T_{rot}/\sigma\theta_{rot} \quad , \qquad (D.5)$$

where σ is the symmetry factor: $\sigma = 1$ for asymmetrical molecules (e.g., diatomic molecules with different nuclei) and $\sigma = 2$ for homonuclear diatomic molecules and symmetrical linear polyatomic molecules, e.g., CO_2. In the rigid rotor approximation this relation is valid for all linear molecules. For nonlinear molecules (H_2O, NO_2, etc.) in the same approximation the expression for Q_{rot} is more complicated (see [2.8, D.3]). Rotational energy e_{rot} is usually calculated in the classical approximation, since $T_{rot} \gg \theta_{rot}$. Thus, for linear molecules we have

$$e_{rot} = kT_{rot} \qquad (D.6)$$

and for nonlinear molecules,

$$e_{rot} = (3/2)kT_{rot} \quad . \qquad (D.7)$$

D.3 Vibrations of Molecules

For the simplest model of a nondegenerate harmonic oscillator (2.1.2), whose energy levels are distributed according to (2.1.3), the population of the vth level by (3.1.1) is

$$N_v = (N/Q_{vib})\exp(-vh\nu_0/kT_{vib}) \quad , \qquad (D.8)$$

where ν_0 is the frequency of vibrations, T_{vib} is the vibrational temperature, and Q_{vib} is the partition function for vibrations. The quantity $h\nu_0/k$

is a constant for a given molecule (or given vibrational mode); therefore it is assumed that $h\nu_0/k = \theta_{vib}$ where θ_{vib} is the so-called characteristic vibrational temperature. The values of θ_{vib} for diatomic molecules are, for example, 465 K (Br_2), 807 K (Cl_2), 2240 K (O_2), 3084 K (CO), 3354 K (N_2), 4160 K (HCl), 6000 K (H_2), and for the three modes of the CO_2 vibrations, θ_{vib} = 1997 K (ν_1), 960 K (ν_2), and 3380 K (ν_3). Thus, for a nondegenerate harmonic oscillator we have

$$N_v = (N/Q_{vib})\exp(-v\theta_{vib}/kT_{vib}) \quad . \qquad (D.9)$$

The vibrational partition function for this model of an harmonic oscillator is

$$Q_{vib} = \sum \exp(-v\theta_{vib}/T_{vib}) = [1 - \exp(-\theta_{vib}/T_{vib})]^{-1} \quad . \qquad (D.10)$$

We shall consider the simplest case of a degenerate harmonic oscillator using the example of a doubly degenerate deformational mode of the CO_2 vibrations. In this case, according to Sect.2.1, every level has a statistical weight $g_{v_2} = v_2 + 1$. When determining the population of the level v_2, according to (3.1.1), this factor should be introduced into (D.9). The partition function for this mode of vibrations in the approximation under consideration is

$$Q_{vib}(\nu_2) = \{1 - \exp[-\theta_{vib}(\nu_2)/T_{vib}(\nu_2)]\}^{-2} \quad . \qquad (D.11)$$

The vibrational energy e_{vib} of one mode of vibrations, which is modeled by a nondegenerate harmonic oscillator, per molecule is obtained by summation

$$e_{vib} = \sum vh\nu_0 N_v/N = k\theta_{vib}[\exp(\theta_{vib}/T_{vib}) - 1]^{-1} \quad . \qquad (D.12)$$

As the temperature rises, e_{vib} grows from zero to its classical limit kT_{vib}, so that, for example, e_{vib}/kT_{vib} = 0.313; 0.582; and 0.771 for T_{vib}/θ_{vib} = 0.5; 1; and 2, respectively. With an increase of temperature, however, the effect of anharmonicity of the molecular vibrations grows, and more complex potentials of intramolecular interaction should be taken into account (see [2.8]).

D.4 Motion of Electrons

Energy levels for the motion of electrons in atoms and molecules are determined by a complex nature of intramolecular interactions; therefore, it is difficult to find a simple model of motion (besides the Coulomb oscillator model, which has a very narrow range of application). The energy of different electronic states as a rule is much higher than the energy of other types of motion considered here. This enables us to consider every electronic state separately, on the basis of electronic spectra of atoms and molecules. The list and the detailed analysis of these states are given, for example, in [2.5,6,8, D.4,5]. The population of electronic states is described by (3.1.1) and the partition function, by (3.1.2). The statistical weight g even for the ground electronic states of different atoms and molecules takes on different values; for example, for the ground state of the N_2 molecule g = 1, for the N atom g = 4, for the O_2 molecule g = 3, and for the O atom g = 9 (see [2.8]).

E. Relations Between Concentrations of Mixture Components

To describe the composition of gas mixtures, various expressions for concentrations are used. The density of particles n_j equals the number of particles of the jth type per unit volume. The molar density of a substance A_j (molar-volume concentration) $[A_j]$ equals the number of moles of this component per unit volume. The molar-mass concentration of the jth component of a gas mixture, c_j, equals the number of moles of this component per unit mass of the mixture. Evidently, they are related as

$$[A_j] = n_j/N_A \quad \text{and} \quad c_j = [A_j]/\rho \tag{E.1}$$

where N_A is the Avogadro number and ρ is the density of a gas mixture. The molar fraction ξ_j equals the ratio of the number of moles of the jth component to the total number of moles per unit volume of the mixture. According to (E.1),

$$\xi_j = n_j/n \quad , \quad \text{where} \quad n = \sum n_j \quad . \tag{E.2}$$

Introducing the partial pressure p_j and taking into account that $p_j = n_jkT$ (see Sect.3.1), we have $\xi_j = p_j/p$, where p is the total pressure of the gas mixture. The mass fraction (or the relative density) α_j equals the ratio of the mass of the component under consideration to the total mass of the mixture or, which is the same, the ratio of the partial density ρ_j to the total density of the mixture ρ. Taking into account that $\rho_j = m_jn_j$, $\rho = \sum \rho_j$, where m_j is the mass of a particle of the jth type, we can write

$$\alpha_j = \rho_j/\rho = m_jn_j/\sum m_jn_j = \mu_jn_j/\sum \mu_jn_j \quad , \tag{E.3}$$

where μ_j is the molecular mass of the jth component. To describe mixtures in a moving medium, sometimes it is convenient to use the normalized molar-mass concentration γ_j,

$$\gamma_j = c_j\mu_\infty \quad , \quad \text{or} \quad \gamma_j = n_j/n_e \quad , \quad \text{where} \quad n_e = n_\infty(\rho/\rho_\infty) \quad . \tag{E.4}$$

The quantities with a subscript ∞ for convenience should be taken in the regions of the flow where their values are known from the conditions of the problem, e.g., in a nonmoving gas (at the nozzle entrance, before the front of a shock wave, etc.). The quantity n_e equals the density of all the particles of the mixture, which is obtained as a result of the variation of the gas density from ρ_∞ to ρ with a constant mixture composition, so that γ_j has the meaning of the relative density of the jth particles.

From (E.1-4) we obtain the following relations for concentrations

$$\begin{aligned}
&[A_j] = \rho_j/\mu_j = \xi_j \sum n_j/N_A = \gamma_j(n_\infty/N_A)(\rho/\rho_\infty) \quad , \\
&\alpha_j = \xi_j \frac{\mu_j}{\mu} \quad , \quad c_j = \rho_j/\mu_j\rho = \alpha_j/\mu_j \quad , \quad \alpha_j = c_j\mu_j = [A_j]\mu_j/\rho \quad , \\
&\xi_j = c_j/\sum c_j = \gamma_j/\sum \gamma_j = \alpha_j/[\mu_j \sum (\alpha_j/\mu_j)] \quad , \quad \gamma_j = \xi_j\mu_\infty/\mu \quad ,
\end{aligned} \tag{E.5}$$

where μ is the molecular mass of the gas mixture,

$$\begin{aligned}
\mu &= \sum \xi_j\mu_j \\
&= \frac{\sum \mu_j[A_j]}{\sum[A_j]} = \frac{\sum \mu_j n_j}{\sum n_j} = \frac{\sum \gamma_j\mu_j}{\sum \gamma_j} = \frac{1}{\sum c_j} = \frac{1}{\sum(\alpha_j/\mu_j)} \quad .
\end{aligned} \tag{E.6}$$

If there are liquid or solid particles of condensed phase in a mixture, their concentration is often described by the mass fractions (α_c') or molar fractions (ξ_c'). In this case the volume occupied by the condensate is usually neglected in comparison with the volume of the gas phase; the mass fraction α_c' is determined by the ratio of the condensed substance mass to the mass of the gas (in a certain volume), and the molar fraction ξ_c' as the ratio of the number of moles of condensate not to the total number of moles, but to the number of moles of the gaseous substance. Then

$$\alpha_c' = \frac{M_c N_A}{V\mu \sum n_j} \quad , \quad \xi_c' = \frac{M_c}{\mu_c V \sum[A_j]} \quad \text{so that} \quad \alpha_c' = \frac{\mu_c}{\mu}\,\xi_c' \quad . \tag{E.7}$$

Here, in addition to the previous notations, V is the volume occupied by the mixture, M_c is the total mass of the condensed phase in this volume, and μ_c is the molecular mass of the substance in this phase (which is not taken into account in the total molecular mass μ of the gas mixture).

F. Material Balance Equations for Mixtures

First of all it should be noted that from Dalton's law (3.1.3) it follows that

$$\sum \xi_j = 1 \quad \text{or} \quad \sum \alpha_j = 1 \ , \tag{F.1}$$

from which

$$\sum \gamma_j = \mu_\infty/\mu \ , \quad \sum c_j\mu_j = 1 \ , \quad \sum \mu_j[A_j] = \rho \ , \quad \sum \mu_j n_j = \rho N_A \ , \tag{F.2}$$

where N_A is the Avogadro number and μ_∞ is the initial molecular mass (see Appendix E).

The material balance equations are obtained from the principle of conservation of chemical elements. Assume that in a molecule s there are r_{qs} atoms of element q. Then in a unit volume the total number of atoms of this element, regardless of the compounds in which they are present, is $\sum_s r_{qs}n_s$. A similar sum can be written for element t. Regardless of the chemical reactions proceeding and the number of particles changing in a unit volume, the ratio of the total number of atoms of the two elements remains constant,

$$\frac{\sum_s r_{qs}n_s}{\sum_s r_{ts}n_s} = C_{qt} \ , \tag{F.3}$$

or, passing over to molar fractions,

$$\frac{\sum_s r_{qs}\xi_s}{\sum_s r_{ts}\xi_s} = C_{qt} \ . \tag{F.4}$$

The constant C_{qt} is determined by the initial conditions of the problem. Similar relations are obtained in describing ionization processes, when the law of charge conservation is taken into account.

With the account of a condensed phase, (F.4) is written in the form

$$\frac{\sum_s r_{qs}\xi'_s + \sum_c r_{qc}\xi'_c}{\sum_s r_{ts}\xi'_s + \sum_c r_{tc}\xi'_c} = C'_{qt} \quad , \tag{F.5}$$

where C'_{qt} is the initial ratio of gram-atoms of elements q and t (regardless the phase of the substance), and the subscripts s and c relate to the gas and condensate, respectively, while the stroke at molar fractions ξ'_c and ξ'_s, as in Appendix E, indicates, that these quantities are related not to the total number of moles, but only to the number of moles of the gaseous component. Relation (F.1) formally does not change,

$$\sum \xi'_s = 1 \quad , \tag{F.6}$$

but now it includes not all the molar fractions, but only that of gaseous substances.

G. Chemical Equilibrium in a Gas Mixture

Let us consider a chemical transformation of molecules A + B = C + D; let the concentration of the components (here the density of particles) in a state of chemical equilibrium be $\bar{n}_A$, $\bar{n}_B$, $\bar{n}_C$, and $\bar{n}_D$; and the reaction proceed in a gas mixture at temperature T in volume V. At equilibrium the number of particles of the jth type in volume V is proportional to the partition function (3.1.2) for this substance for a given temperature: $\bar{n}_j V \propto Q_j(T)$. This assertion reflects the fact that equilibrium is the most probable state of a system, and the number of microscopic ways of realization of this state for the system is proportional to Q. Evidently, in equilibrium the number of particles of the jth type is proportional to the number of possible realizations for a set of particles under consideration.[1] Then for a set of systems of different particles A, B, C, D we obtain the law of mass action,

$$\frac{\bar{n}_C\bar{n}_D}{\bar{n}_A\bar{n}_B} = \frac{(Q_C/V)(Q_D/V)}{(Q_A/V)(Q_B/V)} \equiv K \quad . \tag{G.1}$$

---

1 Similar intuitive considerations in this respect can be found in [3.113]; in the same work, as in many other books on statistical physics, there is a strict derivation of the law of mass action (see, for example, [3.6]).

The quantity K is called the equilibrium constant. According to (D.1), V is canceled out, so that K depends only on temperature. Relations of the type (G.1) can be written for any reaction; for instance, for dissociation $X_2 = X + X$, we have

$$\frac{\bar{n}_X^2}{\bar{n}_{X_2}} = \left(\frac{Q_X}{V}\right)^2 \left(\frac{Q_{X_2}}{V}\right)^{-1} \equiv K \quad , \tag{G.2}$$

or for a more general decomposition reaction $XYZ = X + Y + Z$,

$$\frac{\bar{n}_X \bar{n}_Y \bar{n}_Z}{\bar{n}_{XYZ}} = \left(\frac{Q_X}{V}\right)\left(\frac{Q_Y}{V}\right)\left(\frac{Q_Z}{V}\right)\left(\frac{Q_{XYZ}}{V}\right)^{-1} \equiv K \quad . \tag{G.3}$$

In the Tables and for approximations of K, partial pressures $p_j = n_j kT$ are usually used instead of concentrations (see, for example, [2.8]). In calculations molar fractions $\xi_j = p_j/p$ are often used. Then instead of (G.1-3) we have, respectively,

$$\bar{p}_C \bar{p}_D / \bar{p}_A \bar{p}_B = K \equiv K_{eq} \quad , \quad \bar{\xi}_C \bar{\xi}_D / \bar{\xi}_A \bar{\xi}_B = K_{eq} \quad , \tag{G.4}$$

$$\bar{p}_X^2 / \bar{p}_{X_2} = kTK \equiv K_{eq} \quad , \quad \bar{\xi}_X^2 / \bar{\xi}_{X_2} = K_{eq}/p \quad , \tag{G.5}$$

$$\bar{p}_X \bar{p}_Y \bar{p}_Z / \bar{p}_{XYZ} = (kT)^2 K \equiv K_{eq} \quad , \quad \bar{\xi}_X \bar{\xi}_Y \bar{\xi}_Z / \bar{\xi}_{XYZ} = K_{eq}/p^2 \quad , \tag{G.6}$$

where the bar above the concentration again indicates the equilibrium state. The equilibrium constant K_{eq}, which determines the equilibrium ratio of partial pressures, also depends only on temperature and has different dimensions depending on the difference between the number of particles entering into a chemical reaction and the number of particles obtained as a result of the reaction. So, in (G.4) K_{eq} is a dimensionless quantity, in (G.5) is has the dimension of pressure (most often it is expressed in atm), and in (G.6), the dimension of the square of pressure (atm^2). The values of K_{eq} (T) for decomposition reactions are given, for example, in the second volume of [2.8].

For practical solution of the problem about equilibrium composition of a gas mixture, for certain values of temperature T and pressure p one should write the relations of the type (G.5,6) for decomposition reactions of all the molecules which are present or may form in the mixture under consideration. These relations connect the concentrations of all the molecular components with concentrations of atoms ($\bar{\xi}_X$, $\bar{\xi}_Y$, $\bar{\xi}_Z$, ...). To determine the concentrations of atomic components, use is made of Dalton's law in the form (F.1) or (F.2), and also the material balance relations in the form (F.4) or in some other form. The total number of equations exactly equals the number of required values of concentrations. This procedure takes account of the contribution into equilibrium composition of all the other reactions actually proceeding in the mixture, in addition to the decomposition reactions considered above. The system of nonlinear algebraic equations thus obtained is solved by numerical methods.

If a condensed phase is present in a mixture, the partial pressure of vapor of this substance is taken equal to its saturation pressure $\bar{p}_{sat}(j)$ and, hence, to the indicated relations we must add the balance equation for the condensate and its vapor: $\bar{\xi}'_s(j) = \bar{p}_{sat}(j)/p$ [for $\bar{\xi}'_c(j)$ differing from zero], where p is the overall pressure of the gas in the mixture.

H. Derivation of the Arrhenius Formula for the Rate Constant of a Chemical Reaction

Suppose that a chemical reaction between molecules A and B proceeds only when the relative energy of collision, E_c, along the line connecting the centers of the particles exceeds a certain threshold value Q (activation energy), regardless of the internal state of the colliding particles. In this case the reaction probability is $p = 1$ for $b \le b_{max}$ and $p = 0$ for $b > b_{max}$, where b_{max} is the maximum value of the impact parameter at which the reaction occurs. This value can be easily determined from geometrical considerations of the collision (Fig.H.1), where d is the collision diameter equal to the distance between the centers of the spheres at their closest approach, $\underset{\sim}{v}$ is the vector of velocity of the relative motion of particles A and B, and $\underset{\sim}{v}_c$ is the relative velocity vector along the center line. Since

$$v = \sqrt{2E/\mu} \quad , \quad v_c = \sqrt{2E_c/\mu} \quad , \quad b/d = \sqrt{v^2 - v_c^2}/v$$

(E is the energy of the relative motion, μ is the reduced mass), the maximum impact parameter is $b_{max} = d(1 - Q/E)$. Then, integrating (3.6.2) from 0 to b_{max}, we obtain the expression for the reaction cross section,

$$\sigma(E) = \pi d^2(1 - Q/E) \quad \text{for} \quad E > Q \quad , \quad \sigma(E) = 0 \quad \text{for} \quad E < Q \quad . \tag{H.1}$$

where πd^2 is the elastic scattering cross section.

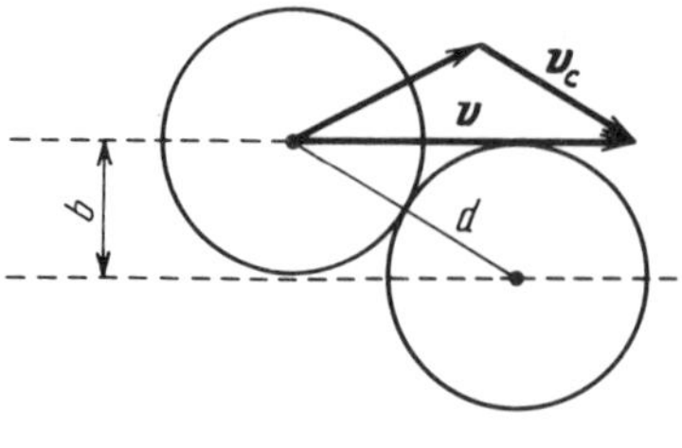

Fig.H.1. Schematic diagram of a molecular collision

Let us in (3.6.6) pass over to the description of the relative motion of A and B; then, using (3.2.6) and (3.6.3), we obtain (omitting the subscript AB)

$$k(T) = \left(\frac{\mu}{2\pi kT}\right)^{3/2} \int v\sigma(v)\exp\left(-\frac{\mu v^2}{2kT}\right)dv \quad . \tag{H.2}$$

Going over to spherical coordinates ($dv = v^2 \sin\theta d\theta d\varphi dv$, where θ is the polar angle and φ is the azimuth, which correspond to the direction of the relative velocity vector during the collision) and to the energy of the relative motion $E = \mu v^2/2$, after integrating with respect to θ (from 0 to π) and φ (from 0 to 2π) we arrive at

$$k(T) = \frac{1}{kT}\left(\frac{8}{\pi\mu kT}\right)^{1/2} \int_0^\infty E\sigma(E)\exp(-E/kT)dE \quad . \tag{H.3}$$

Using the reaction cross section according to (H.1), we get

$$k(T) = \pi d^2\left(\frac{8kT}{\pi\mu}\right)^{1/2} \exp\left(-\frac{Q}{kT}\right) \quad . \tag{H.4}$$

The quantity $(8kT/\pi\mu^{1/2})$ is equal to average velocity of the relative motion of molecules, $\bar{v}$. Since the product $\pi d^2\bar{v}$ determines the collision frequency for unit concentration, it follows from (H.4) that the physical meaning of the quantity k(T) is the fraction of collisions resulting in the reaction. Rewriting (H.4) in the form

$$k(T) = A\sqrt{T}\exp(-Q/kT) \quad , \tag{H.5}$$

where A is a constant, we arrive at the Arrhenius formula for the rate constant of a chemical reaction. For a more general dependence of the reaction cross section σ on the relative energy of collision the pre-exponential factor in the Arrhenius formula is obtained in a generalized form (see [3.8]), so that

$$k(T) = A'T^q \exp(-Q/kT) \quad , \tag{H.6}$$

where q is a constant parameter.

I. Chemical Kinetics Equations

Let the stoichiometric equation for a certain chemical reaction ℓ be of the form

$$\sum_j a_{j\ell} A_j \rightleftharpoons \sum_j a'_{j\ell} A_j \quad , \tag{I.1}$$

where $a_{j\ell}$, $a'_{j\ell}$ are the stoichiometric coefficients, which determine the quantitative relations for substances A_j entering into a chemical reaction. The reaction rate s_ℓ in the forward direction of (I.1), by definition, is

$$s_\ell = \left| \frac{1}{(a'_{j\ell} - a_{j\ell})} \frac{dn_j}{dt} \right| \quad , \tag{I.2}$$

where n_j is the density of particles of the jth component. The quantities on the right-hand side, up to the sign, are the same for any reagent, since the change of the concentration of the components in the reaction is proportional to the corresponding difference in the stoichiometric coefficients [3.2]. On the other hand, the reaction rate is proportional to the product of the concentrations of reagents, so that for (I.1), in the forward and reverse directions, respectively, we have

$$s_\ell = k_\ell \prod_j n_j^{a_{j\ell}} \quad , \qquad s'_\ell = k'_\ell \prod_j n_j^{a'_{j\ell}} \tag{I.3}$$

where k_ℓ and k'_ℓ are the reaction rate constants equal, in particular, to k_f and k_r in (3.6.5) and to k_r and k_d in (3.6.15). In total we obtain

$$S_\ell = |s_\ell - s'_\ell| = \left| k_\ell \prod_j n_j^{a_{j\ell}} - k'_\ell \prod n_j^{a'_{j\ell}} \right| \quad , \tag{I.4}$$

where S_ℓ has the meaning of the total (overall) rate of the ℓth reaction.

In chemical equilibrium $\bar{S}_\ell = 0$, so that

$$k_\ell / k'_\ell = \prod_j \bar{n}_j^{a'_{j\ell}} \Big/ \prod_j \bar{n}_j^{a_{j\ell}} \quad . \tag{I.5}$$

This ratio does not depend on the degree of deviation from chemical equilibrium, i.e., it is valid also in the absence of equilibrium (this is one more example of application of the principle of detailed balancing). The quantity on the right-hand side of (I.5) is called the equilibrium constant; we shall denote it by K_ℓ. Thus,

$$k_\ell/k'_\ell = K_\ell \quad . \tag{I.6}$$

According to Appendix G, the value of K_ℓ is obtained with the help of partition functions of reagents and depends only on temperature. Relation (I.6) is of great practical importance, since it allows us to determine from the experimental value of the reaction rate constant for one direction the reaction rate constant in the opposite direction. In "nonequilibrium" kinetics this relation is valid only when the characteristic time of a chemical reaction greatly exceeds the relaxation time with respect to the internal degrees of freedom of molecules participating in the reaction [3.2].

Usually many reactions occur in a gas simultaneously; in one type of reactions a substance m disappears, in others it is formed. The total rate W_m of the variation of the mth component concentration as a result of all ℓ reactions, each of which is written in the form (I.1), is

$$W_m \equiv (dn_m/dt) = \sum_\ell (a'_{m\ell} - a_{m\ell})S_\ell \quad . \tag{I.7}$$

The sign in front of each term of this sum is determined so that the plus sign corresponds to the appearance of the mth component in the respective reaction and the minus, to its disappearance. The totality of equations of the type (I.7) for different m's gives the system of chemical kinetics equations.

J. Relations for Gasdynamic Parameters of One-Dimensional Steady-State Flow

Let us consider first a perfect gas flow. Passing over from the molar enthalpy H (3.1.8) to the specific enthalpy $h(h = [\gamma/(\gamma - 1)]RT/\mu$, γ is the ratio of specific heats and μ is the molecular mass) and substituting h into (4.1.4), we obtain, taking account of (4.1.3) and (3.1.6),

$$u\,du + \frac{\gamma}{\gamma - 1}\frac{RdT}{\mu} = 0 \quad , \quad \frac{\gamma}{\gamma - 1}\,d\left(\frac{p}{\rho}\right) = \frac{dp}{\rho} \quad \text{or} \quad \frac{dp}{p} = \gamma\frac{d\rho}{\rho} \quad ,$$

and after integration we arrive at the Poisson adiabatic equation

$$\frac{p}{p_0} = \left(\frac{\rho}{\rho_0}\right)^{\gamma} , \quad \frac{T}{T_0} = \left(\frac{p}{p_0}\right)^{(\gamma-1)/\gamma} . \tag{J.1}$$

The velocity of the flow can be found from the energy equation (4.1.5) and the equation of state (3.1.6),

$$\frac{u^2}{2} + \frac{\gamma}{\gamma - 1} \cdot \frac{p}{\rho} = \frac{u_0^2}{2} + \frac{\gamma}{\gamma - 1} \cdot \frac{p_0}{0} . \tag{J.2}$$

From (3.1.13), for the sound velocity in a perfect gas, we have, after differentiating (J.2) and using (3.1.6),

$$a = \sqrt{\gamma RT/\mu} , \tag{J.3}$$

so that introducing the Mach number $\underset{\sim}{M} = u/a$, from (J.2), we obtain (if $u_0 = 0$)

$$\frac{T_0}{T} = 1 + \frac{\gamma - 1}{2} \underset{\sim}{M}^2 . \tag{J.4}$$

Then, according to (J.1), we have

$$\frac{p_0}{p} = \left(1 + \frac{\gamma - 1}{2} \underset{\sim}{M}^2\right)^{\gamma/(\gamma-1)} , \quad \frac{\rho_0}{\rho} = \left(1 + \frac{\gamma - 1}{2} \underset{\sim}{M}^2\right)^{1/(\gamma-1)} . \tag{J.5}$$

For the vibrationally excited gas flow in vibrational equilibrium we obtain similar relations using (3.1.6) in the differential form

$$dp = (R/\mu)(Td\rho + \rho dT) \tag{J.6}$$

and substituting (J.6) into (4.1.3), and then the resulting relation into (4.1.4). Using for the specific enthalpy of this gas equations (3.1.9-11), we get $[\beta = \gamma/(\gamma - 1)]$

$$\left(\beta - 1 + \frac{\mu}{R} \frac{d\xi_{vib}}{dT}\right) \frac{dT}{T} = \frac{d\rho}{\rho} .$$

Integrating from the initial state at the nozzle entrance (T_0, ρ_0), we obtain

$$\frac{\rho}{\rho_0} = \left(\frac{T}{T_0}\right)^{\beta-1} \frac{J}{J_0} , \tag{J.7}$$

$$J = \exp\left[\sum_i \xi_i g_i (\theta_i/T)(e_i + 1)\right] \cdot \prod_i e_i^{\xi_i g_i} ,$$

where all the quantities are determined by (3.1.9,10). According to (3.1.6), from (J.7) we obtain

$$\frac{p}{p_0} = \left(\frac{T}{T_0}\right)^{\beta} \frac{J}{J_0} \quad . \tag{J.8}$$

From the energy equation (4.1.5) it follows that

$$u^2 = 2\beta(R/\mu)(T_0 - T) + 2(\varepsilon_{vib\,0} - \varepsilon_{vib}) \quad . \tag{J.9}$$

Relations (J.7-9) are reduced to (J.4,5) in the absence of molecular vibrations ($e_i = 0$) and for the flow with a constant vibrational energy (vibrationally frozen flow, e_i = const).

For a perfect gas flow in the critical cross section, according to (J.4,5), we have

$$\frac{T_*}{T_0} = \frac{2}{\gamma + 1} \;, \quad \frac{p_*}{p_0} = \left(\frac{2}{\gamma + 1}\right)^{\gamma/(\gamma-1)} \;, \quad \frac{\rho_*}{\rho} = \left(\frac{2}{\gamma + 1}\right)^{1/(\gamma-1)} \quad . \tag{J.10}$$

For an equilibrium flow in a vibrationally excited gas similar relations can be determined as follows. By differentiating (4.1.1), we find that at the nozzle throat the equation

$$\left.\frac{d}{dt}(\rho u)\right|_* = 0 \tag{J.11}$$

is valid. Using (J.6-8), we arrive at the transcendental equation with respect to T_*. Solving it, we obtain T_*, and using (J.7-9), all the other parameters for the critical cross section.

The dependence between the nozzle cross-sectional area S(x) and gasdynamic parameters at any point of the flow (except for $u_0 = 0$) is determined by the continuity equation (4.1.1), so that

$$S = G/\rho u \quad . \tag{J.12}$$

Here the gas flow rate G can be conveniently expressed in terms of the parameters in the critical cross section: $G = S_*\rho_* a_*$. Substituting (J.12) into (J.3-5) and (J.10) in the case of a perfect gas, or (J.7-9) and the result of the solution of (J.11) for a vibrationally excited gas, we can obtain the dependence of the cross-sectional area S on the distribution of any gasdynamic parameter. For example, the dependence of the cross-sectional area on the distribution of pressure for a perfect gas is

$$S = S_*\varphi(\gamma)\,\frac{(p_0/p)^{1/\gamma}}{[1 - (p/p_0)^{(\gamma-1)/\gamma}]^{1/2}} \;, \qquad (J.13)$$

where

$$\varphi(\gamma) = \frac{[(\gamma - 1)/2]^{1/2}}{[(1 + \gamma)/2]^{(\gamma+1)/2(\gamma-1)}} \;.$$

The inverse dependence of gasdynamic parameters on the nozzle cross-sectional area $S(x)$ is obtained only from the solution of the transcendental equation, which in the general case follows from (J.12,13).

K. Continuity Equation for an Individual Component in a Flow

By applying vector analysis to the deformational motion of an elementary volume of a gas medium (see, for example, [4.3]), we can find that the rate of the relative expansion of this volume equals the divergence of the flow velocity vector $\underset{\sim}{u}$. Then the rate of the variation of the gas density ρ in the flow equals the product of ρ and div u with the minus sign, since the positive divergence of velocity leads to the decrease of ρ,

$$\frac{d\rho}{dt} = -\rho\,\mathrm{div}\,\underset{\sim}{u} \;. \qquad (K.1)$$

Besides this, the partial density ρ_j of the jth component varies as a result of chemical reactions, so that

$$\frac{d\rho_j}{dt} = -\rho_j\,\mathrm{div}\,\underset{\sim}{u} + m_jW_j \;, \qquad (K.2)$$

where W_j is the rate of variation of the density of particles n_j as a result of chemical reactions according to (I.7) and m_j is the mass of the jth particle ($m_j = \mu_j/N_A$, μ_j is the molecular mass, and N_A is the Avogadro number). Introducing into (K.2) the mass fraction $\alpha_j = \rho_j/\rho$ and substituting $\rho_j = \alpha_j\rho$, we obtain, with the account of (K.1),

$$\frac{d\alpha_j}{dt} = \frac{\mu_jW_j}{N_A\rho} \;. \qquad (K.3)$$

For a steady-state one-dimensional flow, taking into account that $d/dt = u(d/dx)$, we get

$$u \frac{d\alpha_j}{dx} = \frac{\mu_j W_j}{N_A \rho} \quad . \tag{K.4}$$

L. Implicit Method of the Numerical Solution of Differential Equations

Assume that we have the equation $d\xi/dt = \varphi(\xi,t)$ and that the characteristic time $\tau = (\partial\varphi/\partial\xi)^{-1}$ is determined. Represent this equation in the form of the relation for finite differences

$$\begin{aligned} \Delta\xi/\Delta t &= (\xi_{n+1} - \xi_n)/h \\ &= \sigma\varphi(\xi_n,t_n) + (1-\sigma)\varphi(\xi_{n+1},t_{n+1}) \quad , \end{aligned} \tag{L.1}$$

where σ is the parameter $(0 \leq \sigma \leq 1)$ which determines the contribution of the right-hand sides to computation for n and n + 1 steps. Rewriting (L.1) in the form

$$F(\xi_{n+1}) \equiv (\xi_{n+1} - \xi_n)/h - \sigma\varphi(\xi_n,t_n) - (1-\sigma)\varphi(\xi_{n+1},t_{n+1}) = 0 \quad , \tag{L.2}$$

we obtain the implicit equation for determining the required quantity ξ_{n+1}. The solution of (L.2) is carried out by different methods, most often by the iteration method, in which for the zero approximation we can take $\xi_{n+1}^{(0)} = \xi_n$. The application of implicit methods is also possible for $h \gg \tau$; in this case it is expedient to put $\sigma \leq 0.5$. A simplified method of the solution can be obtained if we add to and subtract from (L.1) the term $(1-\sigma)\varphi_\xi'(\xi_n,t_{n+1})(\xi_{n+1} - \xi_n)h$, where $\varphi_\xi' = (\partial\varphi/\partial\xi)$ and expand $\varphi(\xi_{n+1},t_{n+1})$ into the Taylor series for ξ in the vicinity of ξ_n. Then we have

$$\xi_{n+1} = \xi_n + \frac{\sigma\varphi(\xi_n,t_n) + (1-\sigma)\varphi(\xi_n,t_{n+1})}{1 - (1-\sigma)h\varphi_\xi'(\xi_n,t_{n+1})} h \quad ; \tag{L.3}$$

In this case there is is no problem of convergence of iterations, but it is required that $h \leq \tau$.

M. Relaxation of the System of Harmonic Oscillators in Nonisothermic Conditions

Let us consider a mixture of molecules—harmonic oscillators A and B—with characteristic vibrational temperatures θ_A and θ_B. These can also be two different vibrational modes of a single polyatomic molecule. The microkinetic balance equations for the number of molecules on different vibrational levels in such systems with the account of translation-vibrational (V-T) or vibration-vibrational (V-V) exchange for single-quantum transitions are obtained from the results of Sect.3.4 (see also [1.8]). For the balance of relative population x_n of the nth level of the oscillator (normalizing it for the total number of oscillators $\sum x_n$), we then have

$$\begin{aligned}\frac{dx_n}{dt} = &(Z_{AA}P_{AA} + Z_{AB}P_{AB})\{(n + 1)x_{n+1}\\ &- [(n + 1)\exp(-\Theta_A/T) + n]x_n + n\exp(-\Theta_A/T)x_{n-1}\}\\ &+ Z_{AA}Q_{AA}\{(n + 1)(\alpha_A + 1)x_{n+1} - [(n + 1)\alpha_A + n(\alpha_A + 1)]x_n\\ &+ n\alpha_A x_{n-1}\} + Z_{AB}Q_{AB}\{(n + 1)(\alpha_B + 1)x_{n+1}\\ &- [(n + 1)\alpha_B\exp(\theta_B/T - \theta_A/T) + n(\alpha_B + 1)]x_n\\ &+ n\alpha_B\exp(\theta_B/T - \theta_A/T)x_{n-1})\} ,\end{aligned} \tag{M.1}$$

where P_{AA}, P_{AB} are the probabilities of the V-T deactivation (P_{10}) for the collisions A + A and A + B, respectively, Q_{AA} and Q_{AB} are the probabilities of the single-quantum V-V exchange in the indicated conditions, and α_A is the average vibrational energy of oscillators A, normalized for the total number of these oscillators,

$$\alpha_A = \sum nx_n/\sum x_n \quad . \tag{M.2}$$

The equations and all the remaining quantities for oscillators B can be written similarly. We assume that there is the normalization $\sum x_n + \sum y_n = 1$. Rewrite (M.1) in the form

$$\frac{dx_n}{dt} = (n + 1)G_A x_{n+1} - [(n + 1)F_A + nG_A]x_n + nF_A x_{n-1} \quad , \tag{M.3}$$

where G_A, F_A are the functions whose form becomes clear from the comparison of (M.3) and (M.1). The system of equations (M.3) can be solved using the generating function [3.28, M.1]

$$\pi_A(z,t) = \sum_{n=0}^{\infty} x_n(t) z^n \quad ,$$

so that

$$\frac{\partial \pi_A}{\partial z} = \sum n x_n(t) z^{n-1} \quad , \quad \frac{\partial \pi_A}{\partial t} = \sum z^n \frac{\partial x_n}{\partial t} \quad . \tag{M.4}$$

Then the system of equations (M.3) can be reduced to a single first-order equation in partial derivatives,

$$\frac{\partial \pi_A}{\partial t} = (z - 1)\left\{[F_A(z - 1) - (G_A - F_A)] \frac{\partial \pi_A}{\partial z} + F_A \pi_A\right\} \quad . \tag{M.5}$$

The solution of this equation in variables F_A and G_A can be obtained by the method of characteristics [M.1] and has the form

$$\pi_A(z,t) = [A_A(t)/(z - 1)]\Phi[A_A(t)/(z - 1) - B_A(t)] \quad . \tag{M.6}$$

where

$$A_A(t) = \exp\left[\int_0^t (G_A - F_A)dt'\right] \quad , \quad B_A(t) = \int_0^t A_A(t')F_A dt' \quad .$$

Here Φ is a certain function whose form is determined by the initial conditions $A_A(0) = 1$, $B_A(0) = 0$, $\pi_A(z,0) = \Phi[1/(z - 1)]/(z - 1)$. At first the system under consideration is at equilibrium at temperature T_0 and is characterized by the Boltzmann distribution of molecules over the levels. Then

$$\pi_A(x,0) = [1 - \exp(-\theta_A/T_0)] \sum z^n \exp(-n\theta_A/T_0) \quad ,$$

or

$$\pi_A(z,0) = \frac{1 - \exp(-\theta_A/T_0)}{1 - z\exp(-\theta_A/T_0)} = \frac{1}{1 - (\alpha_A)_0(z - 1)} \quad . \tag{M.7}$$

Denoting the argument of function Φ by η,

$$\eta(z,t) = A_A(t)/(z - 1) - B_A(t) \quad , \quad \text{so that} \quad \eta(z,0) = 1/(z - 1) \quad ,$$

and taking into account the initial conditions, we find that $\Phi = 1/[\eta(z,t) - (\alpha_A)_0]$. Thus,

$$\pi_A(z,t) = \frac{A_A(t)}{(z-1)[\eta(z,t) - (\alpha_A)_0]} \quad , \quad \text{or}$$

$$\pi_A(z,t) = \frac{1}{1 - (z-1)\{[B_A(t) + (\alpha_A)_0]/A_A(t)\}} \quad . \tag{M.8}$$

Comparing (M.7) and (M.8), we arrive at

$$\alpha_A = [B_A(t) + (\alpha_A)_0]/A_A \quad . \tag{M.9}$$

Expanding (M.8) into the Taylor series for z, we get

$$x_n = \frac{1}{1+\alpha_A}\left(\frac{\alpha_A}{1+\alpha_A}\right)^n \quad , \quad \text{or} \quad x_n = \left[1 - \exp\left(-\frac{\theta_A}{T_A}\right)\exp\left(-\frac{n\theta_A}{T_A}\right)\right] , \tag{M.10}$$

where T_A is a certain current value of the vibrational temperature for oscillator system A. Similar mathematical operations can be performed for oscillators of the type B, and generalized for the case of multiquantum exchange. From (M.10) it is clear that in this case the Boltzmann distribution over the levels is retained during the whole period of relaxation irrespective of the change of coefficients A_A and B_A, i.e., irrespective of the variation of temperature and gas pressure.

N. Formulas for the Power of a CO_2 Gasdynamic Laser

Let us consider first the problem for the constant gain coefficient k_* (2.3.23). Assume that the symmetrical modes of the CO_2 vibrations are in equilibrium with the translational motion of molecules ($T_1 = T_2 = T$). Then in the cavity flow the values of $e_2 = \bar{e}_2$, and also of $e_3 = e_3^*$ remain constant, which follows from the expression (5.1.6) for k_*. Integrating the last equation of the system (5.1.2) and ignoring for simplicity the V-T nitrogen deactivation and the difference between θ_3 and θ_4, we obtain the distribution of e_4 along the flow in the cavity. This distribution is used to determine the radiation intensity I from the second equation of (5.3.3) for e_3, which is constant ($de_3/dt = 0$). Taking into account that $T_2 = T$ and (3.5.7), we obtain

$$I = \frac{\xi_{CO_2}(\theta_3 - \theta_1)p}{Tk_*}\left[\xi_{N_2}\, pk_{34}(e_4^0 - e_3^*)\exp\left(-\xi_{CO_2}k_{34}\,\frac{px}{u}\right) - pk_{32}^0(e_3^* - \bar{e}_3)\right] \tag{N.1}$$

(here and further the superscript 0 at the quantities e_i, T_i, ε_i corresponds to the state at the cavity entrance). Hence, the intensity I exponentially decreases downstream and vanishes at the length a_m, when $I \to 0$. The formula for P is obtained from (2.3.30) after integrating (N.1) over the mirror length

$$P = \frac{2\eta_q G\varepsilon_4^0 t}{(1+r)\ln(1/r)}\left\{\left(1 - \frac{e_3^*}{e_4^0}\right)\left[1 - \exp\left(-\xi_{CO_2}k_{34}\frac{pa}{u}\right)\right] - \frac{\xi_{CO_2}}{\xi_{N_2}}k_{32}^0\frac{pa}{u}\left(\frac{e_3^* - \bar{e}_a}{e_4^0}\right)\right\} , \qquad (N.2)$$

where ε_4^0 is the specific vibrational energy of nitrogen at the cavity entrance ($\varepsilon_4^0 = \xi_{N_2}R\theta_4 e_4^0/\mu$), and the mirror length is limited by a_m. The values of k_{32}^0 and k_{34} entering into this expression are given in Figs.3.5 and 4.2 and can also be determined from Table 5.1 with the account of (3.5.7) for k_{32}^0; the value of e_3^* is obtained from k_*, since e_2 is known. The first term in the braces of (N.2) determines the output power connected with the energy transfer from nitrogen to CO_2 (ν_3) and limited by the rate of this process, while the second term corresponds to the power losses due to collisional deactivation of the upper laser level. The effect of the difference between T_2 and T, ignored here, is small. The account of the V-T nitrogen deactivation in these equations presents no problem, but makes them more cumbersome. For example, instead of (N.1) in this case we should write

$$I = \frac{\xi_{CO_2}(\theta_3 - \theta_1)p}{Tk}\left\{\left[\xi_{N_2}pk_{34}(e_4^0 - e_3^*) - \frac{\xi_{N_2}p\bar{e}_4}{\xi_{CO_2}\tau_4}\right] \times \exp\left[-\left(\xi_{CO_2}k_{34} + \frac{1}{\tau_4}\right)\frac{px}{u}\right] + \frac{\xi_{N_2}p\bar{e}_4}{\xi_{CO_2}\tau_4} - pk_{32}^0(e_3^* - e_3)\right\} \qquad (N.3)$$

and correspondingly change (N.2) (here τ_4 is the time of vibrational deactivation of nitrogen).

Let us simplify (N.2), assuming that at the cavity entrance $T_3^0 \approx T_4^0$, and in the cavity $\bar{e}_3 \ll e_3^*$ and r is close to unity. Let e_3 be considerably less than unity, and the value of T_2 be so small that the population of the lower laser level does not affect the gain. Then, according to (5.1.6), the gain coefficient is proportional to e_3. This enables us to transform (N.2) as

$$P = \frac{\eta_q G\varepsilon_4^0 t}{2k_0 d}\left\{\left(\frac{2k_0 d}{\alpha' + 1} - 1\right)\left[1 - \exp\left(-\xi_{CO_2}k_{34}\frac{pa}{u}\right)\right] - \frac{\xi_{CO_2}}{\xi_{N_2}}k_{32}^0\frac{pa}{u}\right\} , \qquad (N.4)$$

where the loss factor α' is calculated for two passes.

This equation takes no account of the contribution from the energy of CO_2 molecules, which is transformed into radiation during the change of the population of laser levels in the transient zone at the cavity entrance, where $k_0 \to k_*$ (see Sect.2.5). The maximal value of this fraction of the power P (CO_2) according to an estimate is $\eta_q GR\theta_3 \xi_{CO_2}(e_3^0 - e_3^*)/\mu$, but actually in the plane-parallel resonator the value of P (CO_2) will be smaller, since its major part is lost due to diffraction on the mirror edge. In unstable resonators of a telescopic type the utilization of this fraction of energy may be more effective due to the increase of the transition region $k_0 \to k_*$ (see Sect.2.5).

The highest power can be extracted with mirrors of maximal length a_m (downstream). Proceeding from (N.1), with the assumptions made, we have

$$a_m = \frac{u}{\xi_{CO_2} p k_{34}} \ln\left[\frac{\xi_{N_2} k_{34}(e_4^0 - e_3^*)}{k_{32}^0(e_3^* - \bar{e}_3)}\right]$$

$$\approx \frac{u}{\xi_{CO_2} p k_{34}} \ln\left[\frac{\xi_{N_2} k_{34}}{k_{32}^0}\left(\frac{2k_0 d}{\alpha' + t} - 1\right)\right] . \qquad \text{(N.5)}$$

Thus, the maximum mirror length a_m is directly proportional to the velocity of the flow and inversely proportional to the probability of the supply of vibrational quanta from nitrogen to CO_2 (ν_3), and it also depends on the state of an active medium in the cavity. Denote the quantity in the square brackets on the right-hand side of (N.5) by β. Substituting (N.5) into (N.3), we arrive at

$$P = \eta_q G \varepsilon_4^0 \frac{t}{2k_0 d}\left(\frac{2k_0 d}{\alpha' + t} - 1\right)\left(1 - \frac{1}{\beta} - \frac{\ln \beta}{\beta}\right) . \qquad \text{(N.6)}$$

Since $k_{34} \gg k_{32}^0$, then β is sufficiently high, and therefore the last bracket of (N.6) is close to unity. As $\beta \to \infty$, we obtain the expression for P when the power does not depend on the energy-exchange kinetics,

$$P = \eta_q G \varepsilon_4^0 (t/2k_0 d)[2k_0 d/(\alpha' + t) - 1] . \qquad \text{(N.7)}$$

Optimizing P (N.7) with respect to the transmission coefficient t, as in Sect.2.3, we obtain

$$P = \eta_q G \varepsilon_4^0 (1 - \sqrt{\alpha'/2k_0 d})^2 . \qquad \text{(N.8)}$$

It should be noted that (N.6-8) are valid for a cavity with the maximum mirror length a_m (N.5). If a_m is unknown, these formulas can be used to estimate power, if we exclude the part of energy ε_4^0 which corresponds to the disappearance of the population inversion (see Sect.5.3). If in this case we also take into account (approximately) the contribution of the energy of CO_2 (ν_3) into the extracted power, then, by analogy with (5.3.4), instead of (N.7,8) we obtain

$$P = \eta_q G[\varepsilon_{II}^0 - \varepsilon_{II}(T_3')](t/2k_0 d)[2k_0 d/(\alpha' + t) - 1] \quad , \tag{N.9}$$

$$P = \eta_q G[\varepsilon_{II}^0 - \varepsilon_{II}(T_3')](1 - \sqrt{\alpha'/2k_0 d})^2 \quad , \tag{N.10}$$

where

$$T_3' = \theta_3 T_1/\theta_1 \quad , \quad \varepsilon_{II}^0 = R(\xi_{CO_2}\theta_3 e_3^0 + \xi_{N_2}\theta_4 e_4^0)/\mu \quad .$$

In the constant intensity approximation, the analytic dependence for P in the same assumptions that in the $k_* = \text{const}$ approximation is more complicated. For this reason, when solving this problem we shall directly consider the simplest scheme of the energy-exchange kinetics (5.1.14), in which it is assumed that $T_2 = T_1 = T$, $T_3 = T_4$. We shall add to (5.1.14) the term accounting for induced transitions and integrate this equation over the distance along the cavity flow. We shall integrate the obtained value of energy of the CO_2 upper laser level [and the respective portion of the total energy of the group of vibrational states of CO_2 (ν_3) + N_2] according to the condition of the steady-level radiation (2.3.28), which in this case with a good approximation is written in the form

$$\frac{1}{ae_3^0}\int_0^a e_3 dx = \frac{\alpha' + t}{2k_0 d} \tag{N.11}$$

(assuming that $\bar{e}_2 \ll e_3$). Then for the output power $P = abtI/2$ we obtain

$$P = \eta_q G\varepsilon_{II}^0 \frac{t}{2k_0 d}\left[\omega - \frac{\xi_{CO_2} k_{32}^0 pa}{(\xi_{CO_2} + \xi_{N_2})u}\right] \quad , \tag{N.12}$$

where ω is determined from the transcendental equation

$$\omega = \frac{2k_0 d}{\alpha' + 1}\left\{(1 - e^{-\omega}) + \frac{\xi_{CO_2} k_{32}^0 pa\bar{e}_3}{(\xi_{CO_2} + \xi_{N_2})ue_3^0}\left[1 - \frac{1}{\omega}(1 - e^{-\omega})\right]\right\} \quad . \tag{N.13}$$

In ordinary conditions, when the gain at the cavity exit considerably exceeds losses and $\bar{e}_3 \ll e_3^0$, the second term in the braces of (N.13) is small. From the remaining part of (N.13) it is clear that already for $A = [2k_0d/(\alpha' + t)] \geq 1.5$ to 2, with a good approximation we can assume that $\omega = A$ (the value of A exceeds ω only by 18% for A = 1.5 and by less than 5% for A = 2). Then

$$P = \eta_q G \varepsilon_{II}^0 \frac{t}{2k_0 d} \left[\frac{2k_0 d}{(\alpha' + t)} - \frac{\xi_{CO_2} k_{32}^0 pa}{(\xi_{CO_2} + \xi_{N_2})u} \right] . \quad (N.14)$$

If in this equation we ignore the collisional deactivation of the upper laser level, we arrive at the result which is close to the expression for power in the $k_* =$ const approximation (N.7,9), if the gain considerably exceeds losses at the cavity entrance ($2k_0 d \gg \alpha' + t$) and the ratio ξ_{CO_2}/ξ_{N_2} is small.

List of the Most Used Symbols

Here, we give the most frequently used notations. Sometimes the same symbol is used for different quantities in different fields. Since these symbols are traditional in usage, it would be difficult and potentially confusing to change them here. However, the reader can easily identify the meaning of the quantity from its context.

Notation	Physical quantity	Page	Equation
a	The ratio of the collisional and Doppler line widths	16	(2.1.21)
a	Velocity of sound	57	(3.1.13)
A_{mn}, a_{mn}	The Einstein integral and spectral coefficients for the probability of spontaneous $m \rightarrow n$ transition	11,12	(2.1.7,10,11, 23)
$[A_j]$	Molar density of a substance A_j	248	(E.1)
B_{nm}, b_{nm}	The Einstein integral and spectral coefficients for the probability of induced absorption in the $n \rightarrow m$ transition	11	(2.1.8,10,11)
B_{mn}, b_{mn}	The Einstein integral and spectral coefficients for the probability of induced radiation in the $m \rightarrow n$ transition	12	(2.1.9-11)
c	Velocity of light		
c_j	Molar-mass concentration of the jth component of a gas mixture	248	(E.1)
C_p	Heat capacity at a constant gas pressure	55	
C_V	Heat capacity at a constant gas volume	55	
d	Resonator width	21	
Da	Damkeler number	137	(4.4.15)

Notation	Physical quantity	Page	Equation
e_i	Average number of vibrational quanta per one mode of molecular vibrations	56	(3.1.9)
E_{rot}	Rotational energy	6,246	(2.11)
E_{vib}	Vibrational energy	6,247	(2.1.3,5)
$\mathcal{E}$	Molar internal energy of a gas	55	(3.1.7)
g	Statistical weight of an energy level, degeneracy	7	
G	Gas flow rate, mass of a gas flowing through a channel per unit time	103	(4.1.1)
g_v	Statistical weight of a mixed mode of CO_2 vibrations	9	
h	Specific enthalpy of a gas	5	
h	Planck constant		
h_*	Throat height of a nozzle	121,133	
H	Molar enthalpy of a gas	55	
H(a, w)	The Voigt function for a spectral line contour	15,243	(2.1.20)
H*	Form factor of the boundary layer in the gas flow	132	
I	Radiation intensity	13	
I	Moment of inertia of a molecule	6,246	
I_r	Radiation intensity for scattering by a turbulent medium	47	(2.5.12)
J	Rotational quantum number	6	
k	Boltzmann constant	54	
K, k_ν	Spectral gain (absorption) coefficient	17	(2.2.3,4, 8-10)
k_*	Saturated gain coefficient	28,29	(2.3.22-24)
k_0	Gain coefficient for a very low light intensity, small-signal gain	28	
K_{eq}	Equilibrium constant for a chemical reaction	252	(G.4-6)
k_j, k_r, k_ℓ	Reaction rate constants	90,91,255	(3.6.5,6), (I.3)
k_{ij}	Rate constant of V-V energy exchange between i^{th} and j^{th} vibrational modes	74	

Notation	Physical quantity	Page	Equation
ℓ	Quantum number for the angular momentum associated with CO_2 deformation vibrations	9	
ℓ_0	Mean free path of a molecule	14	(2.1.17)
M	Linear magnification in unstable resonators	36,37	
$\underline{M}$	Mach number	104	
n, n_j	Density of particles in a gas and density of particles in the jth gas component	54	
N_A	Avogadro number	14,54	
N_F	Fresnel number	25	
N_m, N_n	Population of levels m, n (in Sect.3.4 $N_m \equiv x_m$)	11	
p, p_j	Gas pressure and partial pressure of the jth gas component	54	(3.1.4-6)
P	Laser output power	31,42,264	(2.3.30,32), (2.5.6,7), (N.2,4,6-10, 12,14)
$\underline{Pr}$	Prandtl number	130	
Q	Q-factor of a resonator	25	(2.3.10,11)
Q, Q_t, Q_{rot}, Q_{vib}	Statistical sum (partition function) for translational (t), rotational (rot), and vibrational (vib) degrees of freedom of molecules	53,245-247	(3.1.2), (D.1,5,10,11)
r, r_1, r_2	Reflection coefficients for a resonator mirror	29	
R	Gas constant	14,54	
$\underline{Re}$	Reynolds number	130	
s	Entropy of a gas	107	
S	Area of the mirror in a resonator	30	
S	Cross-sectional area of a channel	103	
S_*	Cross-sectional area of a nozzle throat	105	
t	Time		
t, t_1, t_2	Transmission factors for the mirror in a resonator	29	
T, T_{rot}, T_{vib}	Gas temperature, rotational and vibrational gas temperature	53	

Notation	Physical quantity	Page	Equation
T_i, T_j	Vibrational temperature for the ith and j^{th} vibrational modes	56	(3.1.9)
T_w	Temperature of the nozzle wall	132	
u	Gas flow velocity		
v	Vibrational quantum number	6	
$\bar{v}$	Average velocity of molecules in a gas	14	(2.1.17)
v_1, v_2, v_3	Vibrational quantum numbers for the first, second, and third modes of CO_2 vibrations	9	
v_{AB}	Absolute (scalar) value of relative velocity of molecules A and B	59,60	
V_j	Diffusion rate of the jth component	136	
W_j	Formation rate of the jth component in chemical reactions	256	(I.7)
x_ϵ	Anharmonicity parameter	66	
x_n	Population of the nth level of the molecule (in Sect.3.2 $x_n \equiv n_{A(i)}$, in Sect.2.1 $x_n \equiv N_n$)	69	
Z_{AB}, z_{AB}	Number of collisions of molecule A with all molecules B per unit time and that for one molecule B	60	(3.2.7)
Z^s_{mn}, Z^{in}_{nm}, Z^{in}_{mn}	The number of photons emitted or absorbed per unit gas volume per unit time	11	(2.1.7-9)
α	Loss coefficient for one pass in a resonator	25	
α'	Loss coefficient for a double pass ($\alpha' = 2\alpha$)	29	
α_j	Mass fraction of the jth component	248	(E.3)
β	Loss coefficient per unit length along the optical axis, attenuation factor	41	
γ	Attenuation coefficient due to scattering in a turbulent medium	49	
γ	Ratio of heat capacities	55	

Notation	Physical quantity	Page	Equation
γ_j	Normalized molar-mass concentration of the jth component	249	
δ^*, δ^{**}, δ^{***}	Thickness of displacement, of momentum loss, and of energy loss in a gas flow	131, 132	
Δ_C, Δ_D	Half-widths of the collisional and Doppler broadening of a spectral line		
ε	Specific internal energy of a gas	55	
ε	Vibrational energy	69	(3.4.2)
η_n	GDL nozzle efficiency	203	(5.6.2)
η_q	Quantum efficiency	32	(2.3.36)
η_r	Resonator efficiency	32,50, 206	(2.3.34,35), (2.5.20), (5.6.5)
η_t	Efficiency of thermal excitation of a laser mixture in front of the nozzle	203	(5.6.1)
θ	Angle between the contour and the nozzle axis	133	
θ_d	Diffraction-limited angle of laser beam divergence	34	
θ_r	Scattering angle for light passing through a turbulized medium	47	
θ, θ_{rot}, θ_{vib}	Characteristic temperature, rotational, and vibrational characteristic temperature	246,247	
λ	Wavelength		
Λ	Integral turbulence scale size	47	
μ	Molecular weight of a gas	14,54,249	(E.6)
ν	Kinematic viscosity	130	
ν	Vibrational frequency in a light wave	8	
ν'	Wave number (the number of waves fitting into unit length)	8	
ν_0	Fundamental frequency of vibrations	6	
$\Delta\nu$, $\Delta\nu_C$, $\Delta\nu_D$	Spectral, collisional, and the Doppler broadening line width	13,14	(2.1.16,17,19)

Notation	Physical quantity	Page	Equation
ξ_j	Molar fraction of the jth component of a gas mixture	248	(E.2,5)
ρ	Electromagnetic energy density (light energy per unit volume)	11	
ρ	Gas density		
σ_{mn}, σ_{nm}	Cross section of the induced optical transition $m \rightleftharpoons n$	41	
τ	Relaxation time or any other characteristic time	70,110	(4.1.14)
τ_d	Damping time	24	(2.3.9)
τ_{mn}	Optical lifetime of the mth level with respect to the $m \to n$ transition	11	
ω	Cyclic frequency of molecular vibrations	63	
ω	Frequency difference between the center of the line and frequency under consideration	16	(2.1.21)

References

Chapter 1

1.1 V.N. Karnyushin, R.I. Soloukhin: Combust. Explos. Shock Waves *8*,129 (1972); S.N.B. Murthy: "Gasdynamic and Chemical Lasers: Gas Dynamics", in *High-Power Gas Lasers* (Inst. Phys. Press, Bristol-London 1976); A. Hertzberg: J. Energy *1*,331 (1977)
1.2 J.D. Anderson: *Gasdynamic Lasers: An Introduction* (Academic Press, New York 1976)
1.3 N.G. Basov, A.N. Oraevskii: Sov. Phys. - JETP *17*, 1171 (1963); E.O. Schulz-DuBois, H.E.D. Scovil: "Thermal Maser", Patent USA 3015072 (1961); H.E.D. Scovil, E.O. Schulz-Dubois: Phys. Rev. Lett. *2*, 262 (1959)
1.4 K. Shimoda: Inst. Phys. Chem. Res. Sci. Papers *59*, 53 (1965)
1.5 I.R. Hurle, A. Hertzberg: Phys. Fluids *8*, 1601 (1965)
1.6 L.I. Gudzenko, S.S. Filippov, L.A. Shelepin: Sov. Phys.-JETP *24*, 745 (1967)
1.7 Ya.B. Zel'dovich, Yu.P. Raizner: *Physics of Shock Waves and High-Temperature Hydrodynamic Phenomena* (Academic Press, New York 1966)
1.8 E.V. Stupochenko, S.A. Losev, A.I. Osipov: *Relaxation in Shock Waves*, Applied Physics and Engineering, Vol.1 (Springer, Berlin, Heidelberg, New York 1967)
1.9 A. Kantrowitz: J. Chem. Phys. *14*, 150 (1946)
1.10 I.R. Hurle, A.L. Russo, J.G. Hall: J. Chem. Phys. *40*, 2076 (1964)
1.11 A.N. Terenin, G.G. Neuimin: Izv. Akad. Nauk SSSR, Otd. Khim. Nauk *5*, 242 (1942) (in Russian)
1.12 F. Legay, N. Legay-Sommaire: C.R. Acad. Sci. Paris *259B*, 99 (1964)
1.13 B.S. Patel: Phys. Lett *45*A, 137 (1973)
1.14 L. Allen, D.G.C. Jones: *Principles of Gas Lasers* (Butterworth, London 1967 1967)
1.15 V.K. Konyukhov, A.M. Prokhorov:JETP Lett. *3*, 286 (1966)
1.16 V.K. Konyukhov, A.M. Prokhorov: "Method of Obtaining of Population Inversion", Patent USSR 223954 (November 1966) (in Russian)
1.17 N.G. Basov, A. N. Oraevskii, V.A. Shcheglov: Sov. Phys.-Tech. Phys. *12*, 243 (1967)
1.18 A.S. Biryukov, B.F. Gordiets, L.A. Shelepin: Sov. Phys.- JETP *28*, 762 (1969)
1.19 A.S. Biryukov, L.A. Shelepin: Sov. Phys.-Tech. Phys. *15*, 2018 (1970)
1.20 V.K. Konyukhov, I.V. Matrosov, A.M. Prokhorov, D.T. Shalunov, N.N. Shirokov: JETP Lett. *10*, 53 (1969)
1.21 A.P. Dronov, A.S. D'yakov, E.M. Kudryavtsev, N.N. Sobolev: JETP Lett. *11*, 353 (1970)
1.22 E.T. Gerry: Laser Focus *6*, 27 (1970)
1.23 D.M. Kuehn, D.J. Monson: Appl. Phys. Lett. *16*, 48 (1970)

1.24 V.K. Konyukhov, I.V. Matrosov, A.M. Prokhorov, D.T. Shalunov, N.N. Shirókov: JETP Lett. *21*,321 (1970)
1.25 M.S. Dzhidzhoev, V.V. Korolev, V.N. Markov, V.T. Platonenko, R.V. Khokhlov: JETP Lett. *14*,47 (1971)
1.26 R.L. McKenzie: Appl. Phys. Lett. *17*, 462 (1970)
1.27 V.F. Gavrikov, A.P. Dronov, V.K. Orlov, A.K. Piskunov: Sov. J. Quantum Electron. *3*, 250 (1973)
1.28 Yu.I. Grin' , V.M. Polyakov, V.G. Testov: JETP Lett. *18*, 155 (1973)
1.29 A.Yu. Volkov, A.I. Demin, V.N. Epikhin, E.M. Kudryavtsev: Sov. J. Quantum Electron. *6*, 1001 (1976)
1.30 V.F. Gavrikov, A.P. Dronov, V.K. Orlov, A.K. Piskunov: JETP Lett. *23*, 595 (1976)
1.31 V.M. Gol'dfarb, E.V. Il'ina, I.E. Kostygova, G.A. Luk'yanov, V.A. Silant'ev: Opt. Spectrosc. *20*, 6 (1966)
1.32 A.V. Krauklis, V.N. Croshko, R.I. Soloukhin, N.A. Fomin: Combust. Explos. Shock Waves *12*, 709 (1976)
1.33 R. Borghi, A.F. Carrega, M. Charpenel, J.P.E. Taran: Appl. Phys. Lett. *22*, 661 (1973)
1.34 R. Bailly, M. Pealat, J.P.E. Taran: Rev. Phys. Appl. *12*, 1705 (1977)

Chapter 2

2.1 R.H. Pantell, H.E. Puthoff: *Fundamentals of Quantum Electronics* (Wiley, New York 1969)
2.2 G. Birnbaum: *Optical Masers* (Academic Press, New York, London 1964)
2.3 A.E. Siegman: *An Introduction to Lasers and Masers* (McGraw-Hill, New York 1971)
2.4 B.A. Lengyel: *Lasers*, 2nd ed. (Wiley, New York 1971)
2.5 G. Herzberg: *Molecular Spectra and Molecular Structure*, Vol. 1, Spectra of Diatomic Molecules, 2nd ed. (Van Nostrand, New York 1951)
2.6 G. Herzberg: *Molecular Spectra and Molecular Structure*, Vol. 3, Electronic Spectra and Electronic Structure of Polyatomic Molecules (Van Nostrand, New York 1966)
2.7 M.A. El'yashevich: *Atomic and Molecular Spectroscopy* (Fizmatgiz, Moscow 1962) (in Russian)
2.8 L.V. Gurvich, G.A. Khachkuruzov, V.A. Medvedev: *Thermodynamic Properties of Individual Substances* (NBS, US Dept. of Commerce, Washington 1965)
2.9 S.S. Penner: *Quantitative Molecular Spectroscopy and Gas Emissivities* (Addison-Wesley, London 1959)
2.10 W. Malkmus: J. Opt. Soc. Am. *53*, 951 (1963)
2.11 I.I. Sobelman: *Atomic Spectra and Radiative Transitions*, Springer Series in Chemical Physics, Vol. 1 (Springer, Berlin, Heidelberg, New York 1979)
2.12 A. Mitchell, M. Zemansky: *Resonance Radiation and Excited Atoms* (University Press, Cambridge 1961)
2.13a L.S. Vasilenko, A.A. Kovalev, A.S. Provorov, V.P. Chebotaev: Sov. J. Quantum Electron. *5*, 1382 (1975)
2.13b A.M. Prokhorov: Zh. Eksp. Teor. Fiz. *34*, 1658 (1958) (in Russian)
2.13c A. Javan, W.R. Bennett, D.R. Herriott: Phys. Rev. Lett. *7*, 415 (1961)
2.14 L. Landau, E.M. Lifschiz: *Course of Theoretical Physics* Vol. 2, The Classical Theory of Fields (Pergamon, Oxford 1975)
2.15 M. Born, E. Wolf: *Principles of Optics* (Pergamon Press, Oxford 1975)
2.16 A. Sommerfeld: *Lectures on Theoretical Physics*, Vol. 4, Optics (Academic, New York 1954)
2.17 W.R. Bennett: Phys. Rev. *126*, 580 (1962)
2.18 A.A. Vedenov, A.F. Vitshas, A.P. Napartovich, V.P. Panchenko: Sov. J. Quantum Electron. *5*, 1404 (1975)

2.19 G. Lee: Phys. Fluids *17*, 644 (1974)
2.20 J. Tulip, H. Seguin: J. Appl. Phys. *42*, 3393 (1971)
2.21 A.V. Eletskii, B.M. Smirnov: *Gas Lasers* (Atomizdat, Moscow 1971) (in Russian)
2.22 W.W. Rigrod: J. Appl. Phys. *34*, 2602 (1963)
2.23 N.M. Efremov, B.A. Tikhonov: Fluid Dyn. *12*, 329 (1977)
2.24 E.M. Parmentier, R.A. Greenberg: AIAA J. *11*, 943 (1973)
2.25 P.O. Clark: AIAA Paper 72-708 (1972)
2.26 Yu.A. Anan'ev: Sov. J. Quantum Electron. *1*, 565 (1971)
2.27 A.E. Siegman: Appl. Opt. *13*, 353 (1974)
2.28 A.E. Siegman, R.W. Arrathoon: IEEE J. QE-*3*, 156 (1967)
2.29 Yu.A. Anan'ev, V.E. Sherstobitov: Sov. J. Quantum Electron. *1*, 263 (1971)
2.30 Yu.A. Anan'ev, L.V. Koval'chuk, V.P. Trusov, V.E. Sherstobitov: Sov. J. Quantum Electron. *4*, 659 (1974)
2.31 Yu.A. Anan'ev, V.E. Sherstobitov, O.A. Shorokhov: Sov. J. Quantum Electron. *1*, 65 (1971)
2.32 N.G. Basov, O.N. Krokhin, Yu.M. Popov: Sov. Phys. --Usp. *3*, 702 (1961)
2.33 E.L. Klosterman, A.L. Hoffman: "A High Pressure Shock Tube Driven Gasdynamic Laser", in *Recent Developments in Shock Tube Research*, ed. by D. Bershader, W. Griffith (University Press, Stanford 1973) p. 156
2.34 J. Wilson: Appl. Phys. Lett. *8*, 159 (1960)
2.35 T.A. Cool: J. Appl. Phys. *40*, 3563 (1969)
2.36 A.L. Mikaēlyan, V.P. Minaev, Yu.A. Obod, Yu.G. Turkov: Sov.J. Quantum Electron. *4*, 643 (1974)
2.37 M. Brunne, G. Malaczynski, J. Milewski, J. Stanco: Bull. Acad. Polon. Sci. Ser. Sci. Tech. *18*, 243,251 (1970)
2.38 M. Brunne, G. Malaczynski, J. Stanco: Bull. Acad. Polon. Sci. Ser. Sci. Tech. *18*, 487,497 (1970)
2.39 M. Brunne J. Milewski, J. Stanco, A. Zielinski: Phys. Fluids *18*, 458 (1975)
2.40 N.A. Generalov, G.I. Kozlov, I.K. Selezneva: J. Appl. Mech. Tech. Phys. *13*, 628 (1972)
2.41 S.A. Losev, V.N. Makarov: J. Appl. Mech. Tech. Phys. *16*, 483 (1975)
2.42 Yu.A. Anan'ev, V.P. Trusov, V.E. Sherstobitov: Sov.J. Quantum Electron. *6*, 928 (1976)
2.43 M. Brunne, J. Milewski, J. Stanco, A. Zielinski: Bull. Acad. Polon. Sci. Ser. Sci. Tech. *20*, 395 (1972)
2.44 M. Brunne, A. Zielinski: J. Appl. Phys. *49*, 2288 (1978)
2.45 Yu.A. Dreizin, A.M. Dykhne: JETP Lett. *19*, 371 (1974)
2.46 A.L. Hoffman, T.G. Jones: AIAA Paper 72-217 (1972)
2.47 G.W. Sutton: AIAA J. *7*, 1737 (1969)
2.48 H.C. van de Hulst: *Light Scattering by Small Particles* (Wiley, New York 1957)
2.49 M. Beheshti: AIAA Paper 66-134 (1966)
2.50 V.M. Shmelev, A.D. Margolin: Sov. J. Quantum Electron. *5*, 948 (1975)
2.51 V.V. Appolonov, A.I. Barchukov, A.M. Prokhorov: IEEE J.QE-*10*, 505 (1974)

Chapter 3

3.1 J.O. Hirschfelder, C.F. Curtiss, R.B. Bird: *Molecular Theory of Gases and Liquids* (Wiley, New York 1954)
3.2 V.N. Kondratiev, E.E. Nikitin: *Kinetics and Mechanism of Gas Phase Reactions* (Nauka, Moscow 1974) (in Russian); V.N. Kondratiev: *Chemical Kinetics of Gas Reactions* (Pergamon Press, London, 1964)

3.3 S. Chapman, T.G. Cowling: *The Mathematical Theory of Non-Uniform Gases* (University Press, Cambridge 1952)
3.4 W.G. Vincenti, C.H. Kruger: *Introduction to Physical Gas Dynamics* (Wiley, New York 1965)
3.5 R.C. Millikan: *Molecular Relaxation Processes* (Chem. Soc. and Acad. Press, London, 1966); J. Nicholas: *Chemical Kinetics* (Wiley. New York 1976)
3.6 L.D. Landau, E.M. Lifschiz:*Lehrbuch der theoretischen Physik*, Bd.5 Statistische Physik (Akademie Verlag, Berlin 1966)
3.7 L.D. Landau, E.M. Lifschiz: *Lehrbuch der theoretischen Physik*, Bd.6 Hydrodynamik (Akademie Verlag, Berlin 1966)
3.8 A.R. Hochstim (ed.): *Kinetic Processes in Gases and Plasma* (Academic Press, New York 1969)
3.9 Yu.L. Klimontovich: *Kinetic Theory of Nonideal Gas and Nonideal Plasma* (Nauka, Moscow 1975)(in Russian)
3.10 V.N. Kondratiev, E.E. Nikitin, A.I. Reznikov, S.Ya. Umanskii: *Thermal Bimolecular Reactions in Gases* (Nauka, Moscow 1976)(in Russian)
3.11 E.E. Nikitin: *Theory of Elementary Atomic and Molecular Processes in Gases* (Clarendon Press, Oxford 1974)
3.12 S. Ormonde: Rev. Mod. Phys. *47*, 193 (1975)
3.13 A.B. Petersen, C. Wittig, S.R. Leone: J. Appl. Phys. *47*, 1051 (1976)
3.14 L.D. Landau, E.M. Lifschiz: *Course of Theoretical Physics*, Vol.1, Mechanics (Pergamon, Oxford 1976)
3.15 R.N. Schwartz, Z.I. Slawsky, K.E. Herzfeld: J. Chem. Phys. *20*, 1591 (1952)
3.16 K.F. Herzfeld, T. Litovitz: *Absorption and Dispersion of Ultrasonic Waves* (Academic Press, New York 1959)
3.17 L. Landau, E. Teller: Phys. Z. Sowjetunion *10*,34 (1936)
3.18 B.F. Gordiets, A.I. Osipov, E.V. Stupochenko, L.A. Shelepin: Sov. Phys.-Usp. *15*, 759 (1973)
3.19 A.S. Biryukov: Trans Phys. Inst. Acad. Sci. USSR *83*, 13 (1975)(in Russian)
3.20 W.Q. Jeffers, J.D. Kelley: J. Chem. Phys. *55*, 4433 (1971)
3.21 D. Rapp: J. Chem. Phys. *43*, 316 (1965)
3.22 B.H. Mahan: J. Chem. Phys. *46*, 98 (1967)
3.23 R.D. Sharma: Phys. Rev. A*2*, 173 (1970)
3.24 M. Lev-on, W.E. Palke, R.C. Millikan: Chem. Phys. Lett. *24*, 59 (1974)
3.25 R.C. Millikan, D.R. White: J. Chem. Phys. *39*, 3209 (1963)
3.26 R.L. Taylor, S. Bitterman: Rev. Mod. Phys. *41*, 26 (1969)
3.27 S.A. Losev: Combust. Explos. Shock Waves *9*, 669 (1973)
3.28 E. Montroll, K.E. Shuler: J. Chem. Phys. *26*, 454 (1957)
3.29 S.A. Losev, O.P. Shatalov, M.S. Yalovik: Dokl. Akad. Nauk SSSR *195*, 585 (1970)(in Russian)
3.30 A.I. Osipov: J. Appl. Mech. Tech. Phys. *1*, 32 (1964)
3.31 A.S. Biryukov, B.F. Gordiets: J. Appl. Mech. Tech. Phys. *13*, 787 (1972)
3.32 B.F. Gordiets, A.I. Osipov, L.A. Shelepin: Sov. Phys.-JETP *33*, 58 (1971)
3.33 N.M. Kuznetsov: Sov. Phys.-JETP *34*, 505 (1972)
3.34 J. Teare, R.L. Taylor, C.W. von Rosenberg: Nature *225*, 240 (1970)
3.35 V.F. Gavrikov, A.P. Dronov, V.K. Orlov, A.K. Piskunov: Sov. J. Quantum Electron. *6*, 803, 1000 (1976)
3.36 C.E. Treanor, J.W. Rich, R.G. Rehm: J. Chem Phys. *48*, 1728 (1968)
3.37 J.W. Rich: J. Appl. Phys. *42*, 2719 (1971)
3.38 B.F. Gordietz, Sh.S. Mamedov: J.Appl. Mech. Tech. Phys. *15*, 297 (1974)
3.39 B.F. Gordietz, Sh.S. Mamedov, L.A. Shelepin: Sov. Phys.-JETP *40*,4 (1974)
3.40 R.L. Kerber, N. Cohen, G. Emanuel: IEEE J. QE-*9*, 94 (1973)
3.41 J.A. Blauer, G.R. Nickerson:AIAA Paper 74-536 (1974)
3.42 C.K. Rhodes, M.J. Kelly, A. Javan: J. Chem. Phys. *48*, 5730 (1968)
3.43 R.J. Carbone, W.J. Witterman: IEEE J. QE-*5*, 442 (1969)
3.44 K. Bulthuis, G.J. Ponsen: Chem. Phys. Lett. *14*, 613 (1972)
3.45 W.A. Rosser, E. Hoag, E.T. Gerry: J. Chem. Phys. *57*, 4153 (1972)

(1966)
69)

972)
sfer in CO_2-N_2 Mole-
ymp. (University of

ys. *61*, 1652 (1974)
(1975)
6 (1973)

Pasynkova: Sov. J.

Space Sci. *13*, 57

(1969)

, 368 (1972)
975)
nette, L. Henry: C.R.

s. *56*, 5214 (1972)
ao: "Study of the
as Mixtures", in
. Kamimoto (Shock

56 (1973)
ics, 2nd ed. (Univer-

ons (NBS, US Dept. of

n the Nozzle Gas Flow",
5 (in Russian)
55) (in Russian)
41, 1072 (1961) (in

422 (1959)
53)
(in Russian)
. Mech. Tech. Phys.

ys.-JETP *34*, 299 (1972)
Tech. Phys. *16*,

, 69 (1969) (in

1973)
ork 1955)
th Condensation and
brium Flows (Dekker,

eton University Press,

tron. *4*, 1461 (1975)

39 (1968)

3)
Chem. Phys. *62*, 388 (1975)
: Phys. Rev. A*11*, 54 (1975)

58)
. *16*, 314 (1975)
xplos. Shock Waves *13*, 506 (1977)
967)
. *5*, 1307 (1975)
nergy Transfer in Gases (Butter-

.*57*, 632 (1972)
317, 265 (1970)
J. Chem. Phys. *51*, 2214 (1969)
1969)
51, 844 (1969)
Chem. Phys. *58*, 3837 (1973)
3)
Sci. Paris *267*B, 211 (1973)
hem. Phys. Lett. *8*, 318 (1971)
963)
. *33*, 1120 (1972)
76)
V.A. Pavlov, O.P. Shatalov: Fluid

5)
5 (1976)
em. *76*, 3188 (1972)
971)
969)
1973)
ys. *45*, 3922 (1974)
em. Phys. Lett. *45*, 183 (1977)
ukhov, A.I. Lukovnikov: Trans.
)(in Russian)
skii, V.A. Shcheglov: Sov. Phys.-

ezneva: J. Appl. Mech. Tech. Phys.

. J. Quantum Electron. *4*, 1243

ntum Electron. *4*, 905 (1975)
72)
ukovnikov, R.I. Serikov: Sov. Phys.-

Phys. Lett. *17*, 531 (1972)
s. *51*, 2286 (1969)
e: J. Chem. Phys. *54*, 3097 (1971)
J. Chem. Phys. *50*, 4996 (1969)
yukhov, L.M. Pasynkova: Sov. J.

ys. *54*, 4131 (1971)
Phys. *56*, 1295 (1972)
. J. Chem. Phys. *54*, 1196 (1971)
njukhov, L.M. Pasynkova: Sov. Phys.-

. *60*, 1026 (1974)

3.94 R.D. Sharma, C.A. Brau: Phys. Rev. Lett. *19*, 1273
3.95 R.D. Sharma, C.A. Brau: J. Chem. Phys. *50*, 924 (1
3.96 H.K. Shin: J. Chem. Phys. *57*, 3484 (1972)
3.97 P. Borrel, G.E. Milward: J. Chem. Phys. *57*, 462 (
3.98 R.L. Taylor, S. Bitterman: "Vibrational Energy Tra cular Systems", in *Shock Tubes*, Proc 7th Intern. Toronto Press, Toronto 1970)
3.99 R.C. Benson, D.J. Benard, R.E. Walker: J. Chem. P
3.100 M.I. Buchwald, G.J. Wolga: J. Chem. Phys. *62*, 282
3.101 J.W. Cramp, J.D. Lambert: Chem. Phys. Lett. *22*, 1
3.102 J.D. Seery: J. Chem. Phys. *56*, 631 (1972)
3.103 A.N. Vargin, V.V. Gogokhiya, V.K. Konyukhov, L.M. Quantum Electron. *6*, 119 (1976)
3.104 G.Kamimoto, H. Matsui: Trans. Jpn. Soc. Aeronaut. (1970)
3.105 R.L. Taylor, S. Bitterman: J. Chem. Phys. *50*, 172
3.106 K. Bulthuis: J. Chem. Phys. *58*, 5786 (1973)
3.107 B.M. Christophe, A.A. Offenberger: Can. J. Phys.
3.108 G. Inoe, S. Tsuchida: J. Phys. Soc. Jpn. *39*, 479 (
3.109 H. Guegen, J. Ardati, M. Margottin-Maclou, L. Doye Acad. Sci. Paris *272*B, 1139 (1971)
3.110 J.C. Stephenson, J. Finzi, C.B. Moore: J. Chem. Ph
3.111 I.V. Rao, M. Tyagaraju, Y.V.C. Rao, S.V. Babu, V. Energy Transfer Mechanism in CO_2-C_2H_6 and CO_2-SO_2 *Modern Development in Shock Tube Research*, ed. by Tube Research Soc., Japan 1975) p.504
3.112 T.A. Dillon, J.C. Stephenson: J. Chem. Phys. *58*, 2
3.113 R. Fowler, E.A. Guggenheim: *Statistical Thermodyna* sity Press, Cambridge 1949)
3.114 V.N. Kondratiev: *Rate Constants of Gas Phase React* Commerce, Washington 1972)
3.115 S.A. Losev: "Energy Exchange and Chemical Kinetics in *Combustion and Explosion* (Nauka, Moscow 1977) p
3.116 N.M. Kuznetsov: Dokl. Akad. Nauk SSSR *164*, 1097 (1
3.117 S.A. Losev, N.A. Generalov: Dokl. Akad. Nauk SSSR Russian)
3.118 P. Hammerling, J. Teare, B. Kivel: Phys. Fluids *2*,
3.119 P.V. Marrone, C.E. Treanor: Phys. Fluids *6*, 215 (1
3.120 E.E. Nikitin: Dokl. Akad. Nauk SSSR *119*, 526 (1958)
3.121 B.F. Gordiets, A.I. Osipov, V.Ya. Panchenko: J. App *15*, 441 (1974)
3.122 B.F. Gordiets, A.I. Osipov, L. A. Shelepin: Sov. Ph
3.123 M.B. Zheleznyak, A.Kh. Mnatsakanyan: J. Appl. Mech. 309 (1975)
3.124 B.V. Kuksenko, S.A. Losev: Dokl. Akad. Nauk SSSR *18* Russian)
3.125 O.P. Shatalov: Combust. Explos. Shock Waves *9*, 610
3.126 J. Frenkel: *Kinetic Theory of Liquids* (Dover, New
3.127 P.P. Wegener (ed.): "Gasdynamics of Expansion Flow w Homogeneous Nucleation of Water Vapor", in *Nonequil* New York 1969)
3.128 H. Emmons (ed.): *Fundamentals of Gas Dynamics* (Prin Princeton 1958)
3.129 V.K. Konyukhov, V.N. Faizulaev: Sov. J. Quantum Ele
3.130 H.K. Shin: J. Chem. Phys. *55*, 5233 (1971)
3.131 M. Kovacs, D.R. Rao, A. Javan: J. Chem. Phys. *48*, 3

3.132 L. Doyennette, M. Margottin-Maclou, H. Guegen, A. Carion, L. Henry: J. Chem. Phys. *60*, 697 (1974); M. Margottin-Maclou, L. Doyennette, L. Henry: Appl. Opt. *10*, 176 (1971)
3.133 Yu.M. Gershenzon, V.B. Rosenshtein, S.Ya. Umanskii: *Chemistry of Plasma*, Vol.4(Atomizdat, Moscow 1977) (in Russian)
3.134 T.F. Hunter: J.Chem. Phys. *51*, 2641 (1969)
3.135 N. LrGay-Sommairy, F. LeGay: Canad. J. Phys. *48*, 1966b (1970)
3.136 G. Black, H. Wise, S. Schechter, R. Sharpless: J. Chem. Phys. *60*, 3526 (1974)
3.137 R. Abouaf, F. LeGay: J. Chem. Phys. *63*, 1393 (1966)
3.138 V.K. Konyukhov, V.N. Faizulaev: Sov. J. Quantum Electron. *8*, 851 (1978)
3.139 Yu. M. Gershenzon, V.I. Egorov, V.B. Rosenshtein, S.Ya. Umanskii: Chem. Phys. Lett. *20*, 77 (1973)
3.140 V.K. Konyukhov, V.N. Faizulaev: Sov. J. Quantum Electron. *8*, 297 (1978)
3.141 W.H. Dorrance: *Viscous Hypersonic Flow* (McGraw-Hill, New York 1962)

Chapter 4

4.1 G.N. Abramowitsch: *Angewandte Gasdynamik* (Technik, Berlin 1958)
4.2 J.F. Clarke, M. McChesney: *The Dynamics of Real Gases* (Butterworth, London 1964)
4.3 L.G. Loitsanskii: *Mechanics of Liquids and Gases* (Nauka, Moscow 1973) (in Russian)
4.4 G.I. Kozlov, D.I. Roitenburg, E.I. Stupitskii: J. Appl. Mech. Tech. Phys. *4*, 18 (1969)
4.5 V.M. Khailov: *Chemical Relaxation in Rocket Nozzles* (Mashinostroenie, Moscow 1975) (in Russian)
4.6 V.N. Kamzolov, U.G. Pirumov: Fluid Dyn. *1*, 6,14 (1966)
4.7 A.Q. Eschenroeder, D.W. Boyer, J.G. Hall: Phys. Fluids *5*, 615 (1962)
4.8 S.W. Bowen, C. Park: AIAA J. *9*, 493 (1971)
4.9 J.D. Anderson: AIAA J. *8*, 545,2280 (1970)
4.10 V.M. Shmelev, N.Ya. Vasilik, A.D. Margolin: Sov. J. Quantum Electron. *4*, 944 (1975)
4.11 J.D. Anderson: Phys. Fluids *13*, 1983 (1970)
4.12 S. Godunov (ed.): *Resolutions Numeriques des Problems Multidimensionnels de la Gazodynamique* (Mir, Moscow 1979)
4.13 V.M. Vasil'ev, M.N. Safaryan, O.V. Skrebkov: Fluid Dyn. *6*, 872 (1971)
4.14 V.K. Dushin: USSR J. Comp. Math. Math Phys. *9*, 5,184 (1969)
4.15 G.A. Moretti: AIAA J. *3*, 223 (1965)
4.16 N.M. Reddy, F.L. Daum: AIAA J. *9*, 1622 (1971)
4.17 K.N.C. Bray: J. Fluid Mech. *6*, 1 (1959)
4.18 I.R. Hurle: *Shock Tube Research* (Chapman and Hall, London 1971)
4.19 A.L. Russo: J. Chem. Phys. *47*, 5201 (1967)
4.20 T.I. McLaren, J. Appleton: J. Chem. Phys. *53*, 2850 (1970)
4.21 K. Sato, T. Sekiguchi: J. Phys. Soc. Jpn. *39*, 1351 (1975)
4.22 C.W. von Rosenberg, R. Taylor, J. Teare: J. Chem. Phys. *54*, 1974 (1971)
4.23a N.V. Evtyukhin, S.A. Losev, V.N. Makarov, V.A. Pavlov, M.S. Valovik: J. Appl. Mech. Tech. Phys. *14*, 763 (1973)
4.23b V.N. Makarov, O.P. Shatalov: Fluid Dyn. *9*, 323 (1974)
4.24 J.R. McDonald: J. Chem. Phys. *57*, 1016 (1972)
4.25 K. Nanbu: J. Phys. Soc. Jpn. *40*, 1391 (1976)
4.26 L.C. Franciscus, E.A. Lezberg: AIAA J. *1*, 2077 (1973)
4.27 K.A. Wilde: AIAA J. *3*, 1846 (1965)
4.28 V.M. Vasil'ev, S.V. Kulikov, O.V. Skrebkov: J. Appl. Mech. Tech. Phys. *18*, 437 (1977)

4.29 R.A. Greenberg, A.M. Schneiderman, D.R. Ahouse, E.M. Parmentier: AIAA J. *10*, 1494 (1972)
4.30 V.N. Croshko, R.I. Soloukhin, N.A. Fomin: Combust. Explos. Shock Waves *9*, 306 (1973)
4.31a R. Sauer: *Ecoulementes des Fluides Compressibles* (Betanger, Paris, Liege 1951)
4.31b A. Ferri: *Elements of Aerodynamics of Supersonic Flows* (MacMillan, New York 1949)
4.32 V.P. Verkhovskii: Trans. CAGI 1680 (1975) (in Russian)
4.33 J.D. Anderson, E.L. Harris: Laser Focus *8*, 32 (1972)
4.34 M.N. Director: AIAA Paper 73-626 (1973)
4.35 D.M. Kuehn: Appl. Phys. Lett. *21*, 112 (1972)
4.36 D.A. Russell: AIAA Paper 74-223 (1974)
4.37 G.M. Ryabinkov: CAGI Sci. Paper *1*, 34 (1970) (in Russian)
4.38 U.G. Pirumov, G.S. Roslyakov: *Nozzle Gas Flow* (Springer, Berlin, Heidelberg, New York 1981)
4.39 M.Ya. Ivanov, A.N. Kraiko, N.B. Mikhailov: USSR J. Comp. Math. Phys. *12*, 2,196 (1972)
4.40 V.A. Levin, Yu.V. Tunik: Fluid Dyn. *11*, 101 (1976)
4.41 H. Schlichting: *Boundary Layer Theory*, 4th ed. (McGraw-Hill, New York 1960)
4.42 P.F. Massier, L.H. Black, M.B. Noel, F. Sahel: AIAA J. *8*, 605 (1970)
4.43 A.N. Sekundov: Fluid Dyn. *6*, 828 (1971)
4.44 D.B. Spalding, B.E. Launder: *Mathematical Models of Turbulence* (Academic Press, New York 1972)
4.45 E.J. Hopkins, E.R. Keener, H.A. Dwyer: AIAA Paper 71-167 (1971)
4.46 A.F. Burke, K.D. Bird: "The Use of Conical and Contoured Expansion Nozzles in Hypervelocity Facilities", in *Advances of Hypervelocity Techniques* (Plenum Press, New York 1963)
4.47 G.A. Simons: AIAA Paper 72-709 (1972)
4.48 J.C. Williams: AIAA J. *1*, 186 (1963)
4.49 N.K. Mitra, M. Fiebig:" Low Reynolds Number Vibrational Nonequilibrium Flow in Laval Nozzle", in *Rarefield Gas Dynamics*, Proc. 9th Intern. Symp. (DFVLR-Press, Porz-Wahn 1974)
4.50 V.V. Mikhailov: Fluid Dyn. *4*, 46 (1969)
4.51 J.C. Hyde, G.A. Hosack, G.S. Osugi: AIAA Paper 73-641 (1973)
4.52 M.J. Monsler, R.A. Greenberg: AIAA Paper 71-24 (1971)
4.53 G.A. Simons: AIAA J. *10*, 1037 (1972)
4.54 A.N. Sekundov, O.V. Yakovlevskii: Fluid Dyn. *7*, 513 (1972)
4.55 A. Demetriades: Phys. Fluids *12*, 24 (1969)
4.56 C.W. Peterson: AIAA Paper 78-1214 (1978)
4.57 A. Demetriades: J. Energy *1*, 73 (1977)
4.58 D.A. Russell, S.E. Neice, P.H. Rose: AIAA J. *13*, 593 (1975)
4.59 A.G. Gaydon, I.R. Hurle: *The Shock Tube in High-Temperature Chemical Physics* (Chapman and Hall, London 1963)
4.60 T.V. Bazhenova, L.G. Gvozdeva: *Nonstationary Interaction of Shock Waves* (Nauka, Moscow 1977) (in Russian)
4.61 T.V. Bazhenova: *Shock Waves in Real Gases* (Springer, Wien, New York 1973)
4.62 C.E. Wittliff, M.R. Wilson, A. Herzberg: J. Aeron. Sci. *26*, 219 (1959)
4.63 T.V. Bazhenova, I.M. Naboko, A.V. Eremin, V.A. Kochnev:"Test Time Behind Reflected Wave in a Shock Tube with Nozzle", in *Recent Development in Shock Tube Research*, Proc. 9th Intern. Symp. (Stanford Press, Stanford 1973)
4.64 L. Davis, D.H. Edwards: ARC Rep. Memor. 3446 (1964)
4.65 D.W. Holder, D.L. Schultz: ARC Rep. Memor. 3265 (1962)
4.66 M.G. Dunn: AIAA J. *7*, 1553 (1969)
4.67 D.J. Monson: AIAA J. *9*, 1872 (1971)

4.68 D.C. Bull, D.H. Edwards: AIAA J. *6*, 1549 (1968)
4.69 S.A. Losev, V.N. Makarov, V.A. Pavlov, O.P. Shatalov: Combust. Explos. Shock Waves *9*, 401, (1973)
4.70 A.I. Demin, E.M. Kudryavtsev: J. Phys. Soc. Jpn. *38*, 1222 (1975)
4.71 K. Sato, T. Sekiguchi: J. Phys. Soc. Jpn. *35*, 315 (1973)
4.72 K. Sato, Z. Sekiguchi: J. Phys. Soc. Jpn. *35*, 630 (1973)
4.73 J.A. Ackroyd: ARC Current Paper 883 (1967)
4.74 H.O. Amann: Phys. Fluids *12*, Suppl. 1, 150 (1969)
4.75 R.F. Chisnel: J. Fluid Mech. *2*, 286 (1957)
4.76 P.H. Rose:"Advanced Laser Technology Development in Shock Tubes", in *Shock Tube and Shock Wave Research*, Proc. 11th Intern. Symp.(University of Washington Press, Seattle 1978) p. 508
4.77 K. Sato, J. Yamashita, Y.Iida, T. Sekiguchi:"Studies on CO_2 Gas-Dynamic Laser by Means of Shock Tube", in *Modern Development in Shock Tube Research*, ed. by G. Kamimoto (Shock Tube Research Soc., Japan 1975) p. 520
4.78 R.I. Soloukhin:"Shock Tubes in Flow Laser Research; Modeling and Application", in *Shock Tube and Shock Wave*, Proc. 11th Intern. Symp. (University of Washington Press, Seattle,London 1978)

Chapter 5

5.1 J.D. Anderson: AIAA Paper 74-176 (1974)
5.2 L.B. Evans: JASA *51*, 409 (1972)
5.3 C.W. von Rosenberg, K.N.C. Bray, N.H. Pratt: J. Chem. Phys. *56*, 3230 (1972)
5.4 E.T. Gerry, D.A. Leonard: Appl. Phys. Lett. *8*, 227 (1966)
5.5 A.S. Birjykov, A.Yu. Volkov, A.I. Demin, E.M. Kudryavtsev, R.I. Serikov: Sov. J. Quantum Electron. *6*, 946 (1976)
5.6 V.V. Danilov, E.P. Kruglyakov, E.V. Shun'ko: J. Appl. Mech. Tech. Phys. *13*, 783 (1972)
5.7 C. Cousin, C. Rossetti, C. Mever: C. R. Acad. Sci. Paris *262B*, 1640 (1969)
5.8 T.K. McCubbin, R.R. Mooney: JQSRT *8*, 1255 (1968)
5.9 R.R. Patty, E.R. Manring, J.A. Gardner: Appl. Opt. *7*, 2241 (1968)
5.10 A.R. Strilchuk, A.A. Offenberger: Appl. Opt. *13*, 2643 (1974)
5.11 C. Young, R.E. Chapman: JQSRT *14*, 679 (1974)
5.12 R. Ely, T.K. McCubbin: Appl. Opt. *9*, 1230 (1970)
5.13 R.L. Leonard: Appl. Opt. *13*, 1920 (1974)
5.14 S.A. Munjee, W.H. Christiansen: Appl. Opt. *12*, 999 (1973)
5.15 J.D. Anderson, R.L. Humphrey, J.S. Vamos, M.J. Plummer, R.E. Jensen: Phys. Fluids *14*, 2620 (1971)
5.16 R.I. Soloukhin, Yu.A. Yakobi: J. Appl. Mech. Tech. Phys. *15*, 289 (1974)
5.17 G.V. Gembarzhevskii, N.A. Generalov, G.I. Kozlov: J. Appl. Mech. Tech. Phys. *14*, 448 (1973)
5.18 G.V. Gembarzhevskii, N.A. Generalov, G.I. Kozlov, D.I. Roitenburg: Sov. Phys.-JETP *35*, 447 (1972)
5.19 E.M. Kudryavtsev, V.N. Faizulaev: J. Appl. Tech. Phys. *14*, 757 (1973)
5.20 W.H. Christiansen, A. Hertzberg: Proc. IEEE *61*, 1060 (1973)
5.21 W.H. Christiansen, G.A. Tsongas: Phys. Fluids *14*, 2611 (1971)
5.22a S.S.R. Murthy: AIAA Paper 74-226 (1974)
5.22b G. Lee, F.E. Gowen, J.R. Hagen: AIAA J. *10*, 65 (1972)
5.23 A.T. Jones: J. Phys. D: Appl. Phys. *9*, 1193 (1976)
5.24 D.J. Monson: AIAA J. *15*, 1660 (1977)
5.25 J.D. Anderson: AIAA J. *12*, 1699 (1974)
5.26 J.I. McManus, J.D. Anderson: AIAA J. *14*, 1770 (1976)

5.27 A.S. Biryukov, A.P. Dronov, E.M. Kudryavtsev, G.A. Rainin, N.N. Sobolev: High Temp. High Pressures *5*, 389 (1973)
5.28 A.Yu. Volkov, A.I. Demin, A.N. Logunov, E.M. Kudryavtsev, N.N. Sobolev: Trans. Phys. Inst. Acad. Sci. USSR *113* (1979) (in Russian)
5.29 N.N. Kudryavtsev, S.S. Novikov, I.B. Svetlichnyi: Combust. Explos. Shock Waves, *12*, 653 (1976)
5.30 M. Thomas-Andraud, A. Carrega, O. Leuchter, J. Taran: Rech. Aerosp. *6*, 325 (1972)
5.31 P.V. Belkov, V.P. Vakatov, B.K. Tkachenko, N.N. Shirokov: Combust. Explos. Shock Waves *12*, 515 (1976)
5.32 A.B. Britan, S.A. Losev, O.P. Shatalov: Sov. J. Quantum Electron. *4*, 1459 (1975)
5.33 A.I. Demin, E.M. Kudryavtsev, N.N. Sobolev, V.N. Faizulaev, N.A. Shubina: Sov. J. Quantum Electron. *4*, 403 (1974)
5.34 J.D. Anderson, E.M. Winkler: Proc. IEEE *59*, 651 (1971)
5.35 J. Tulip, H. Seguin: Appl. Phys. Lett. *19*, 263 (1971)
5.36 J.S. Vamos: AIAA Paper 74-177 (1974)
5.37 S. Yatsiv: Appl. Phys. Lett. *19*, 65 (1971)
5.38 V.N. Croshko, N.A. Fomin, R.I. Soloukhin: Acta Astron. *2*, 929 (1975)
5.39 V.N. Croshko, R.I. Soloukhin, N.A. Fomin: Combust. Explos. Shock Waves *10*, 410 (1974)
5.40 V.B. Buonadonna, W.H. Christiansen:"Gain Measurements of High Temperature CO_2 Laser Mixtures in a Shock Tube Driven Flow", in *Recent Development in Shock Tube Research*, Proc. 9th Intern. Symp. (Stanford Press, Stanford 1973)
5.41 G.I. Kozlov, V.N. Ivanov, A.S. Korablev: Sov. Phys.-JETP *38*, 41 (1974)
5.42 S.A. Losev, V.N. Makarov: Sov. J. Quantum Electron. *5*, 780 (1975)
5.43 S.A. Losev, V.N. Makarov: Sov. J. Quantum Electron. *6*, 514 (1976)
5.44 D.J. Wilde: *Optimum Seeking Methods* (Prentice-Hall, Engelewodd Cliffs, New Jersey 1964)
5.45 V.N. Makarov: Combust. Explos. Shock Waves *12*, 659 (1976)
5.46 S.A. Losev, V.N. Makarov: J. Appl. Mech. Tech. Phys. *18*, 292 (1977)
5.47 G.I. Kozlov, E.L. Stupitskii: Sov.Phys.-Tech. Phys. *20*, 226 (1975)
5.48 R.I. Soloukhin, N.A. Fomin: Sov. Phys.-Dokl. *21*, 281 (1976)
5.49a A.I. Anan'kin, V.M. Khailov, Yu.M. Shirkhman: High Temp. *14*, 5 (1976)
5.49b A.S. Biryukov, R.S. Serikov, A.I. Starik, L.A. Shelepin: Sov. J. Quantum Electron. *7*, 438 (1977)
5.50 M.G. Ktalkherman, V.A. Levin, V.M. Mal'kov, Yu.V. Tunik: Combust. Explos. Shock Waves *15*, 1 (1979)
5.51 N.K. Mitra, M. Fiebig: "Effect of Vibrational Wall Accommodation on Small Signal Gain in CO_2 -N_2-H_2O Gasdynamic Lasers", in *Rarefied Gas Dynamics*, Proc 10th Intern. Symp., ed. by J. Potter (AIAA, New York 1977)
5.52 J.H. Lundell, L.J. Otten, R.R. Dickey: AIAA Paper 75-177 (1975)
5.53 R.A. Meinzer: AIAA J. *10*, 388 (1972)
5.54 K. Sato, T. Sekiguchi: J. Phys. Soc. Jpn. *36*, 808 (1974)
5.55a H. Hara: J. Appl. Phys. *48*, 5341 (1977)
5.55b G.B. Abrosimov, A.A. Vedenov, A.F. Vitshas, A.P. Napartovich, V.F. Sharkov: High Temp. *13*, 4 (1975)
5.56 G.I. Kozlov, V.N. Ivanov, A.S. Korablev, I.K. Selezneva: Sov. Phys.-JETP *41*, 825 (1976)
5.57 S. Yatsiv: IEEE J. QE- *8*, 161 (1972)
5.58 E.V. Locke, R. Hella, L. Westra: IEEE J. QE-*7*, 581 (1971)
5.59 V.N. Makarov, Yu.V. Tunik: J. Appl. Mech. Tech. Phys. *19*, 5 (1978)
5.60 A.I. Anan'kin: Sov. J. Quantum Electron. *5*, 89 (1975)
5.61 P. Klass: Aviat. Week *97*, 8,32 (1972)

5.62 N.N. Kudryavtsev, S.S. Novikov, I.B. Svetlichnyi: Dokl. Akad. Nauk SSSR *219*, 844 (1974) (in Russian)
5.63 G.P. Pallay, C.T. Zovko: AIAA Paper 73-1233 (1973)
5.64 R.J. Hill, N.T. Jewell, A.T. Jones, R.B. Price: J. Energy *1*, 125 (1977)
5.65 N.V. Evtyukhin, A.P. Genich, G.B. Manelis: Combust. Explos. Shock Waves *14*, 4, (1978)
5.66 N.V. Evtyukhin, A.P. Genich, A.A. Yudanov, G.B. Manelis: Sov. J. Quantum Electron. *8*, 580 (1978)
5.67 M.G. Ktalkherman, V.M. Mal'kov, A.V. Petukhov, Ya.I. Kharitonova: Combust. Explos. Shock Waves *12*, 524 (1976)
5.68 J.D. Anderson: Acta Astronaut. *2*, 911 (1975)
5.69 A.P. Genich, N.V. Evtyukhin, G.B. Manelis: Combust. Explos. Shock Waves *11*, 644 (1975)
5.70 M.G. Ktalkherman, V.M. Mal'kov, A.V. Petukhov, Ya.I. Kharitonova: Sov. J. Quantum Electron. *7*, 97 (1977)
5.71 L.K. Beletskii, V.A. Vazyulin, V.F. Gavrikov, A.I. Didyukov, A.P. Dronov, V.K. Orlov, A.K. Piskunov, E.M. Cherkasov: Sov. J. Quantum Electron *4*, 250 (1974)
5.72 A.P. Genich, N.V. Evtyukhin, S.V. Kulikov, G.B. Manelis, M.E. Solov'eva: J. Appl. Mech. Tech. Phys. *20*, 1 (1979)
5.73 A.P. Genich, S.V. Kulikov, G.B. Manelis: J. Appl. Mech. Tech. Phys. *20*, 4 (1979)
5.74 A. Gabai, M. Rokni, E. Greenfield: IEEE J. QE-*8*, 619 (1972)
5.75 J.D. Anderson: J. Appl. Phys. *43*, 435 (1972)
5.76 R. Tennant, R. Vargas, S. Hadley: AIAA Paper 74-178 (1974)
5.77 J.B. Watterson, G.S. Knoke: "Effects of Oxygen Contamination on Gas-dynamic Laser Performance", in *Shock Tube and Shock Wave Research*, Proc. 11th Intern. Symp. (University of Washington Press, Seattle 1978) p. 527
5.78 B.V. Egorov, V.N. Komarov: J. Appl. Mech. Tech. Phys. *16*, 172 (1975)
5.79 V.N. Makarov, S.A. Losev: Combust Explos. Shock Waves *11*, 689 (1975)
5.80 N.Ya. Vasilik, V.M. Shmelev, A.D. Margolin: Sov. J. Quantum Electron. *6*, 1181 (1976)
5.81 J.P.E. Taran, M. Charpenel, R. Borghi: AIAA Paper 73-622 (1973)
5.82 A.I. Demin, E.M. Kudryavtsev, N.N. Sobolev, V.N. Faizulaev: Sov. J. Quantum Electron. *4*, 293 (1974)
5.83 A.B. Britan, N.M. Kortsenshtein: Sov. J. Quantum Electron. *5*, 1388 (1975)
5.84 N.G. Basov, A.N. Oraevskii, V.A. Shcheglov: Sov. Phys.-Tech. Phys. *15*, 126 (1970)
5.85 B.R. Bronfin, L.R. Boedeker, J.P. Cheyer: Appl. Phys. Lett. *16*, 214 (1970)
5.86 A.N. Sekundov: Industrial Aerodynamics *34* (Mashinostroenie, Moscow 1979) (in Russian)
5.87 D.Vandromme, W. Kollmann, J.F. Wendt: "Turbulent Mixing in the Downsteam-Mixing Gasdynamic Lasers, Von Karman Inst. Fluid Dyn., Belgum, Preprint 1978-4 (1978)
5.88 J.D. Anderson: " Navier-Strokes Solutions of High Energy Laser Flows", in 2nd Intern. Symp. Gas-Flow and Chem. Lasers Proc. (Hemisphere Publ. Corp. 1978)
5.89 J. Milewski, M. Brunne, J. Stanco, A. Zielinski, M. Irczuk, J. Kusmierek: Bull. Acad. Polon. Sci. Ser. Sci. Tech. *20*, 313 (1972)
5.90 J.P.E. Taran, M. Charpenel, R. Borghi: AIAA Paper 70-622 (1970)
5.91 K. Kasuya, T. Kumasaka, T. Murasaki: Jpn. J. Appl. Phys. *12*, 771 (1973)
5.92 V.N. Croshko, R.I. Soloukhin: Dokl. Akad. Nauk SSSR *211*, 829 (1973) (in Russian)
5.93 A.S. D'yakov, A.I. Didyukov, B.K. Tkachenko, E.M. Cherkasov: Sov. J. Quantum Electron. *8*, 669 (1978)
5.94 L.I. Kiselevskii, D.K. Skutov, S.A. Sokolov: J. Appl. Spectrosc. (Minsk) *20*, 35 (1974) (in Russian)

5.95 J.A. Stregack, B.L. Wexler, G.A. Hart: Appl. Phys. Lett. *27*, 670 (1975)
5.96 P. Cassady, J. Newton, P. Rose: AIAA J. *16*, 305 (1978)
5.97 R. Borghi, M. Charpenel: Astron. Acta *17*, 833 (1972)
5.98 W. Schall, P. Hoffmann, H. Hügel: J. Appl. Phys. *48*, 688 (1977)
5.99 O.V. Achasov, R.I. Soloukhin, N.A. Fomin: Sov. J. Quantum Electron. *8*, 1312 (1978)
5.100 P. Hoffmann, H. Hügel, W. Schall: AIAA J. *15*, 1527 (1977)
5.101 N.N. Ostroukhov, B.K. Tkachenko: Sov. J. Quantum Electron. *8*, 531 (1978)
5.102 A.S. Biryukov, V.M. Marchenko, A.M. Prokhorov: Sov. Phys.- JETP *44*, 905 (1976)
5.103 K. Kasuya, M. Migamete, K. Herioka, K. Fujiwara, T. Murasaki: Jpn. J. Appl. Phys. *15*, 1397 (1976)
5.104a V.K. Konyukhov: Sov. J. Quantum Electron. *7*, 568 (1977)
5.104b V.V. Breev, A.I. Kukharenko, A.P. Napartovich, V.F. Sharkov: High Temp. *12*, 6 (1974)
5.105 W.H. Christiansen, D.A. Russel, A.Hertzberg: Annu. Rev. Fluid Mech. *7*, 115 (1975)
5.106 J. Wieder: Phys. Lett *24*A 759 (1967)
5.107 V.A. Batanov, F.V. Bunkin, A.M. Prokhorov, V.B. Fedorov: Dokl. Akad. Nauk SSSR *191*, 1267 (1970) (in Russian)
5.108 V.M. Shmelev, A.D. Margolin: Sov. J. Quantum Electron. *5*, 776 (1975)
5.109 N.G. Basov, V.V. Gromov, E.L. Koshelev, E.P. Markin, A.N. Oraevsky: JETP Lett. *10*, 2 (1969)
5.110 M.S. Dzhidzhoev, M.I. Pimenov, V.T. Platonenko, Yu.V. Filippov, R.V. Khokhlov: Sov. Phys.- JETP *30*, 225 (1970)
5.111 V.V. Kovtun, N.N. Kudryavtsev, S.S. Novikov, I.B. Svetlichnyi, P.N. Shagov: Sov. Phys.- Dokl. *23*, 503 (1978)
5.112 N.N. Kudryavtsev, S.S. Novikov, I.B. Svetlichnyi: Sov. Phys.- Dokl. *21*, 748 (1976)
5.113 I.S. Zaslonko, S.M. Kogarko, E.V. Mozzhukhin, V.N. Smirnov: Dokl. Akad. Nauk SSSR *212*, 1138 (1973) (in Russian)
5.114 A.E. Hill: Appl. Phys. Lett. *18*, 194 (1971)
5.115 A.S. Biryukov, V.M. Marchenko, L.A. Shelepin: Trans. Phys. Inst. Acad. Sci. USSR *83*, 87 (1975) (in Russian)
5.116 V.K. Konyukhov, V.N. Faizulaev: Sov. J. Quantum Electron. *8*, 1473 (1978)
5.117 A.A. Vedeneev, A.Yu. Volkov, A.I. Demin, A.N. Logunov, E.M. Kudryavtsev, N.N. Sobolev: Sov. Tech. Phys. Lett. *4*, 275 (1978)
5.118 T.J. Manuccia, J.A. Stregack, N.W. Harris, B.L. Wexler: Appl. Phys. Lett. *29*, 360 (1976)
5.119 B.L. Wexler, T.J. Manuccia, R.W. Waynant: Appl. Phys. Lett. *31*, 730 (1977)
5.120 N.V. Karlov, A.N. Orlov, Yu.N. Petrov, A.M. Prokhorov: Sov. Tech.Phys. Lett. *3*, 49 (1977)
5.121 E. Wasserstrom, J. Stricker: Phys. Fluids *17*, 1335 (1975)
5.122 V.M. Marchenko, A.M. Prokhorov: JETP Lett. *14*, 76 (1971)
5.123 V.A. Levin, Yu.V. Tunik: Combust. Explos. Shock Waves *13*, 380 (1977)
5.124 J.H. Lee, T.D. Bui, R. Knystautas: Appl. Phys. Lett. *22*, 434 (1973)
5.125 M.I. Podduev: Sov. J. Quantum Electron. *9*, 225 (1979)
5.126 V.M. Kuznetsov: J. Appl. Mech. Tech. Phys. *16*, 177 (1975)
5.127 R.W.F. Gross, R.R. Ciedt, T.A. Jacobs: J. Chem. Phys. *51*, 1250 (1969)
5.128 A.S. Biryukov, R.I. Serikov, A.I. Starik: Sov. J. Quantum Electron. *8*, 736 (1978)
5.129 V.A. Levin, A.I. Starik: Fluid Dyn. *14*, 1 (1979)

Chapter 6

6.1a Yu.I. Grin', V.G. Testov: Sov. Phys.-Tech. Phys. *19*, 975 (1974)
6.1b A.S. Biryukov, A.Yu. Volkov, A.I. Demin, E.M. Kudryavtsev, Yu.A. Kulagin, N.N. Sobolev, L.A. Shelepin: Sov. Phys.-JETP *41*, 834 (1976)
6.2 A.S. Biryukov, A.Yu. Volkov, A.I. Demin, E.M. Kudryavtsev, Yu.A. Kulagin, N.N. Sobolev: Sov. J. Quantum Electron. *5*, 1408 (1975)
6.3 A.S. Biryukov, Yu.A. Kulagin, L.A. Shelepin: Sov. Phys.-Tech. Phys. *21*, 199 (1976)
6.4 Yu.I. Grin', V.G. Testov: Sov. Phys.-Dokl. *21*, 200 (1976)
6.5 V.G. Testov, Yu.I. Grin', V.V. Golub, V.M. Polyakov, I.M. Naboko: Sov. Phys.-JETP *44*, 46 (1976)
6.6 V.G. Testov, Yu.I. Grin', G.I. Mishin, R.L. Petrov, V.D. Menenkov: Sov. Tech. Phys. Lett. *4*, 449 (1978)
6.7 N. Djeu, T.Kan, C. Wolga: IEEE J. QE-*4*, 783 (1968)
6.8 M.C. Lin, S.H. Bauer: J. Chem. Phys. *50*, 3377 (1969)
6.9 J. Milewski, M. Brunne, M. Irczuk, A.I. Demin, E.M. Kudryavtsev, A.Yu. Volkov, N.N. Sobolev: Appl. Phys. Lett. *30*, 342 (1977)
6.10 J. Milewski, M. Brunne, J. Stanco, A. Zielinski, M. Irczuk, G. Rabczuk, A.I. Demin, E.M. Kudriavtsev, A.Yu. Volkov, N.N. Sobolev: Appl. Phys. Lett. *32*, 642 (1978)
6.11 W.S. Watt: Appl. Phys. Lett. *18*, 487 (1971)
6.12 V.F. Gavrikov, A.P. Dronov, V.K. Orlov, A.K. Piskunov: Sov. J. Quantum Electon. *3*, 437 (1974)
6.13 V.F. Gavrikov, A.P. Dronov, V.K. Orlov, A.K. Piskunov: Sov. J. Quantum Electron. *5*, 69 (1975)
6.14 R.L. McKenzie: Phys. Fluids *15*, 2163 (1972)
6.15 B.F. Gordiets, A.I. Osipov, L.A. Shelepin: Sov. Phys.- JETP *32*, 334 (1971)
6.16 E.R. Fisher, R.H. Kummler: J. Chem. Phys. *49*, 1085 (1968)
6.17 G. Hancock, I.W.M. Smith: Appl. Opt. *10*, 1827 (1971)
6.18 A. J. Lightman, E.R. Fisher: Appl. Phys. Lett. *29*, 593 (1976)
6.19 R.L. McKenzie: J. Chem. Phys. *64*, 1498 (1976)
6.20 V.F. Gavrikov, A.P. Dronov, V.K. Orlov, A.K. Piskunov: Sov. J. Quantum Electron. *6*, 803 (1976)
6.21 N.N. Sobolev, V.V. Sokovikov: Sov. Phys.-Usp. *16*, 350 (1973)
6.22 W.J. Hooker, R.C. Millikan: J. Chem. Phys. *38*, 214 (1963)
6.23 D.F. Starr, J.K. Hancock, W.H. Green: J. Chem. Phys. *61*, 5421 (1974)
6.24 M.R. Verter, H. Rabitz: J. Chem. Phys. *64*, 2939 (1976)
6.25 J.C. Stephenson, E.R. Mosburg: J. Chem. Phys. *60*, 3562 (1974)
6.26 M.C. Gower, G. Srinivasan, K.W. Billman: J. Chem. Phys. *63*, 4206 (1975)
6.27 H.T. Powell: J. Chem. Phys. *63*, 2635 (1975)
6.28 P.F. Zittel, C.B. Moore: Appl. Phys. Lett. *21*, 81 (1972)
6.29 C.K.N. Patel: Phys. Rev. Lett. *13*, 617 (1964)
6.30 V.A. Balanin, G.V. Abrosimov, G.A. Andronov, A.G. Armer, B.M. Dymshits, Ya.P. Koretskii, V.F. Sharkov: Sov. Phys.-Tech. Phys. *22*, 364 (1977)
6.31 J.P. Martin: Acta Astronaut. *3*, 839 (1976)
6.32 N.Ya. Vasilik, V.A. Vakhenenko, A.D. Margolin, V.M. Shmelev: J. Appl. Mech. Mech. Tech. Phys. *19*, 5 (1978)
6.33 T. Kan, J.A. Stregak, W.S.Watt: Appl. Phys. Lett. *20*, 137 (1972)
6.34 M.M. Mann: AIAA Paper 75-34 (1975)
6.35 J.W. Rich, R.C. Bergman, J.A. Lordi: AIAA Paper 74-179 (1974)
6.36 A.N. Vargin, N.A. Ganina, V.K. Konyukhov, A.I.Lukovnokov, V.I. Selyakov: Sov. J. Quantum Electron. *5*, 338 (1975)
6.37 S.A. Reshetnyak: Trans Phys. Inst. Acad. Sci. USSR *83*, 146 (1975) (in Russian)
6.38 V.K. Konyukhov, A.M. Prokhorov: JETP Lett. *13*, 153 (1971)

6.39 V.K. Konyukhov, A.M. Prokhorov, V.I. Tikhonov, V.N. Faizulaev: Sov. J. Quantum Electron. *5*, 1131 (1975)
6.40 A.N. Oraevskii: Sov. Phys.-JETP *32*, 856 (1971)
6.41 S.I. Pekar: Dokl. Akad. Nauk SSSR *187*, 555 (1969) (in Russian)
6.42 C.H. Bamford, C.F.H. Tipper (eds.): *Comprehensive Chemical Kinetics*, Vol. 3 (Elsevier, Amsterdam 1969)
6.43 B.P. Levitt (ed.) *Physical Chemistry of Fast Reactions*, Vol. 1 (Plenum Press, London New York 1973)
6.44 S.I. Pekar, V.A. Kochelap: Dokl. Akad. Nauk SSSR *196*, 808 (1971) (in Russian)
6.45 V.A. Kochelap Yu.A. Kukibnyi, S.I. Pekar: Sov. J. Quantum Electron. *4*, 155 (1974)
6.46 R.A. Young: J. Chem. Phys. *40*, 1848 (1964)
6.47 A.S. Bashkin, A.N. Oraevskii: Sov. J. Quantum Electron. *6*, 14 (1976)
6.48 A.S. Biryukov, A.M. Prokhorov, L.A. Shelepin, N.N. Shirokov: Sov. Phys.-JETP *40*, 1025 (1974)
6.49 I.A. Izmailov, V.A. Kochelap, Yu.A. Kukibnyi: Ukr. Phys. J. *21*, 508 (1976) (in Russian)
6.50 A.S. Bashkin, A.N. Oraevskii: Sov. J. Quantum Electron. *3*, 1 (1973)
6.51 A.S. Bashkin, A.N. Oraevskii, N.N. Yuryshev: Sov. J. Quantum Electron. *1*, 630 (1971)
6.52 V.A. Kochelap: Ukr. Phys. J. *15*, 1213 (1970) (in Russian)
6.53 V.A. Kochelap: Ukr. Phys. J. *16*, 396 (1971) (in Russian)
6.54 V.A. Kochelap,Yu.A. Kukibnyi: Sov. J. Quantum Electron. *5*, 790 (1975)
6.55 V.A. Kochelap, Yu.A. Kukibnyi: Quantum Electron. (Kiev) *10*, 27 (1976) (in Russian)
6.56 V.L. Tal'rose, E.B.Gordon, Yu.L.Moskvin, A.P. Kharitonov: Dokl. Akad. Nauk SSSR *216*, 146 (1974) (in Russian)
6.57 B. Fontaine, B. Forestier: Sov. J. Quantum Electron. *6*, 486 (1976)
6.58 B. Forestier, B. Fontaine, V. Valensi: "Experimental Investigation on Recombination Molecular Gasdynamic Lasers Using the Exitation of Electronic Levels", in *Modern Development in Shock Tube Research*, ed, by G. Kamimoto (Shock Tube Research Soc., Japan 1975) p. 496
6.59 V.A. Kochelap, Yu.A. Kukibnyi: Fluid Dyn. *12*, 476 (1977)
6.60 S.R. Leone, K.L. Kosnik: Appl. Phys. Lett. *30*, 346 (1977)
6.61 A.S. Bashkin, N.L. Kupriyanov, A.N. Oraevskii: Sov. J. Quantum Electron. *8*, 518 (1978)
6.62 A.Yu. Volkov, A.I. Demin, E.M. Kudryavtsev, N.N. Sobolev: "Recombinational Gasdynamic Visible O_2 Laser: Experimental and Theoretical Feasibility Study", In *Electronic Transition Lasers*, ed. By J.I. Steinfeld (MIT-Press), Cambridge, Mass., London 1976)
6.63 A.Yu. Volkov, A.I. Demin, E.M. Kudryavtsev, N.N. Sobolev: Sov. Phys.-JETP *43*, 259 (1976)
6.64a A.S. Biryukov, L.A. Shelepin: Sov. Phys.-Tech. Phys. *19*, 775 (1974)
6.64b E.B. Gordon, Yu.L. Moskvin, S.A. Sotnichenko: Sov. J. Quantum Electron. *6*, 1425 (1976)
6.65 I.A. Izmailov, V.A.Kochelap, Yu.A. Kukibnyi: Ukr. Phys. J. *21*, 371 (1976) (in Russian)
6.66 V.A. Kochelap, Yu.A. Kukibnyi, I.A. Izmailov: Opt. Commun. *16*, 380 (1976)
6.67 S.I. Pekar, I.A. Izmailov, V.A. Kochelap, Yu.A. Kukibnyi: Sov. Phys.-Dokl. *23*, 488 (1978)
6.68 I.A. Izmailov, V.A. Kochelap, Yu.A. Kukibnyi, S.I. Pekar: Sov. Tech. Phys. Lett. *5*, 2 (1979)
6.69 N.M. Kuznetsov, Yu.P. Raizer: J. Appl. Mech. Tech. Phys. *4*, 6 (1965)
6.70 L.I. Gudzenko, D.I. Yakovlenko: *Plasma Lasers* (Atomizdat, Moscow 1977) (in Russian)

6.71 L.I. Gudzenko, L.A. Shelepin, S.I. Yakovlenko: Trans. Phys. Inst. Acad. Sci. USSR *83*, 100 (1975) (in Russian)
6.72 G.A. Luk'yanov: Sov. Phys.-Tech. Phys. *21*, 440 (1976)
6.73 M. Nishida: Z. Naturforsch. *30*a, 1563 (1975)
6.74 W.L. Bohn, M.U. Beth, P. Hoffmann: AIAA Paper 74-565 (1974)
6.75 A.W. Ali, W.W. Jones: Phys. Lett. *55*A, 462 (1976)
6.76 L.I. Gudzenko, Yu.I. Sytsko, S.S. Filippov, S.I. Yakovlenko: J. Appl. Mech. Tech. Phys. *14*, 605 (1973)
6.77 E.L. Stupitskii, G.I. Kozlov: Sov. Phys.-Tech. Phys. *18*, 482 (1973)
6.78 V.M. Gold'farb, E.V. Il'ina, G.A. Lukyanov: AIAA Paper 68-682 (1968)
6.79 V.M. Gold'farb, E.V. Il'ina, I.E. Kostygova, G.A. Luk'yanov: Opt. Spectrosc. *27*, 2 (1969)
6.80 A. Czernichowski, J. Chapelle, F. Cabannes: C. R. Acad. Sci. Paris *270*B, 54 (1970)
6.81 P. Hoffmann, W.L. Bohn: Z. Naturforsch. *27*A, 878 (1972)
6.82 G.A. Luk'yanov, V.V. Nazarov, N.O. Pavlova: Opt Spectrosc. *44*, 26 (1978)
6.83 G.M. Zhinzhikov, G.A. Luk'yanov, V.V. Nazarov, N.O. Pavlova: Sov. Phys.-Te Tech. Phys. *23*, 554 (1978)
6.84 S.W. Bowen, C. Park: AIAA J. *10*, 522 (1972)
6.85 E.M. Campbell, W.F. von Jaskowsky, K.E. Clark: AIAA Paper 75-852 (1975)
6.86 E.M. Campbell, R.G. John, W.F. von Jaskowsky, K.E. Clark: AIAA Paper 77-242 (1977)
6.87 D.L. Murphree, R.P. Carter, D.R. Smith: AIAA J. *12*, 1732 (1974)

Appendixes

B.1 V.S. Matveev: J. Appl. Spectrosc. (Minsk) *16*, 228 (1972) (in Russian)
B.2 W. Lochte-Holtgreven (ed): *Plasma Diagnostics* (North Holland, Amsterdam (1968)
D.1 J. Mayer, M. Goeppert-Mayer: *Statistical Mechanics* (Wiley, New York 1977)
D.2 N. Davidson: *Statistical Mechanics* (McGraw-Hill, New York 1962)
D.3 G. Herzberg: *Molecular Spectra and Molecular Structure*, Vol.2, Infrared and Raman Spectra of Polyatomic Molecules (Van Nostrand, New York 1945)
D.4 G. Herzberg: *The Spectra and Structures of Simple Free Radicals* (Cornell University Press, Ithaca, London 1971)
D.5 K.P. Huber, G. Herzberg: *Molecular Spectra and Molecular Structure*, Vol. 4, Constants of Diatomic Molecules (Van Nostrand, New York 1979)
M.1 C.C. Rankin, J.C. Light: J. Chem. Phys. *46*, 1305 (1967)

Bibliography

Abramovich, G.N. (ed.): *Turbulent Mixing of Gas Jets* (Nauka, Moscow 1974) (in Russian)
Bond, J.W., Watson, K.M., Welch, J.: *Atomic Theory of Gas Dynamics* (MIT-Press, Massachusetts 1965)
Cason, C., Horton, T.E.: Phys. Lett *36*A, 303 (1971)
D'yakov, A.S., Piskunov, A.K., Cherkasov, E.M.: Sov. J. Quantum Electron. *5*, 762 (1975)
Dzhidzhoev, M.S., Platonenko, V.T., Khokhlov, R.V.: Sov. Phys.-Usp. *13*, 247 (1970)
Gurvich, L.V., Rtishcheva, N.P.: High Temp. *3*, 1 (1965)
Hausenblas, H.: Z. Flugwiss. *6*, 84 (1958)

Hertzberg, A., Christiansen, W.H., Johnston, E.A., Ahlstrom, H.G.:AIAA J. *10*, 394 (1972)
Hoffman, A.L., Vlases, G.C.: IEEE J. QE-*8*, 46 (1972)
Lapworth, K.C., Townsend, J.E.G.: ARC Rep. Memor. 3479 (1967)
Legay-Sommaire, N., Henry, L., Legay, F.: C.R. Acad. Sci. Paris *260*, 3339 (1965)
Simpson, C., Simmie, J.M.: Proc. R. Soc. A*325*, 197 (1971)
Takayanagi, K.: Progr. Theor. Phys. Suppl. *25*, 1 (1963)
Yushchenkova, N.I., Kalenkov, Yu.A.: J. Appl. Spectrosc. (Minsk) *16*, 39 (1972) (in Russian)

Subject Index